Functional Properties of Bio-Inspired Surfaces

Characterization and Technological Applications

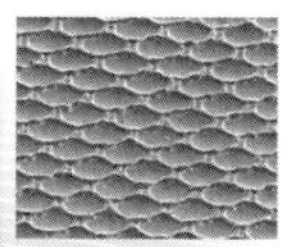

Functional Properties *of* Bio-Inspired Surfaces

Characterization and Technological Applications

Editors

Eduardo A Favret
Instituto Nacional de Tecnología Agropecuaria, Argentina

Néstor O Fuentes
Comisión Nacional de Energía Atómica, Argentina

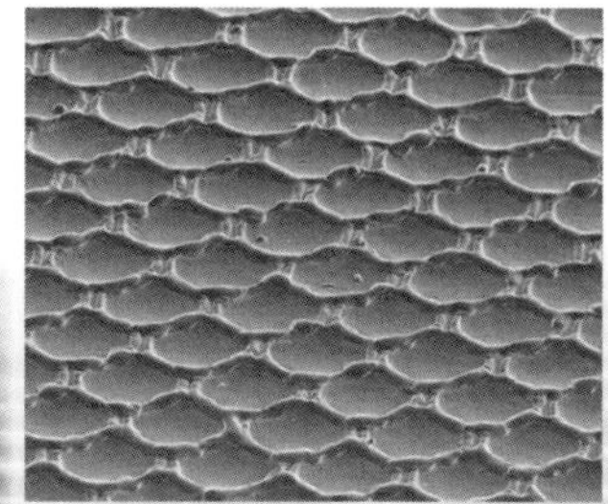

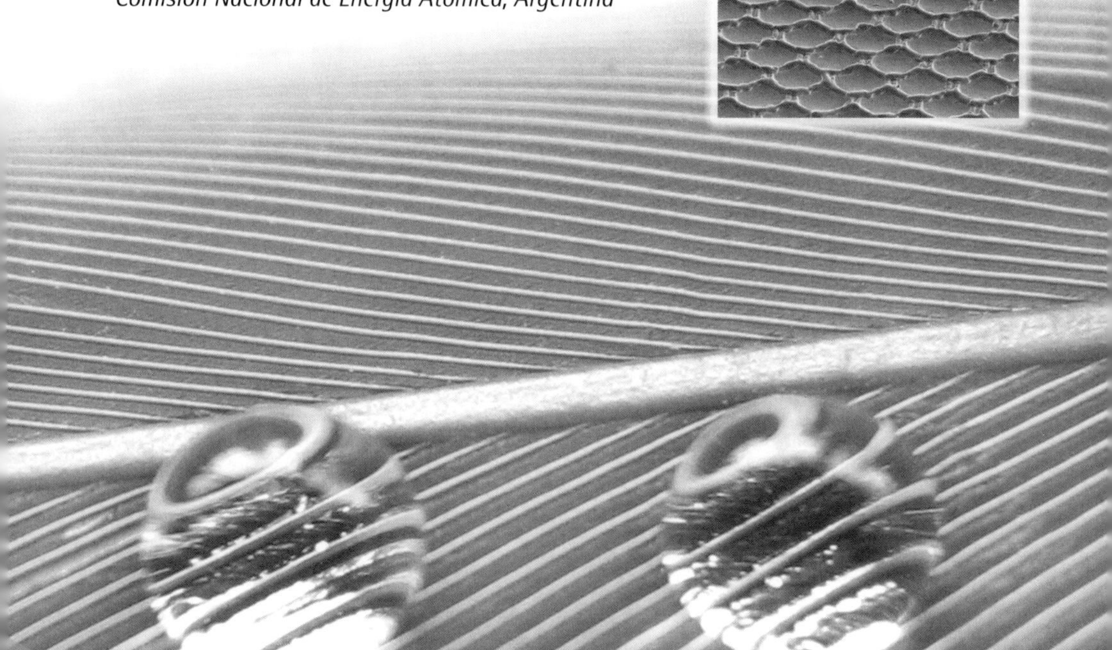

World Scientific

NEW JERSEY • LONDON • SINGAPORE • BEIJING • SHANGHAI • HONG KONG • TAIPEI • CHENNAI

Published by

World Scientific Publishing Co. Pte. Ltd.
5 Toh Tuck Link, Singapore 596224
USA office: 27 Warren Street, Suite 401-402, Hackensack, NJ 07601
UK office: 57 Shelton Street, Covent Garden, London WC2H 9HE

British Library Cataloguing-in-Publication Data
A catalogue record for this book is available from the British Library.

FUNCTIONAL PROPERTIES OF BIO-INSPIRED SURFACES
Characterization and Technological Applications

ISBN-13 978-981-283-701-1
ISBN-10 981-283-701-9

Typeset by Stallion Press
Email: enquiries@stallionpress.com

Printed in Singapore by Mainland Press Pte Ltd.

"God made solids, but surfaces were the work of the devil"

Wolfgang Pauli

(1900–1958)

To the memory of my ***father Ewald*** (1921–1992) whose enthusiasm for natural sciences fueled my interest in physics and to my ***mother Betty*** and my ***sisters Graciela*** and ***Estela*** who encouraged and stimulated me in the writing of this book

Eduardo A. Favret

To my ***wife Hebe*** and to my ***daughter Natalia***, in gratitude for their endless love and support.

Néstor O. Fuentes

Preface

Many good books have been written recently on this new field called biomimetics or bionics, but only few explore simultaneously the characterization and technological processes to produce man-made surfaces with similar properties as the biological ones. Bio-inspired surface structures offer significant commercial potential for the creation of antireflective, self-cleaning, and drag reducing surfaces, as well as new types of adhesive systems. This review book explores how the current knowledge of the biological structures occurring on the surface of moth eyes, leaves, shark skin, and the feet of reptiles can be transferred to functional technological materials. It analyses how such surfaces can be described and characterized using microscopic techniques and thus reproduced, encompassing the important areas of current surface replication techniques and the associated acquisition of good master structures.

It is well known that biological systems have the ability to sense, react, regulate, grow, regenerate, and heal. Recent advances in materials chemistry, micro and nanoscale fabrication techniques have enabled biologically inspired material systems that mimic many of these remarkable functions. Nature is the best technologist to overcome any adaptation for survival difficulty. The field of bio-inspiration is emerging as one of the most innovative areas of science today. Researchers and engineers are studying nature and its novel design solutions, using its principles to engineer technical breakthroughs. Self-cleaning surfaces, powerful LEDs, adhesive tape, and even novel computer algorithms are just some of the benefits of this research partnership with the natural world. The advent of bio-inspiration has not only brought nature closer to high technology engineering, but has also brought together scientists of complementary specialties. Bio-inspiration means that not only scientists and engineers must look to nature, but also be experts in other fields. In many aspects of science and technology, collaboration between nature and science is really the key to achieve breakthroughs in technological materials.

The book is divided into three sections, (1) an introduction of the skin functions and four functional properties of biological surfaces; (2) physical, chemical, and microscopy techniques for describing and characterizing the surfaces, and (3) replication techniques for modifying non-natural surfaces.

The book begins with a chapter that describes certain functions of the skin which contributes to the new area of bio-inspired design. The next four chapter have more specific subjects with its technological applications. Chapter 2 analyzes the shark skin effect or the commonly known antifriction surfaces. Chapter 3 discusses the Lotus effect or the usually known self-cleaning surfaces. Chapter 4 analyzes the Moth-eye effect or the commonly known antireflection surfaces and Chap. 5 describes the Gecko effect or the usually known dry adhesive surfaces.

The second part of the book consists of four chapters. Chapter 6 discusses the principles and techniques used for observing biological surface structures, including optical microscopy (light microscopy, laser scanning confocal microscopy), electron microscopy (scanning electron microscopy, transmission electron microscopy), and scanning probe microscopy. Chapter 7 describes two image analyses techniques (RIMAPS and Variograms) used for topographical characterization in technological surfaces and recently in biological surfaces. Chapter 8 discusses interfacial and capillary phenomena present in multiple biological processes. Chapter 9 describes the chemical characterization of biological and technological surfaces, such as optical, electronic, and particle beams spectroscopies.

The third section has three chapters. Chapter 10 analyzes the Laser Interference Metallurgy which allows the creation of periodic patterns with a well-defined long-range order at the scale of typical microstructures (from the sub micrometer level up to micrometers). Chapter 11 discusses general phenomena involved in electrodeposition process, commonly used techniques and application of electrodeposition in different areas of research and industry. Lastly, Chap. 12 describes the plasma state and its application to modify surfaces in order to obtain a desired functional property in the biomedical field.

We hope that this book will be useful for undergraduate and graduate students from materials science and biology as well as for researchers working in the field of biomimetics, materials science, and engineering.

We are especially indebted to all of the contributing authors for their dedication in writing the chapters. We gratefully acknowledge the useful work done by Estela Favret and Lorena La Fuente, who helped to contact the chapter authors and in the organization of the book. Finally, we also wish to thank Marta Colon, Hwee Yun Tan and Chian Yang Tan (WS) for their support during the preparation of the book. This book has been possible because of all of them.

Eduardo A. Favret
Néstor O. Fuentes
Buenos Aires, April 2009

Contents

List of Contributors

Bläsi, Benedikt
Fraunhofer-Institut für Solare Energiessysteme ISE
Heidenhofstr. 2, 79110 Freiburg, Germany
benedikt.blaesi@ise.fraunhofer.de

Bormashenko, Edward
Head of the Laboratory of Polymers
College of Judea & Samaria
P.O.B. 3, Ariel 44837, Israel
edward@ariel.ac.il

Braceras Izaguirre, Iñigo
Fundación INASMET — Tecnalia
Mikeletegi Pasealekua 2, Parque Tecnológico
Donostia San Sebastián, Provincia de Gipuzkoa
E-20009 España
ibracera@inasmet.es

Brankovic, Stanko
Electrical and Computer Engineering Department
Cullen College of Engineering, University of Houston
N308 Engineering Bldg. 1, Houston TX 77204-4005, USA
SRBranko@mail.uh.edu

Calvo, Adriana
Grupo de Medios Porosos
Departamento de Física. Facultad de Ingeniería
Universidad de Buenos Aires, Paseo Colón 850
1063 Buenos Aires, Argentina
acalvo@fi.uba.ar

Callegari, Gerardo
TRI/Princeton, 601 Prospect Ave.
Princeton, NJ 08536, USA
gcallegari@triprinceton.org

Corengia, Pablo
Instituto Nacional de Tecnología Industrial (INTI)
Av. Gral. Paz 5445, B1650WAB San Martín
Buenos Aires, Argentina
corengia@inti.gob.ar

Fundación INASMET — Tecnalia
Mikeletegi Pasealekua 2, Parque Tecnológico
Donostia San Sebastián, Provincia de Gipuzkoa
E-20009 España
pablo.corengia@inasmet.es

De Las Heras, Evangelina
Instituto Nacional de Tecnología Industrial (INTI)
Av. Gral. Paz 5445, B1650WAB San Martín
Buenos Aires, Argentina
evangedlh@hotmail.com

Favret, Eduardo
Instituto de Suelos
Centro de Investigaciones de Recursos Naturales (CIRN)
Instituto Nacional de Tecnología Agropecuaria (INTA)
De Los Reseros y N. Repetto s/n, 1686 Hurlingham
Buenos Aires, Argentina
eafavret@cnia.inta.gov.ar

Consejo Nacional de Investigaciones Científicas y Técnicas (CONICET),
Av. Rivadavia 1917, C1033AAJ
Ciudad de Buenos Aires, Argentina

Instituto de Tecnología "Prof. J. Sábato"
UNSAM — CNEA, Av. Gral Paz 1499
B1650KNA — San Martín. Buenos Aires, Argentina

Fuentes, Néstor Osvaldo
Head Advanced Materials Division
Comisión Nacional de Energía Atómica (CNEA)
Gerencia de Area Energía Nuclear
GIDAT — U. A. Materiales, Av. Gral. Paz 1499
B1650KNA — San Martín,
Buenos Aires, Argentina
nestor.fuentes@cnea.gov.ar

Instituto de Tecnología "Prof. J. Sábato"
UNSAM — CNEA, Av. Gral Paz 1499
B1650KNA. San Martín. Buenos Aires, Argentina

Gombert, Andreas
Concentrix Solar GmbH
Bötzinger Str 31
79111 Freiburg, Germany
andreas.gombert@concentrix-solar.de

Johnson, Megan
Department of Biological Sciences
University of Calgary
2500 University Dr. NW, Calgary, Alberta
Canada T2N 1N4
mkjohnso@ucalgary.ca

Kruse, Peter
Department of Chemistry
McMaster University
1280 Main St.W., Hamilton, Ontario
Canada L8S 4M1
pkruse@mcmaster.ca

Lang, Amy
Aerospace Engineering & Mechanics Department
University of Alabama,
Box 870280, Tuscaloosa, AL 35487-0280, USA
alang@eng.ua.edu

Lasagni, Andrés
Universität des Saarlandes
Gebäude C6.3, 7 Etage
66123 Saarbrücken, Germany
a.lasagni@mx.uni-saarland.de

Fraunhofer Institute for Material and Beam Technology
Winterbergstr. 28, D-01277 Dresden, Germany
Andres-Fabian.Lasagni@iws.fraunhofer.de

Müklich, Frank
Lehrstuhl für Funktionswerkstoffe
Universität des Saarlandes
Gebäude C6.3, 7 Etage, Postfach 151150
66041 Saarbrücken, Germany
muecke@matsci.uni-sb.de

Nosonovsky, Michael
Department of Mechanical Engineering
EMS Building, Room E371G
College of Engineering and Applied Science
University of Wisconsin-Milwaukee
Milwaukee, WI 53201-0413, USA
nosonovs@uwm.edu

Ren, Qun
Microscopy Core Facility
Department of Zoology and Physiology
University of Wyoming
Laramie, WY 82071, USA
qren@uwyo.edu

Russell, Anthony
Department of Biological Sciences
University of Calgary
2500 University Dr. NW
Calgary, Alberta, Canada T2N 1N4
arussell@ucalgary.ca

Vincent, Julian
Director, Centre for Biomimetic and Natural Technologies
Department of Mechanical Engineering
University of Bath, Bath BA2 7AY, UK
j.f.v.vincent@bath.ac.uk

Ybarra, Gabriel
Instituto Nacional de Tecnología Industrial (INTI)
Av. Gral. Paz 5445, B1650WAB San Martín
Buenos Aires, Argentina
gabriel@inti.gob.ar

Zhang, Zhaojie
Director, Microscopy Core Facility
Department of Zoology and Physiology
University of Wyoming
Laramie, WY 82071, USA
zzhang@uwyo.edu

Part I

Functional Properties of Biological Surfaces

"Nature does nothing in vain"
Aristotle
(384 BC – 322 BC)

Chapter 1

Biomimetics of Skins

Julian F. V. Vincent

The skin is the interface between an organism and its world. It has many functions, some of which are of current interest (e.g. superhydrophobicity, color, rheological effects). It is probably the most versatile of all organ systems, having mechano-sensory and skeletal functions as well, whose relative importance is different in different groups of animals. These functions can all be incorporated into our technology, contributing to the new area of bio-inspired design.

1. Introduction

Life is made possible by membranes. Part of their function is to provide a surface on or from which interactions and reactions can occur and be controlled, and part is to provide selective barriers to keep reactants together and the rest of the world out. These membranes can occur at any level within the hierarchy of the organism. Within the cell, certain organelles such as mitochondria, the endoplasmic reticulum, and the Golgi are made of, recycle, or produce, membranes. Then there are the membranes around the nucleus and around the cell itself. In plants, the hierarchy almost stops here, since all cells are surrounded by a stiff cell wall that, in various guises, supplies the skeletal structure of the entire plant. Size for size, plants are much simpler than animals, requiring only a tenth or a hundredth of the number of cell types.[1] In plants, the cells then form sheets and tubes which gain rigidity from internal pressure, commonly about 1 MPa; the forces so generated are largely resisted by the skin (epidermis) which is made of thick cell walls and is stiff. Whilst multicellular plants are relatively rigid and stationary, most animals are mobile, so the skin has to be compliant, either locally (hinges) or globally. Rigidity is selectively adaptive and supplied by muscular tissues; internal pressures are rarely more than 0.05 MPa. In all cases, both plant and animal, the skin is specialized not only as a covering but also as a selective barrier to passage in both directions of mechanical, physical, and chemical stimuli such as force, heat, water, and volatiles. Since this single layer has to perform many different functions, it is inevitable that there will be conflict between the various requirements. We make the assumption that in biology these conflicts have been largely resolved by evolution, and that we can benefit from the abstraction of concepts from these natural structures into a form that can be integrated into our own technology. This transfer of technologies is known as biomimetics (or, equally, biomimicry, bionics, bionique, bio-inspiration, bio-inspired design, biognosis, etc.). The variety of applications described in this book both underline the variety of functions in skins and demonstrates how concepts can be shifted from one context to another, thus solving design problems and increasing the effectiveness of our technology.

The definition of a problem as a conflict in design requirements, and the solution of the problem as the resolution of this conflict, has been pointed out several times, apparently first by the ancient Greeks. More recently, Genrich Altshuller used it in the development of one of the main components in his system of inventive problem solving ("TRIZ" — *Theoriya Resheniya Izobreatatel'skih Zadatch*) that he and others in Russia have developed as a design tool for engineering. His approach was to derive a set of standardized factors (weight, speed, ease of use, etc. — 39 in all) and to select from these, the properties that the problem definition would bring into conflict. Thus the problem is then defined as a pair of properties that are apparently mutually incompatible: for instance, if one parameter (e.g. strength) is improved it will probably compromise another (e.g. lightness). The properties are listed along orthogonal sides of a matrix (known as the "Contradiction Matrix") so that in the square where the two properties coincide is the solution for resolving the conflicts (known as "Inventive Principles" in TRIZ; there are about 40 of them). The Inventive Principles cover all possible manipulations of the system, context, and components under examination. The properties and the Inventive Principles were derived from successful patents and so represent a collection of best engineering practices.[2,3] Thus in comparing the resolutions provided from TRIZ with the resolutions to similar problems provided from biology, we are able to provide a measure of the similarity between technology and biology, and thus to test (a) whether the emerging study of biomimetics is likely to provide novel practical solutions to technical problems and (b) how those solutions can be expressed in a technical environment.[4] We can then express the various morphological and physiological mechanisms found in biology in the same way as Altshuller did with engineering — what is the novelty that the mechanism requires, and what is the adjustment that will deliver that novelty? In this way we have quantified adaptations as solved technical problems in the same way that inventive innovations in engineering can be defined in patents. There are differences. In very general terms, the similarity between biological and technological solutions to problems is only about 12%; biology solves problems in very different ways to ourselves. This may be an important insight, since it suggests that we are looking to biology for solutions to technical problems, we may quite frequently be looking in the wrong place since we have mis-identified the problem! More important for the development of our technology (which is presumably what biomimetics is about) is the realization that where technology uses energy as the main controlling parameter, especially in the processing of materials, biology uses the information from DNA embedded in those materials when they were synthesized (Fig. 1). In fact, living organisms rely very little on energy to solve technical problems, and it should be the *main* objective of biomimetics to show how we can similarly rid our technology of the energy requirement, thus giving us a more sustainable future on this planet.

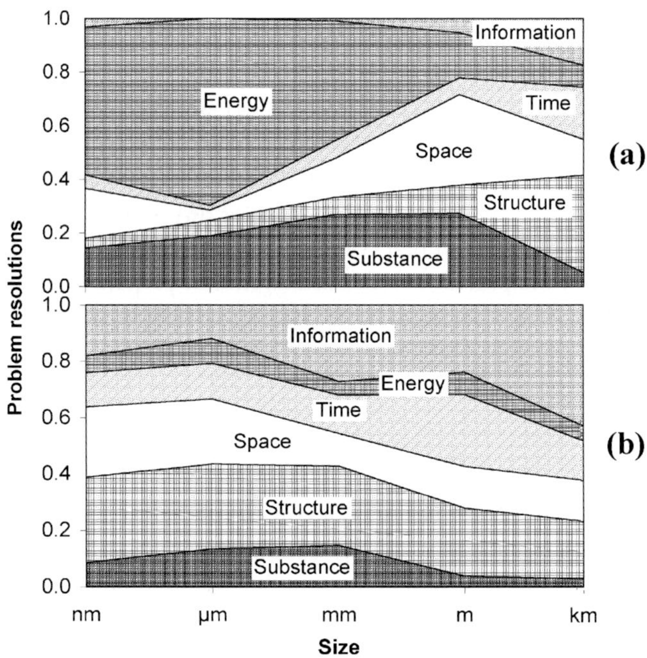

Fig. 1. These two graphs show the main factor which is changed in order to solve a problem over a range of size scales. For instance energy can be increased, decreased, the source can be changed, it can be localized or spread out, etc. The upper graph shows the way technology uses these factors while the lower graph shows how biology does it. Energy is not an important variable in biology; information (more or less ignored by technology) is paramount.[4]

In a recent case study,[5] the conflicts inherent in the functional requirements (about 20 of them) of insect cuticle, a relatively simple but well-studied "skin," were defined and resolved using the TRIZ system. This showed that most of the functions of cuticle are provided by detailed control of properties over a very short distance at a chemical and morphological level, summarized by the Inventive Principle "Local Quality," a principle which is rarely used in technology. The manipulations which "Local Quality" suggests are:

- Change an object's structure, or its environment, from homo- to heterogeneous; use gradients instead of uniformity.
- Make each part of an object more adapted to its own purpose; compartmentalize.
- Make each part of an object fulfill a different function, e.g. pencil with eraser; hammer with nail-puller; Swiss army knife.

The other Principles most used by insect cuticle were "spheroidality," "flexible shell," and "composite material." These three are related to the skeletal function of the insect cuticle, being concerned with providing a good supportive structure that is strong and light.[6] But this reflects the predominance of skeletal functions in insect cuticle when it is compared with vertebrate skin (Fig. 2). None of these principles would be so commonly used in our technology, which tends to use a blunter, more global approach, the commonest resolution involving changing a parameter such as temperature or pressure. Thus the main outcome of this study was that biology and technology "solve" problems in design in very different ways, and that most of the functions of cuticle are provided by detailed control of properties over a very short distance. The only area where solutions seemed to be held in common were in the control of spectral response, both in the production of physical and pigmentary colors and the use of UV filters to protect underlying tissues.

2. Surface Hardening

These are examples of "Local Quality" in insect (arthropod) cuticle.

Wear resistance is required in the tips of ovipositors, claws, and mouth-parts of insects, all of them localized areas or layers. Most work has been on the mandibles of herbivorous insects. Wood, plant leaves, and seeds are commonly hard, either because they are dense and dry or because they are reinforced by silica or calcium oxalate monohydrate crystals.[7] The cuticle of the cutting and grinding surfaces of the mandibles of these insects (Fig. 3) is reinforced.[8,9] The Vicker's hardness of the cutting edge of dry mandibles compares to that of enamel and is about twice as high as that of the sheared face of the dry mandible and several times that of wet cuticle which has a hardness in the range of that of dentine. Studies on a number of insect orders (locusts,

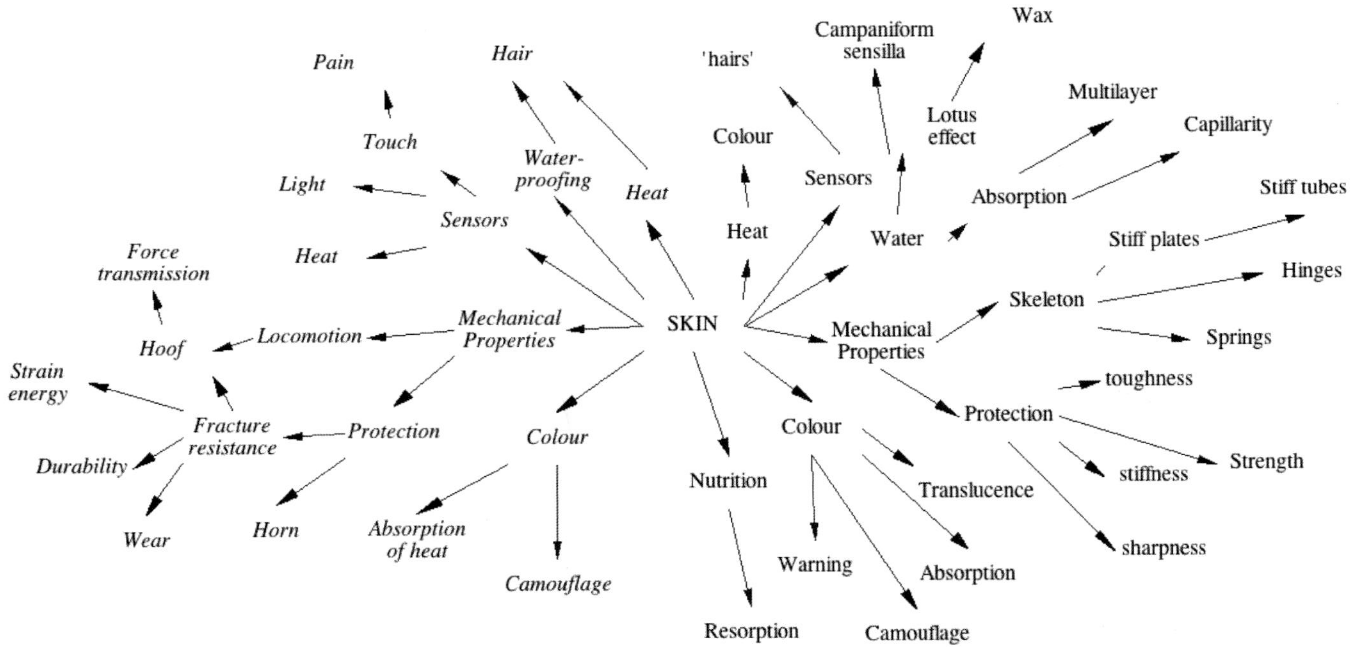

Fig. 2. The functions of skin (left, italic) and insect cuticle (right, roman). Not an accurate or detailed diagram, merely an indicator of relative importance of the various functions.

Fig. 3. Surface hardening of insect cuticle, showing how the hard outer surface (A) of the right mandible (B) of locust wears away (D) the soft outer surface (C) of the left mandible (E). The inner surfaces are soft and hard respectively, so the hard surfaces are continually being undercut, break away, and so leave a sharp edge (A), rather like the break-off blade of a hobby knife.[10]

stick-insects, butterflies, ants, wasps, and beetles)[10–12] revealed that the reinforcement against wear and tear is achieved by impregnating the hardened cuticle of the mandible with heavy metals (form unknown) such as Zn, Mn, or occasionally Fe. These metals are present in relatively large amounts ranging from a few percent to up to 16% of dry mass of the mandibular cutting edges and increase their hardness significantly, trebling it from about 25 to 80 $kgf\,mm^2$.[9] How the incorporation of metals into the cuticle hardens and stiffens the cuticle is not yet understood, nor why metals are preferred to the more common mineralization with Ca-salts.

3. Strain Sensors

An external skeleton is a barrier to transmission of sensory information about the external environment. The resolution here is achieved by having holes through it (an Inventive Principle of "porosity"), although the holes have to be carefully designed if they are to give reliable information, be sensitive and not weaken the material (control of local quality again).

Insects strain sensors are basically oval holes in the cuticle with a domed bell shaped cap, hence called campaniform sensilla. These organs allow the animal to measure displacements in the plane of the cuticle, and they do it by introducing compliance due to the hole. The geometry and mechanical properties of the suspension cause the cap to move up and down as the hole changes its dimensions when the plate is stretched, compressed, bent or twisted. Thus the cap system rotates deformation in the plane of the plate through 90°, allowing the deformation to be detected out of the plane.[13,14] In insects the sensillum, together with its associated sensory and nerve cells, forms a simple yet sensitive mechanism[15,16] which is capable of detecting displacements of the order of a nanometer.[17] The plate in which the sensillum is formed can be anisotropic depending on the orientation of the chitin fibers and heterogeneous depending on the volume fractions of the components, degree of cross-linking of the matrix, etc. There is therefore, considerable freedom in the design parameters. In most arthropods (including insects and crustaceans) the sensillum hole is oval, so it can be "aimed" to pick up and amplify displacements in defined directions.[14,18,19] These organs are always placed in areas where the load is likely to be the greatest. These are also the places where the organ in which the load is being measured — the wing (Fig. 4) or the leg — seems most likely to break.[20] The animal seems to be weakening those areas that we would expect it would most want to strengthen. The design conflict seems to be met

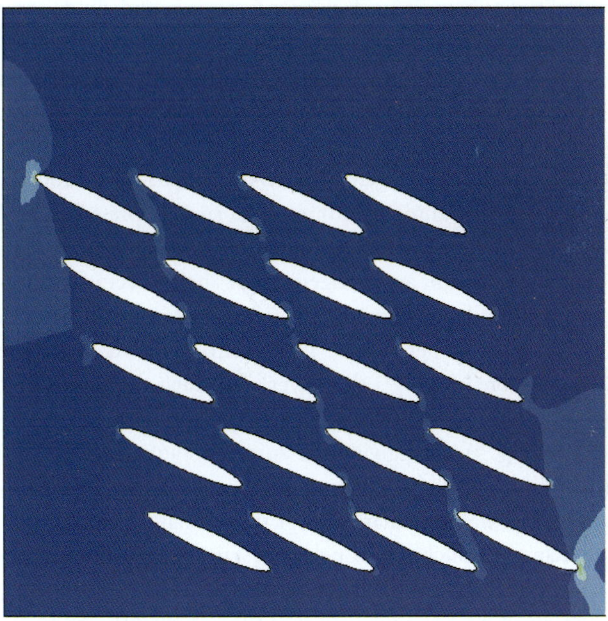

Fig. 4. Campaniform sensilla — a finite elements model of part of the array of sensilla at the base of the haltere (the balancing organ) of a fly. The oval holes are the sensilla and the only significant stress concentrations are top left and bottom right, indicated by a lighter shade of blue.

by two factors, neither of them yet adequately investigated. These are the orientation of the chitin fibers which reinforce the cuticle, and the clustering of the holes, which evens out the stress concentrations[21] or confines the stress concentrations to specific areas where a small amount of reinforcement associated with two holes at opposite corners of the array seems to be sufficient without compromising the performance of the remainder.[22] In spiders the strain sensors have a much higher aspect ratio and are called slit sensilla. They work on the same design principle, but also seem to be frequency analyzers, allowing the animal not only to detect vibration, as in a spider's web, but to differentiate the frequencies of vibration peripherally, and thus reduce the amount of neural processing of information. Once again, "local quality" is the main design consideration.

4. Water Repellence

One of the major surface effects of skins found over the last 20 or more years is related to the concept of superhydrophobicity. This has been reinforced by the development of self-cleaning surfaces based on the "Lotus effect" (Chap. 3 pp. 23 and 24). Such surfaces are found not only on leaves but on insect wings and, no doubt, will be found on many other surfaces. The concept is a product of two effects: a hydrophobic surface created by a layer of hydrocarbons, usually crystalline, and a rugose morphology of bumps at a spacing of about 10 μm. There is no indication how many times such surfaces have been evolved. However, research on these surfaces is not new.[25] At one end of the scale is lumpy surfaces that Chinese researchers have dubbed "nonsmooth" and are covered in small bumps or domes. The surface presented by the denticles on shark skin come into this category (Chap. 2, pp. 26–28). At the other end of the scale is the insect plastron, an arrangement of "hairs" (actually outgrowths of the cuticle, and so very different from mammalian hairs) a few micrometers long arranged at a density of 10^7 or more per cm^2.[29] The plastron is a well-researched system for repelling water, therefore keeping a thin layer of air next to the surface of the animal and providing an air–water interface at the tips of the hairs across which exchange of gases can occur (Fig. 5), thus giving the animal an aquatic gill.[30] This is not only a common adaptation of insects living on and around water, but is obviously also a technique for keeping the immediate skin of the animal dry and insulated.[31] Thus a plastron in the shape of a thick pile or pelage of hairs gives a waterproof surface, since water cannot penetrate between the hairs. In the way TRIZ is normally expressed, the conflict being resolved here is obviously that the animal is living beneath the water surface and yet it is breathing air — there is air yet there is no air! Such a conflict (categorized as "A and not A" where "A" is a thing, property or function) is called a "Physical Conflict." Classical (engineering) TRIZ does not indicate an equivalent in

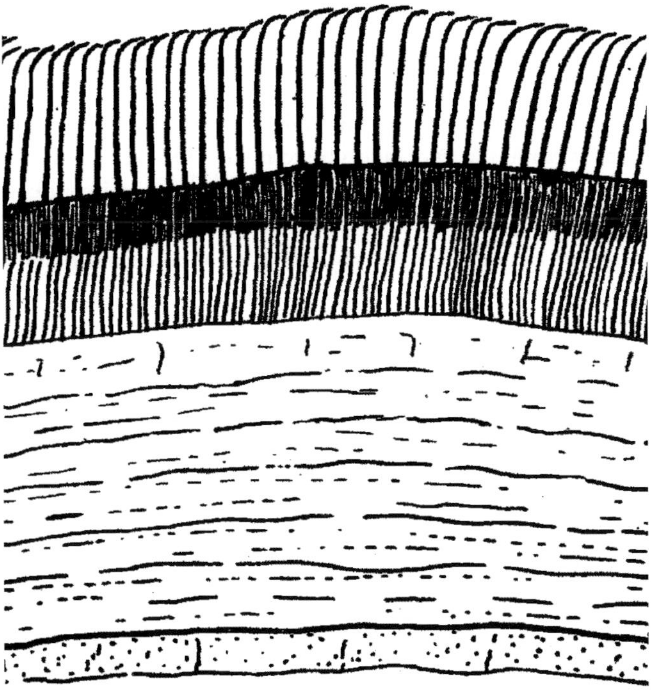

Fig. 5. A section through the plastron of Aphelocheirus showing the densely arranged hairs on the surface of the cuticle. Reproduced with permission from.[40]

technology, so this device is therefore a novel biomimetic solution. A textile similar to velvet has been developed by a sportswear company in the UK as a breathable and warm waterproof material for surfers to wear. It is very successful. The inventor did not realize he was reinventing the plastron, although he based his initial concepts on the waterproof pelage of seals and otters. Another technical version of the same concept is provided by the surface of an extremely water-repellent foam which mimics this mechanism and allows direct extraction of oxygen from aerated water[32] and so, presumably, the evaporation of water from the skin surface providing the equivalent of Goretex as a 3D material rather than a single sheet. Water-repellent surfaces are also being developed to make devices which can walk on water.[33] This is clearly a rich area for development.

The "non-smooth" surfaces found on the elytra of many soil-burrowing beetles have also been shown to be self-cleaning, but in a totally different way from the lotus effect. The mechanism appears to be due to the high local shear stresses developed at the top of the domes. Experiments showed that the main characteristic is simply the unevenness of the surface, which reduces friction against the soil by up to 40%. The morphology has been used in the design and development of new mould-boards on ploughs and new bulldozing plates in China[34,35] and has shown significant fuel savings. In TRIZ terms, the conflict is again physical — the surface should be there to contain the object or transmit the force, and should not be there in order to reduce the friction of plate and soil. The solution is separation in space — small contact areas (bumps) and air between them. An equivalent in engineering would be a ball or roller bearing, where two surfaces are held apart by rotating components with minimal contact area. Ultimately, of course, one would want a fluid bearing, which would be a plain bearing (filled with the fluid, oil) or an air bearing. An agricultural version might use something like a cartilaginous (e.g. knee) bearing which "weeps," producing a liquid (hyaluronic acid or synovial fluid) when it is under load.

5. Color

Coloration can be physical or chemical (pigmentary) or a combination, and biomimetic versions of the principles involved are contributing to textiles, plastics, and camouflage systems. There are several types of physical coloration, loosely classified as scattering (coherent and incoherent), diffraction, and interference. They can be produced by close-packed arrays of spheres or platelets, by surface lines, by layered materials with varying refractive index and responses to polarized light, and by 3D structures. The reflection of colors is probably the best known from the physics point of view, and was investigated by Newton. Biology has managed to produce systems which not only produce physical colors, but can do so over a much wider range of viewing angles.[36] The details of this form of color production are still unknown, but some of the mechanisms are used in textiles. In some scarab beetles the cuticle is made of structures which look like liquid crystals, mainly nematic and cholesteric. Of the incident light on the cuticle, the right circularly polarized component can be reflected and the left circularly polarized light can penetrate the helicoidally structured cuticle. However, at a certain depth there is a layer of nematic structure which acts as a half-wave plate, reversing the sense of polarization of the light, which is then reflected when it reaches the next layer of helicoidal structure, is reflected, has its sense reversed again by the nematic layer, and continues back out through the helicoidal cuticle with very little loss. The refractive index of the cuticle is increased by the addition of uric acid. Thus the cuticle is an almost perfect reflector, making the beetle appear the same green as its surroundings. This system will only work when the color and light intensity are the same in all directions.[37] In some cephalopods there is a very bright white reflection, which cannot be due to a mirror because there is no layer of silvering which could be the reflecting layer. Stacks of platelets in the iridophores contain reflectins, proteins rich in aromatic and sulfur-containing amino acids.[38] The photonic structures reflect light through alternating layers of high and low index materials, with the high-index layer comprising reflectins. The refractive index of recombinant reflectin was calculated to be 1.591 ± 0.002 which is the highest reported refractive index for a naturally occurring protein. These proteins self-assemble into a variety of structures. Another way of producing total reflection is a stack of platelets of different thicknesses. Each will refract light of a different wavelength, so reflecting the entire spectrum of colors as discrete wavebands. Since the platelets can be suspended in soft tissue such as skin, and the orientation and order (in terms of graded thickness) of the platelets is not important, this system produces a flexible mirror. It is common in "silvery" fish, where the platelets are made of guanine.[39]

TRIZ does not have much specifically to say about color and its generation or control, although the general trend within biology to control color by splitting it up into its components and manipulate each one separately is rather different to technical systems and can be much more effective. Once again biology maximizes the possibilities inherent in self-assembly systems and the production of small repeating structures.

6. Envoi

The rest of this book deals with a number of surface functions of skins that have been well researched in recent years, and I have touched on some of them above. But anyone should realize that not only is this a small selection of what skin can do, and what we can use for inspiration and innovation, but that the engineering environment into which we are transferring these functions may not be the ideal. In making the technological transfer, we strip away the uniqueness of biology since that is the only way the transfer is currently possible. Perhaps we should spend more time understanding how the function is generated and made adaptive and make that part of the biomimetics. I have the feeling biomimetic materials processing would be far more important than the functions which are currently being copied.

References

[1] McMahon TA, Bonner JT. (1983) On size and life. Freeman, NY.

[2] Altshuller G. (1999) The innovation algorithm, TRIZ, systematic innovation and technical creativity. Technical Innovation Center Inc, Worcester, Massachusetts, USA.

[3] Altshuller G. (1988) Creativity as an exact science. Gordon and Breach, NY.

[4] Vincent JFV, Bogatyreva OA, Bogatyrev NR, Bowyer A, Pahl A-K. (2006) Biomimetics — Its practice and theory. *J R Soc Interface* **3**: 471–482.
[5] Vincent JFV. (2005) Deconstructing the design of a biological material. *J Theor Biol* **236**: 73–78.
[6] Vincent JFV, Wegst UGK. (2004) Design and mechanical properties of insect cuticle. *Arth Struct Devel* **33**: 187–199.
[7] Lucas PW, Turner IM, Dominy NJ, Yamashita N. (2000) Mechanical defences to herbivory. *Ann Bot* **86**: 913–920.
[8] Hillerton JE, Reynolds SE, Vincent JFV. (1982) On the indentation hardness of insect cuticle. *J Exp Biol* **96**: 45–52.
[9] Schofield RMS, Nesson MH, Richardson KA. (2002) Tooth hardness increases with zinc-content in mandibles of young adult leaf-cutter ants. *Nat Wiss* **89**: 579–583.
[10] Hillerton JE, Vincent JFV. (1982) The specific location of zinc in insect mandibles. *J Exp Biol* **101**: 333–336.
[11] Robertson B, Hillerton JE, Vincent JFV. (1984) The presence of zinc or manganese as the predominant metal in the mandibles of adult stored product beetles. *J Stored Prod Res* **20**: 133–137.
[12] Fawke JD, McClements JG, Wyeth P. (1997) Cuticular metals: Quantification and mapping by complementary techniques. *Cell Biol Int* **21**: 675–678.
[13] Skordos A, Chan C, Jeronimidis G, Vincent JFV. (2002) A novel strain sensor based on the campaniform sensillum of insects. *Phil Trans R Soc A* **360**: 239–254.
[14] Pringle JWS. (1938) Proprioception in insects: II. The action of the campaniform sensilla on the legs. *J Exp Biol* **15**: 114–131.
[15] Zill SN, Moran DT. (1981) The exoskeleton and insect proprioception I Responses of tibial campaniform sensilla to external and muscle-generated forces in the american cockroach, *Periplaneta americana*. *J Exp Biol* **91**: 1–24.
[16] Zill SN, Moran DT. (1981) The exoskeleton and insect proprioception III Activity of tibial campaniform sensilla during walking in the American cockroach, *Periplaneta americana*. *J Exp Biol* **94**: 57–75.
[17] Dickinson MH. (1990) Linear and nonlinear encoding properties of an identified mechanoreceptor on the fly wing measured with mechanical noise stimuli. *J Exp Biol* **151**: 219–244.
[18] Keil TA. (1997) Functional morphology of insect mechanoreceptors. *Microsp Res Tech* **39**: 506–531.
[19] Keil TA. (1997) Comparative morphogenesis of sensilla: A review. *Int J Insect Morphol Embryol* **26**: 151–160.
[20] Gnatzy W, Gruenert U, Bender M. (1987) Campaniform sensilla of *Calliphora vicina* (Insecta: Diptera) I. Topography. *Zoomorphol* **106**: 312–319.
[21] Barth FG. (2004) Spider mechanoreceptors. *Curr Opin Neurobiol* **14**: 415–422.
[22] Vincent JFV, Clift SE, Menon C. (2007) Biomimetics of campaniform sensilla: Measuring strain from the deformation of holes. *J Bionic Eng* **4**: 63–76.

[23] Barthlott W, Neinhuis C. (1997) Purity of the sacred lotus, or escape from contamination in biological surfaces. *Planta* **202**: 1–8.

[24] Bechert DW, Bruse M, Hage W, Meyer R. (2000) Fluid mechanics of biological surfaces and their technological application. *Nat Wiss* **87**: 157–171.

[25] Cassie ABD, Baxter S. (1944) Wettability of porous surfaces. *Trans Farad Soc* **40**: 546–551.

[26] Ball P. (1999) Engineering shark skin and other solutions. *Nature* **400**: 507–509.

[27] Bechert DW, Bruse M, Hage W. (2000) Experiments with three-dimensional riblets as an idealized model of shark skin. *Expts Fluids* **28**: 403–412.

[28] Koeltzsch K, Dinkelacker A, Grundmann R. (2002) Flow over convergent and divergent wall riblets. *Expts Fluids* **33**: 346–350.

[29] Thorpe WH, Crisp DJ. (1949) Studies on plastron respiration: IV. Plastron respiration in the Coleoptera. *J Exp Biol* **26**: 219–260.

[30] Rahn H, Paganelli CV. (1968) Gas exchange in gas gills of diving insects. *Resp Physiol* **5**: 145–164.

[31] Bush JWM, Hu DL, Prakash M, Casas J, Simpson SJ. (2007) The integument of water-walking arthropods: Form and function. In *The Integument of Water-Walking Arthropods: Form and Function*, pp. 117–192. Academic Press.

[32] Shirtcliffe NJ, McHale G, Newton MI, Perry CC, Pyatt FB. (2006) Plastron properties of a superhydrophobic surface. *Appl Phys Lett* **89**: 104106.

[33] Hu D, Prakash M, Chan B, Bush J. (2007) Water-walking devices. *Expts Fluids* **43**: 769–778.

[34] Han ZW, Xu XX, Qiu ZM, Ren LQ. (2005) Investigation of micro-wear and micro-friction properties for bionic non-smooth concave components. *J Bionic Eng* **2**: 63–67.

[35] Li J-Q, Sun J-R, Ren L-Q, Chen B-C. (2004) Sliding resistance of plates with bionic bumpy surface against soil. *J Bionic Eng* **1**: 207–214.

[36] Vukusic P, Sambles JR. (2003) Photonic structures in biology. *Nature* **424**: 852–855.

[37] Caveney S. (1971) Cuticle reflectivity and optical activity in scarab beetles: The role of uric acid. *Proc R Soc B* **178**: 205–225.

[38] Kramer RM, Crookes-Goodson WJ, Naik RR. (2007) The self-organizing properties of squid reflectin protein. *Nat Mater* **6**: 533–538, advanced online publication.

[39] McKenzie DR, Yin Y, McFall WD. (1995) Silvery fish skin as an example of a chaotic reflector. *Proc R Soc A* **451**: 579–584.

[40] Thorpe WH, Crisp DJ. (1947) Studies on plastron respiration: I. The biology of *Aphelocheirus* [Hemiptera, Aphelocheiridae (Naucoridae)] and the mechanism of plastron retention. *J Exp Biol* **24**: 227–269.

Chapter 2

The Shark Skin Effect

Amy W. Lang

This chapter reviews the research and current state of knowledge with respect to the particular morphology associated with shark skin. Shark skin has a unique microstructure that appears to serve multiple functions for the shark. In the case of fast-swimming sharks, arguments can be made that the microgeometry associated with the shark skin reduces the drag allowing for higher swimming speeds and increased turning ability on some species. The majority of the research has involved deducing the turbulent skin-friction drag-reduction capabilities of the shark skin microgeometry consisting of streamwise riblets, although the separation control mechanisms that may be present have also been considered. Finally, current biomimetic technological applications of shark skin microstructure are reviewed, including shark skin swimsuits to anti-fouling surfaces for ocean-going vessels.

1. Introduction

Elasmobranch fishes (sharks, rays, and skates) have a unique skin structure which has adapted for different species to serve various purposes. Their skin is covered with miniscule scales that are extremely similar to teeth (hard enamel outside with a pulpy inside) in which the base of each scale is embedded in the surface of the skin. These scales are known as placoid scales or denticles. Sharks (Fig. 1), in particular, have a chain of evolutionary development dating back 400 million years, with some of the youngest species of highly adapted fast-swimming sharks (e.g. the shortfin mako shark *Isurus oxyrinchus)* dating back approximately 8 million years. To meet the needs of various species, the microstructure of the denticles has adapted to serve four essential functions.[1] The primary purpose for this bony skin is protection from predators and or prey as well as ectoparasites. A secondary purpose for most species is incorporation

Fig. 1. A sand tiger or grey nurse shark (*Carcharias taurus*) in captivity (Image courtesy of Fundación Temaikèn, Escobar, Buenos Aires, Argentina).

of sensory organs, and even a few species have developed bioluminescence. Tertiary functions of the shark skin are protection from abrasion (e.g. sharks that live near reefs) and control of the fluid flow around their body, which has been theorized to lead to drag reduction on fast-swimming sharks.[1–6]

For several thousand years, man has found shark skin to be a valuable resource to meet various needs. From the early use as sand paper by ancient Greeks[1,7] to use on weapons and the soles of shoes,[1,8] mankind has found the durable and abrasive nature of the shark skin extremely functional. However, it was not until the late 1800s that the first scientific observations concerning the structure of the skin and scales were carried out, with some of the earliest and most complete work being that of Hertwig.[9] Work that followed has been numerous, documenting the shape of the scales for most species of sharks, with some of the more recent studies relying on modern microscopic techniques for more precise and accurate measurements.[5,10] It was not until the latter half of the 20th century that the shark skin was first considered for its drag-reducing capabilities.[11,12] Several theories have been posited over the years including separation control[13] and turbulent skin-friction reduction.[14,15] An overview of these theories is provided in this chapter, with conclusions as to the most probable mechanisms based on the latest work performed in this area. Finally, modern day applications of the shark skin geometry to serve various purposes are discussed along with future applications that may still yet be achieved.

2. Shark Skin Structure

The skin of sharks has evolved into a highly complex morphology, which as explained previously, has the potential to serve a variety of purposes. The scales, or denticles, strongly resemble tiny teeth that cover the shark's body. These denticles, just as teeth, have a hard enamel outside and a pulpy inside. They are attached to the skin at the base by a collagen fiber, which is an elastic-type tendon that can be found in the ligaments of most animals. In fact, the base allows for the crown to sit above the skin and is orientated approximately orthogonal to the crown. The top surface of the denticle, or crown (Fig. 2), represents the surface exposed to the water, and a staggered, overlapping array of these crowns constitutes the complete shark skin. As observed in Fig. 3, the denticles can vary by size and shape according to the position on the shark's body. Denticles also vary greatly by size and shape for various species. However, there are some common characteristics that have been observed in faster swimming sharks.[1,5] In particular, faster sharks tend to have smaller denticles (averaging a crown length of 160–400 microns in size) and a well-defined set of ribs. For example, three ribs are consistently observed in the case of one of the fastest species, the shortfin mako *I. oxyrinchus*, which also has the smallest denticles (Figs. 2 and 3). It is this staggered array of ribs along the surface (aligned in the streamwise direction of the flow) that has been long associated with the turbulent skin-friction reduction studies inspired by the shark skin.

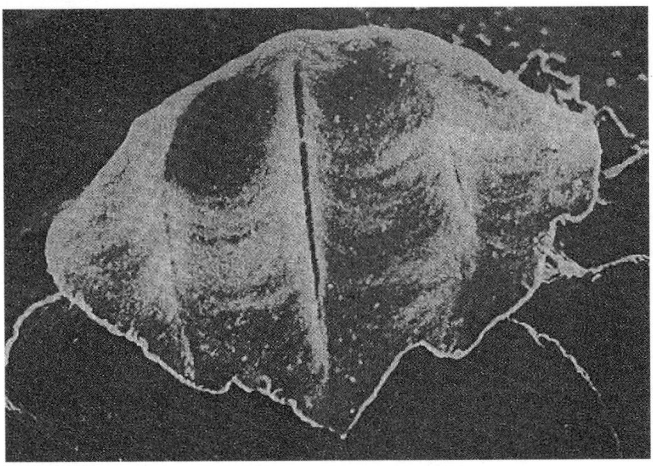

Fig. 2. Scanning electron micrograph of the posterior margin of a shortfin mako shark scale from a 132 cm long male specimen.[5] Crown length (along the rib direction) is about 160 microns.

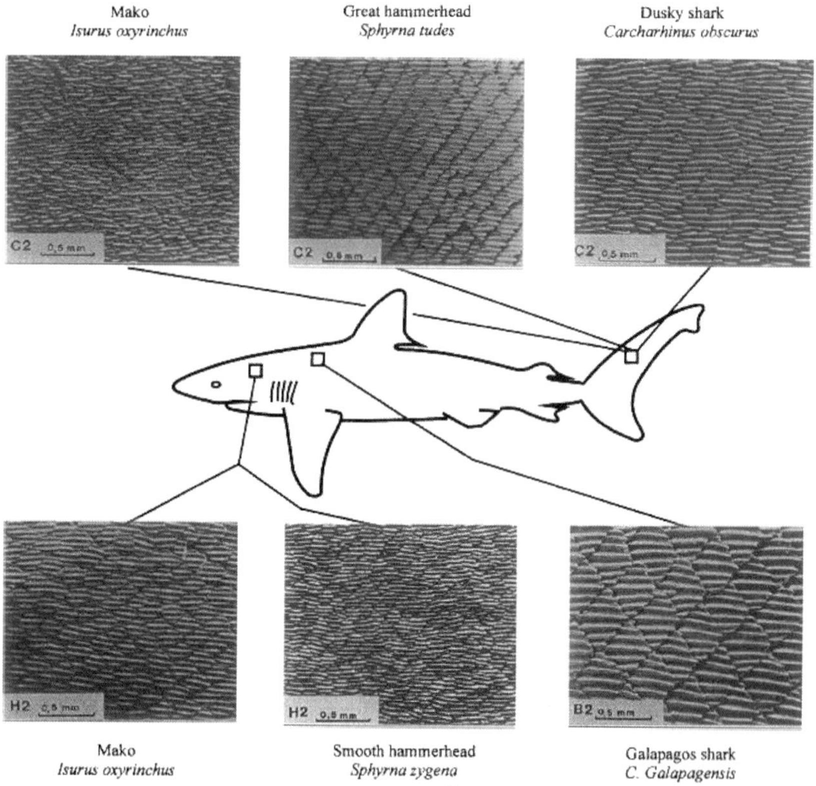

Fig. 3. Scale patterns of various species of fast sharks. Reproduced with permission by Springer from Bechert *et al.*[16] with scale images originally obtained by Reif.[3]

The scales of the fastest swimming ($Re = UL/\nu \sim 10^7$ based on shark length (L) corresponding to $U \sim 20$ m/s[17]) pelagic sharks, such as the smooth hammerhead *Sphyrna zygaena* and shortfin mako shark *I. oxyrinchus*, apparently can change the angle of attack of the crown to the streamwise flow such that the back tips of the scales bristle.[2,4,18] It has also been observed on several slower species that movement of the scales on dead shark skin is confined to the rear part of the body and fins such as on the Galapagos shark[18,19] and blacktip shark (Fig. 4). It is notable that these are the areas which are more prone to flow separation.

Several mechanisms leading to denticle bristling have been theorized. Besides possible bristling induced by a localized pressure gradient in the flow, a morphological mechanism resulting from an increase in skin tension is also considered to be feasible. The ability to achieve denticle bristling on some species appears to have resulted in

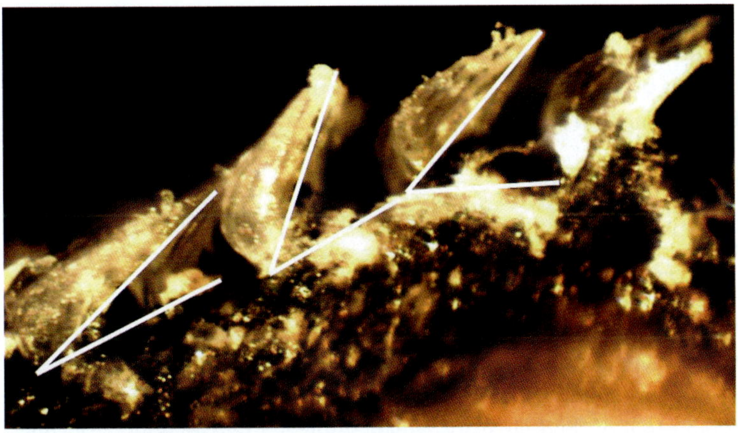

Fig. 4. A coronal section through the skin and scales anterior to the first dorsal fin of a blacktip shark *C. limbatus*. Anterior is towards the left. The scales have been manually manipulated by a probe so that the three scales towards the right are erected to greater than 30°, whereas the scale on the extreme left is in the resting position. The picture was taken with a dissecting microscope at 400× magnification and in-line digital camera. Image courtesy of Dr. Phil Motta (Professor of Biology, University of South Florida).

Fig. 5. "Hand-like" scales at approximate angle of bristling as that observed on dead sharks.

a reduction in the size of the base of the scales. While scales on the nose are densely packed, immoveable, with thick bases, those that have been observed to be moveable to the touch on some species have a "hand-like" appearance with a small base that appears to be loosely anchored to the skin on dead specimens (Fig. 5). The skin of sharks, just underneath of this denticle outer-shell, is made of interwoven collagen fibers, or bundles, that branch off to smaller fiber sizes as the surface of the skin is approached from beneath; where each denticle is anchored into the skin by one of these fibers. It has been purported that as a shark swims faster, the skin as a whole is pressurized, thereby putting collagen bundles under tension, resulting in whole-field denticle bristling. This hypothesis is backed by the measurement of an increase in subcutaneous pressure from 7 to 14 Pa in a resting lemon shark *Negaprion brevirostris* to 200 kPa during fast swimming.[20]

3. Drag Reduction

3.1. *Marine Animal Locomotion*

An animal swimming through a fluid experiences three types of drag that impedes its passage. These include (i) form drag due to a difference in pressure around the body, (ii) drag-due-to-lift, and (iii) skin friction (τ) due to boundary layer formation.[6] Skin friction is a result of fluid viscosity, which enforces a no-slip boundary condition on the flow known such that a swimming body will impart momentum to the water molecules in the region close to the body (boundary layer), and the energy expended in pulling these molecules with the body results in drag. At low Reynolds numbers (*Re*) skin-friction predominates, while at higher *Re* pressure drag can dominate if not minimized. Drag-due-to-lift, or induced drag, is associated with the generation of longitudinal tip vortices formed primarily from the thrust producing caudal fin and is often grouped with form drag, such that the summation is referred to as pressure drag. Figure 6 exhibits the estimated distribution of drag depending on whether the flow remains attached or separated around the body, where flow separation is the determining factor in causing high-pressure drag. It is not surprising then to consider the fact that aquatic organisms have evolved to minimize drag,[21] with the primary decrease coming from a streamlined body shape to reduce form drag.

Aquatic organisms that swim at high *Re* ($>10^3$) have a variety of shapes and structures to reduce drag, which we often attempt to duplicate.[22,23] It has been deduced that a crescent tail design could decrease induced drag on the order of 8%, and not surprisingly this lunate tail design is found on many marine animals.[21,24] Several researchers, e.g. Anderson *et al.*,[25] have observed that a swimming fish experiences more friction drag than the same body towed. This higher friction is attributed to motion of the body as it produces thrust inducing thinner boundary layers, and may result in a five-fold increase in skin-friction drag.[23] The argument has been made that because of this higher power output requirement, to overcome drag and maintain a certain speed, swimming drag reduction due to various morphological mechanisms is extremely probable.[26] In addition, skin friction can increase by as much as 5–10 times if the boundary layer transitions from laminar to turbulent flow. However, two arguments could be made that a natural delay of transition (and consequent drag reduction from lower skin friction) does indeed occur over a large portion of the body

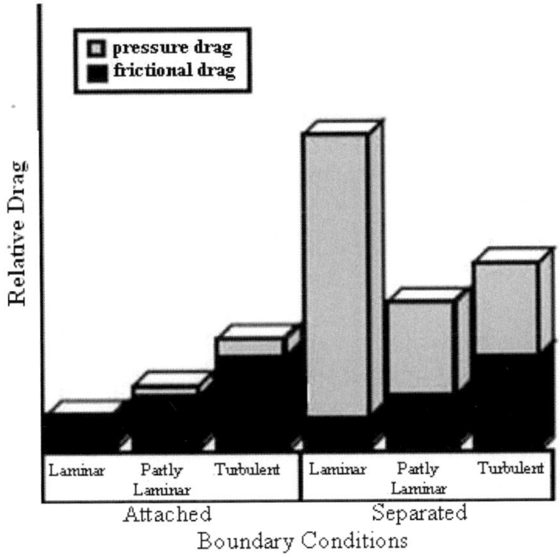

Fig. 6. Relative drag associated with boundary conditions.[21] Reproduced by permission from Dr. Frank Fish.

of many streamlined fish. The curvature effect on the anterior part of the body, which results in convective flow acceleration and a favorable pressure gradient, is known to delay boundary layer transition.[27,28] Second, during acceleration, a key maneuver in the predator–prey interaction, the boundary layer is subjected to a time varying favorable pressure gradient additionally contributing to transition delay.[28,29]

To give an estimate of the range of acceleration experienced over fast-swimming sharks, a recent observation of a shortfin mako (estimated to be capable of reaching speeds upwards of 20 m/s) is herein described. Starting from almost a stand still a small (1.2 m long) shortfin mako in 3 s covered a distance of roughly 44 m to reach a towed bait, which corresponds to an average acceleration over the maneuver of 5 m/s^2.[30] In addition, induced thinner boundary layers from transverse motion may result in local relaminarization and thus lower skin friction.[31] As seen in Fig. 6, laminar flow results in lower skin friction, but potentially higher pressure drag as it is more prone to separation. Thus, a turbulent boundary layer can reduce pressure drag if it helps to delay flow separation from the body.[28,32] Finally, Blake[33] concluded that, "further experimental studies on the possible role on drag reduction in thunniform swimming fishes (i.e. shortfin mako shark and tuna) of structures such as keels, finlets, and scale corselets would be worthwhile". As described above, the trade-off to achieve maximum drag reduction is complex.

3.2. *Skin-Friction Reduction*

Many researchers would contend that the best method to achieve skin-friction reduction would be a unique surface geometry, where no additional energy is required. In the case of a fully turbulent boundary layer, researchers have demonstrated drag reduction on the order of 10% through unique man-made surface geometries. The most well-known of these is riblets (ridges) or streamwise grooves, considered to be first inspired by the ridges found on shark denticles (Fig. 3). Bechert *et al.*[34] performed exhaustive experiments with adjustable geometries in optimizing both riblet spacing and cavity geometry. The earliest success with riblet designs reported that sawtooth riblets with a triangular cross-section achieved a 5.1% reduction in drag. Bechert *et al.*[34] further altered the shape of the riblets and concluded that slit/blade riblets were the optimal geometry, with a record value of 9.9% reduction in turbulent skin-friction drag (Fig. 7).

How is it that a rough surface, consisting of streamlined grooves, can result in a reduction in skin-friction drag? To answer that question, one must comprehend the flow field present in a turbulent boundary layer, and an understanding of the coherent vortical structures forming within the flow over a smooth surface is critical.

Drawing on the results of numerous experimental and computational results, an overall model for the coherent vortical structures and their formation/disruption within a turbulent wall-bounded flow is now available. Perhaps the most in-depth look at such an up-to-date model is given by Bernard and Wallace.[36] A wall-bounded turbulent flow is made up of a series of horseshoe or hairpin shaped vortices in various stages of growth/decay. As these vortices grow and interact, one may find vortex disconnection/re-connection and amalgamation occurring with neighboring vortices resulting in what may appear as disjointed or only half-horseshoe vortices as has been observed in numerical simulations of turbulent channel flows.[37] Predominantly, negative vorticity is formed in a boundary layer (i.e. fluid further away from the surface is moving faster than that closer to the surface due to the no-slip condition, and this results in an overall rotational characteristic of the flow in the clockwise direction for flow moving from left to right). The intricacies of this flow model are still an area of debate, but there are some key characteristics that are widely agreed upon and have been observed theoretically, computationally, and experimentally.

First, if y is defined as the distance away from the wall, all dimensions within the boundary layer can be nondimensionalized by a viscous length scale given as ν/u_* where u_* is the friction velocity and ν is the kinematic viscosity of the fluid. This leads to a length scale away from the wall defined as $y^+ = u_* y/\nu$. The turbulent boundary layer flow can now be divided into three main regions. These are the viscous layer (which is also made up of both the viscous sublayer ($0 < y^+ < 5$) closest

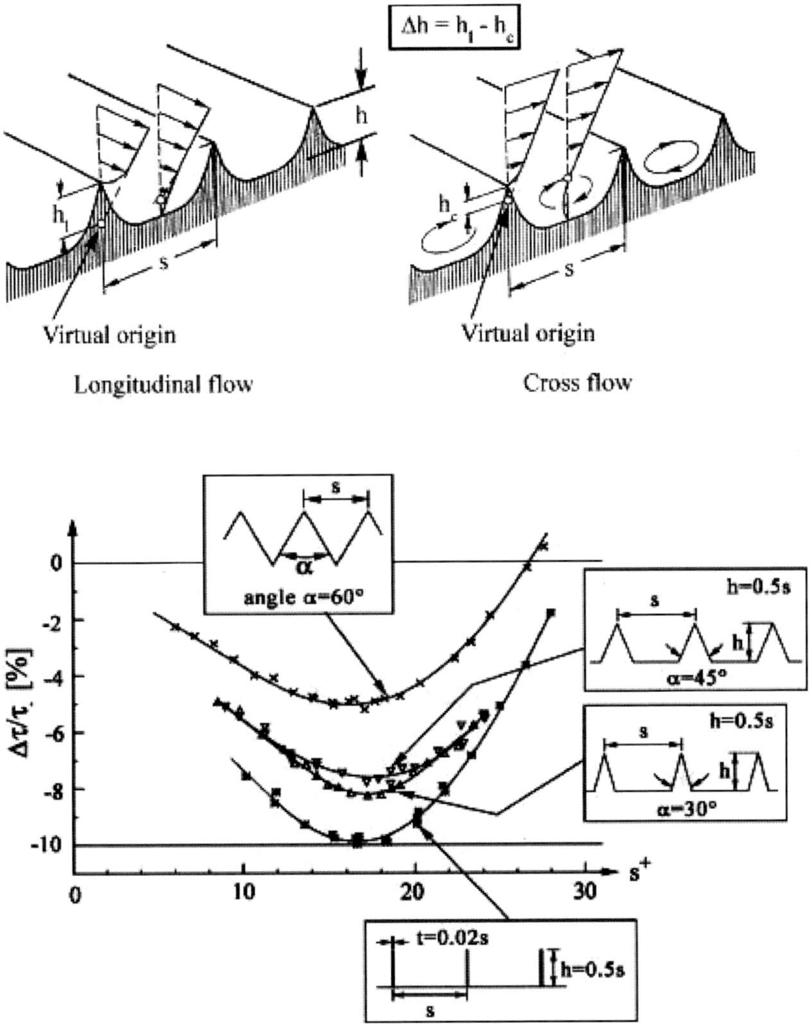

Fig. 7. Above: longitudinal and cross-flow on a ribbed surface; below: drag-reduction performance of various rib geometries. Reprinted with permission from Bechert *et al.*[35] by Springer.

to the wall and the buffer layer ($5 < y^+ < 30$)), the intermediate layer, and the outer layer (sizes of the latter two dependent on Reynolds number (*Re*)). It has been shown that the viscous sublayer has a mean linear velocity distribution throughout most of the region, and it is within this region that viscous forces are dominant and flow fluctuations leading to turbulent stresses (Reynolds stresses) are negligible. The buffer layer is categorized by a rising Reynolds stress away from the wall (with the peak *Re* stress occurring at about $y^+ = 14$), and it is within this region that both viscous and Reynolds stresses contribute to the flow. At the outer end of the buffer layer, the mean velocity has typically only reached approximately 55% of the mean flow magnitude (U). Within the intermediate (fully turbulent) layer, the flow is dominated more by Reynolds stresses while viscous stress is negligible. Finally, the outer layer is categorized by intermittent regions of turbulent and laminar flow and extends out to the thickness of the boundary layer, δ. This is also the region where mixing of the outer flow with the turbulent eddies primarily takes place, leading to the rapid growth of turbulent boundary layers compared to the laminar ones. The buffer layer is considered to be home by many to the most interesting dynamical processes within the flow, including turbulence creation.[36] Therefore, roughness elements that have a variation in height which penetrate into the buffer layer could certainly affect the process by which turbulence is produced.

The edge of the viscous layer (at $y^+ = 5$) is home to a key characteristic of wall-bounded flows; these are low-speed streaks orientated in the streamwise direction. These streaks are typically spaced apart with an average of 100 viscous length scales and have a length of about 1000 viscous length scales. They have also been observed to migrate away from the wall, and oscillate in the buffer region at $8 < y^+ < 12$, and break up when penetrating deeper into the buffer layer for $10 < y^+ < 30$. This process of lifting low-speed streaks is known as a turbulent burst. Nearly all the turbulent kinetic energy production is assumed to be a result of this burst process which can penetrate as high as $y^+ = 100$.[36] The converse event, which is intricately connected with the turbulent burst, is the turbulent sweep. If fluid is moving away from the wall in a burst, due to conservation of mass one could argue (assuming the flow is statistically on average self-similar in the spanwise direction) that somewhere nearby higher momentum fluid must move towards the wall. Sweeps have been observed to penetrate deep into the viscous sublayer, and it is this process that is primarily responsible for the high-skin friction associated with turbulent boundary layer flows.

The formation of the low-speed streaks and the lifting of the streaks away from the wall is attributed to the presence of streamwise vortices (also with a typical streamwise extent of 1000 viscous length scales), that when paired produce an upwash that may pull low-speed fluid away from the wall. These pairs, that have been observed to form in the region of $14 < y^+ < 50$ (with the most energetic at $14 < y^+ < 25$), are also orientated at a slight angle (3°–7°) to the wall with their downstream end further off the wall. The success to date with riblets appears to be due to inhibiting their interaction with the surface (drag reduction). However, if not spaced optimally then drag enhancement can occur.

A recent experimental study on the formation of streamwise vortices in the case of both drag-enhancing and drag-reducing riblets was performed by Lee and Lee.[38] They measured the flow field inside both types of riblet spacings, and used smokewire visualizations to observe the streamwise vortex formation within a turbulent boundary layer. In the case of drag enhancing riblets (achieved experimentally by increasing the velocity of the flow which increased the shear stress and thus decreased the size of a viscous length scale), on average stable streamwise vorticity was found to persist within the riblet valleys. These vortices caused a downwash of high-speed fluid into the valleys increasing the skin friction. Their case of drag-reducing riblets (although not the optimal spacing as reported by Bechert *et al.*[34] of $s^+ = 15$ but rather slightly higher at $s^+ = 25.2$) showed that most of the streamwise vortices were observed to stay above the riblets, with secondary vorticity often forming off of the riblet tips with the interaction of the primary streamwise vortices. The riblet valleys saw a decrease in the rms velocity fluctuations compared to the smooth case near the surface. Thus, even though the riblets result in an increase in surface area a net reduction in drag results. It should be noted that optimal riblet heights are on the order of half the spacing, which for the optimal spacing of $s^+ = 15$ results in the riblet tips just penetrating into the buffer layer.

The other method, by which a reduction in skin friction over a surface may be achieved, is to delay the transition of the boundary layer from the laminar to turbulent state, better known as laminar flow control (LFC). A thorough review of many of the attempts to date has been given by Joslin.[39] The primary methods whereby laminar flow is maintained are through the use of body shaping, extending the region of favorable pressure gradient and suctioning off the low-speed flow in the boundary layer. There are other possible passive mechanisms that show potential to delay boundary layer transition. Compliant walls, perhaps similar to the dolphin skin, have been shown theoretically to dampen the Tollmein–Schlichting (TS) waves leading to transition.[28,40,41] Grek *et al.*[42] found that riblets caused a delay in the late stages of the transition process by suppressing high-frequency traveling waves occurring in the later stages of transition. Although the TS waves are excited at a slightly lower critical *Re* over a riblet surface, riblets have been found to cause a delay in the later stages of transition for both two-dimensional and three-dimensional boundary layers.[43] Neumann and Dinkelacker[44] observed up to 13% drag reduction on a body of revolution with the application of riblets leading to observable transition delay. Thus, the riblets found on shark denticles may also act as a transition delay mechanism.

3.3. *Separation Control*

Separation of the boundary layer from the body typically occurs in vicinities where the flow is decelerating due to change in body curvature, resulting in an adverse pressure gradient. As a result, separation typically occurs in areas posterior of the maximum body thickness. Incipient separation is characterized by regions of decreasing skin friction approaching zero, and consequent reversal of the flow at the surface.[45] A similar process, known as dynamic stall, characterizes unsteady separation from a moving surface producing lift (i.e. such as a pitching airfoil) or thrust (i.e. oscillating caudal fin). Unsteady separation is characterized by a locality where both the shear stress (or skin friction) and velocity approach zero as seen by an observer moving with the separation point, known as the MRS criterion.[28] In this case, a separated region is most likely to occur near the point of highest curvature (typically near the leading edge) prior to blending with the wake near the trailing edge. If such separation occurs in the latter case, it has been found that lower propulsive efficiencies result.[46] However, if the unsteady separation process can be controlled, such that the leading edge separation bubble remains disconnected with the wake, then an unsteady high-thrust (or high-lift) generation mechanism could occur.[28]

When three-dimensionality is added to the separation flow kinematics, boundary layer separation does not always coincide with a point of zero shear stress at the wall. In fact, the shear stress may vanish only at a limited number of points along the separation line, and a convergence of skin-friction lines onto a particular separation line is required for separation to occur. As a result, 3D boundary layers can be more capable of overcoming an adverse pressure gradient without separating.[28] Flow over a shark's body is largely three-dimensional in nature.[3] If the denticles on shark skin prevent this required convergence of skin-friction lines, they might passively act to keep the flow attached thereby reducing pressure drag.

When considering drag reduction for aquatic locomotion, the issue of body curvature cannot be neglected. Swimming kinematics in thunniform fish are characterized by cyclically repeating motions and small linear and angular accelerations.[33] Most fast-swimming sharks, such as the shortfin mako, are thunniform swimmers where oscillations are limited to the posterior end of the body. In reporting studies of swimming styles, Anderson[25] makes the statement that, "the occasional appearance of incipient separation (in carangiform swimmers) and subsequent stabilization may be evidence of complex flow manipulation on the part of the fish ... Avoiding separation, a fish essentially eliminates form drag and increases the effectiveness of the caudal fin in thrust production ... Perhaps fish tune their swimming movements to take advantage of the lowered shear stress of a nearly separating boundary layer while simultaneously benefiting from the reduced form drag and increased lift of fully attached flow. The inflected boundary layer profiles observed may be an example of the fish 'pushing the envelope'..." The unique experimental measurements made by Anderson[25] of the boundary layers forming on real (carangiform and anguilliform) swimming fish were performed at fairly low-swimming speeds. The boundary layer characteristics observed may thus be accentuated at higher swimming speeds and higher frequency posterior oscillations. It is feasible that thunniform swimming sharks need a mechanism to "push the envelope" as described by Anderson[25] to control flow separation and decrease drag; and, it is a realistic hypothesis that this boundary layer control mechanism resides in the skin microstructure. In addition, video evidence of a swimming shortfin mako[30] pursuing bait show the shark's ability to turn in one direction and then reverse direction before the entire body completed the initial turn, a turning behavior not previously analyzed which is labeled *contragility*. This is a maneuver that requires low form drag and large muscular effort and control (personal communication with Dr. Frank Fish, West Chester University).

Delaying separation of the flow from a solid boundary not only results in reduced pressure drag, but also decreased pressure losses in ducted flows such as through diffusers and turning elbows. As a result, various mechanisms by which separation can be controlled have been investigated and successfully applied. A key to successful separation control is imparting momentum to the very near-wall region of the flow, thus preventing flow reversal.[28] One of the most successful passive means to date has been the use of vortex generators, or small typically v-shaped protrusions with profiles less than half the boundary layer thickness. It has been speculated that bristled shark skin denticles could act as vortex generators.[35] Man-made vortex generators have been shown to produce a system of streamwise vortices which mix high- and low-momentum fluid, thus energizing the flow close to the surface. Vortex generators need to be placed at a specific downstream location within a turbulent boundary layer for maximum performance such that the streamwise vortices affect the region where separation would normally occur.[47,48] Patterned surfaces can also result in separation control and golf ball dimples present one of the most well-known illustrations of surface patterning resulting in separation control with a subsequent reduction in drag. However, the dimples do more than just trip the boundary layer to the turbulent state. It has been shown that the formation of embedded cavity vortices, or small localized regions of separation within the surface allowing the outer boundary layer flow to skip over the dimples, is a crucial aspect to this mechanism of separation delay.[49] Thus, it can readily be perceived that the use of patterned surfaces, capable of imposing partial-slip flow conditions at the wall due to the formation of embedded vortices, is a viable approach for achieving drag reduction via separation control. One hypothesis explored in the following section is that a bristled shark skin results in the formation of an array of embedded vortices which may result in a means for the shark to inhibit or delay separation.

4. Drag-Reducing Capabilities of the Skin on Fast-Swimming Sharks

Studies of real shark skin have shown that faster shark species have riblet spacing consistent with experimental observations leading to turbulent skin-friction reduction[1,5] if burst velocities are considered. Over the years there have been several attempts to test various models of shark skin experimentally. Gren[50] built one of the first shark skin plastic models for experimental testing in an oil channel, allowing for change in angle of attack of the denticles. However, he used the scales from a rather slow shark, the spiny dogfish, and determined shear stress from the velocity measurements. About the same time, Bechert *et al.*[51] built shark skin models for wind tunnel testing with various angles of attack for the denticle crowns. Shear stress measurements for the various models, using a wind tunnel balance, were obtained. However, they admit that their plastic cast replicas had various limitations, including the inability to fully model the gaps between the scales. They observed that at higher angles of attack, the shear stress increased dramatically.

The most recent experimental study performed by Bechert and colleagues[35,52] attempted to account for denticle bristling with compliant anchoring of the denticle scales. In this study, a fully turbulent oil channel facility was used based on the assumption that fully turbulent flow takes place over a large majority of the shark's skin. They meticulously built 800 small replicas of a hammerhead (*S. zygaena*) shark denticle scaled up 100 times (Fig. 8). They also accounted for various tensions by which the scales were held by changing the rigidity of the springs attached to the scales, and shear stress measurements were obtained with a specially designed force balance.[53] Their replica skin could only achieve at best, a 3% reduction in skin-friction (τ_w) drag with the denticles laid flat, and reported an increase in drag when the denticles were bristled (Fig. 8). In this case, the denticles laid flat corresponded to a less efficient riblet surface and worked only in the drag reduction range of approximately 10–20s^+, consistent with all previously reported work on riblets. Work was also performed to consider whether a system of staggered, discrete riblets (as found on shark skin) would perform better than single, long riblets extending the entire surface. However, they reported no significant improvement with this design.[54]

More recent work by the author and colleagues,[55] has looked closer at the flow field induced inside a bristled shark skin model. To date, most of the work on shark skin models has been confined to force (shear) measurements when submerged in a fully turbulent boundary layer. However, if shark skin denticles are truly capable of bristling, what is the flow induced by such a geometry? To answer this question, a series of experiments, using a water tunnel facility and the Digital Particle Image Velcocimetry (DPIV) measurement technique, were conducted. A bristled shortfin

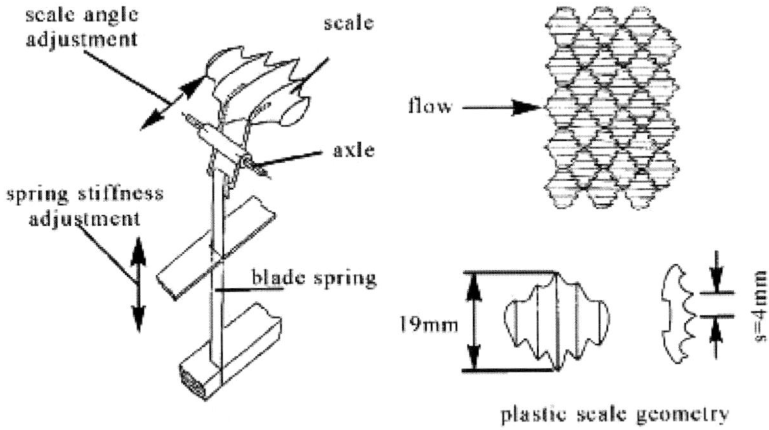

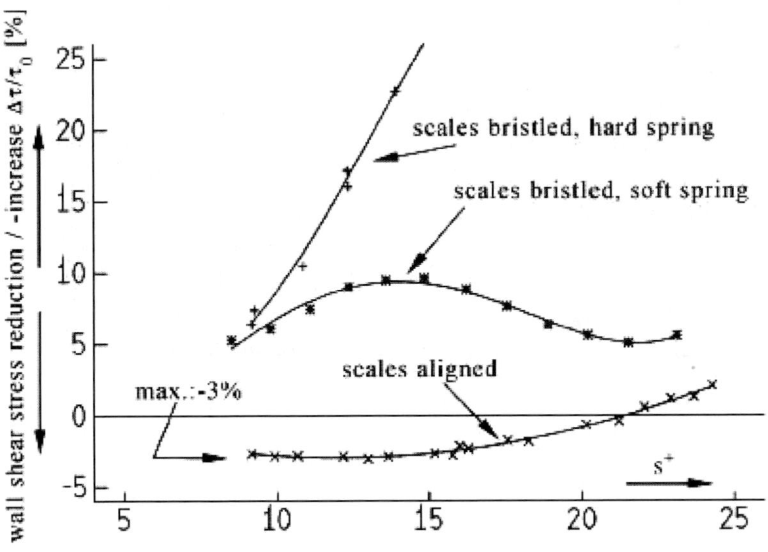

Fig. 8. Above: plastic scale and suspension mechanism; below: shear stress data obtained over the shark skin model. Reprinted with permission from Bechert *et al.*[35] by Springer.

mako shark skin was modeled; this choice was based on its status as the fastest species of shark[3,30] as well as evidence suggesting that its scales are moveable to the touch on dead specimens.[52] Using a rapid-prototype machine, shortfin mako denticles forming a 16 × 24 array were scaled up from 0.2 mm on the shark to 20 mm for the model (Fig. 9). The model was based on measurements reported in literature of short fin mako

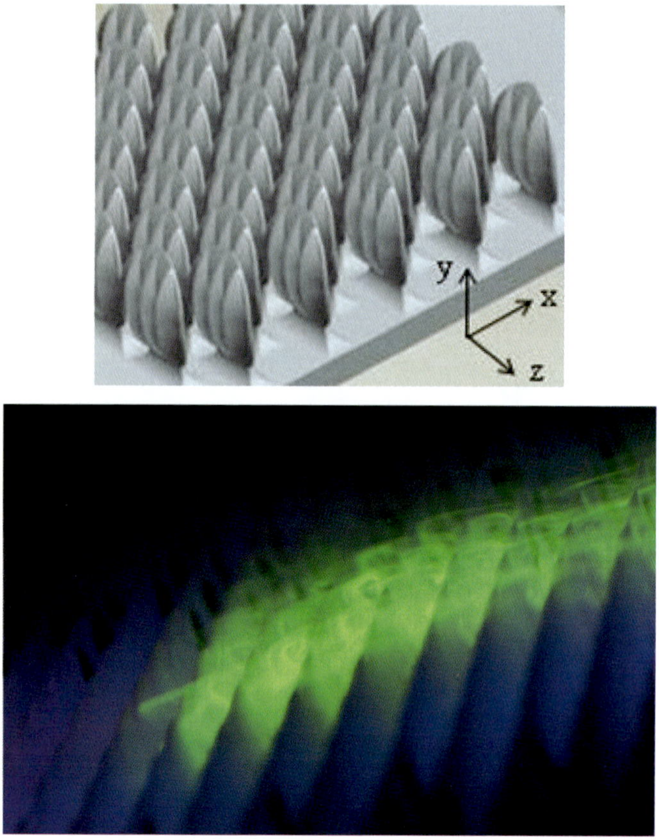

Fig. 9. Top: rendering of the bristled shark skin model; bottom: dye visualization of embedded vortices forming within the bristled shark skin model (flow left to right).

denticles,[3,5] and an extreme angle of bristling was chosen for these initial experiments to aid in gaining maximal optical access to the model for cavity flow measurements. Similarity of the cavity flow was achieved by matching the $Re \sim 3000$ based on crown length of the shark skin and model; the scale up in size of 100 times is countered by a scale down in velocity of 100 times with negligible change in viscosity. A long flat plate was used to grow the boundary layer sufficiently thick such that shear layer instabilities over the cavity vortices were not observed to develop, as would occur in the case of the real shark skin (nose and leading edges of fins have smooth, immoveable denticles). Flow visualization (Fig. 9) confirmed the presence of embedded cavity vortices within the model.

It was observed that the cavity vortices generated within the shark skin microgeometry represent a controlled formation of primary and secondary vorticity within the model. First, as shown in Fig. 10, a primary cavity vortex forms in the area ahead of each denticle peak. This occurs for both laminar and turbulent boundary layer conditions above the model, but for the turbulent boundary layer conditions, the vortex is more deeply embedded within the geometry. The ribs on the face of the denticles resulted in the formation of secondary vorticity which was predefined by the geometry (Fig. 11). After DPIV data was acquired for various measurement planes within the model cavity, we reconstructed the dominant flow field as shown in Fig. 12. There exists an interlocking web of vorticity embedded within the scale surface, with the strongest vorticity forming in front of each scale peak.

Of particular interest as well is that, as the flow accelerates as it passes over the cavity between the denticles, it reaches speeds on the order of 5–20% of the freestream flow (U) and has an average velocity in the $y = 0$ plane of 0.03 U for laminar boundary layer conditions. In the purely flat surface case, the no slip condition at $y = 0$ enforces a zero velocity boundary condition to the boundary layer flow. It was also observed that the flow velocity at the streamline separating the cavity flow from the outer boundary layer flow further increases concomitantly with a decrease in the boundary layer thickness, δ (in the case presented $\delta = 21$ mm, or roughly the same size as the crown length and thus a fairly thick boundary layer is used which is close to mimicking δ/d for real shark skin). In the case where the boundary layer is tripped prior to the model, this increases to an average velocity in the $y = 0$ plane of $0.10U$ owing to the higher momentum closer to the surface from the presence of the turbulent boundary layer above the denticle model. Peak velocities at $y = 0$ measured on the order of 28% of the freestream flow. Periodic exchange of fluid was observed in the turbulent boundary layer case between the cavity flow and boundary flow, but on average the flow displays the primary embedded cavity vortex. These results indicate that a bristled shark skin microgeometry leads to higher momentum in the fluid closer to the wall for both laminar and turbulent boundary layer conditions making the surface a good candidate as a mechanism for separation control.

It is interesting to note that the bristled shark skin surface as tested in the above study appears to provide a different mechanism for separation control than the previously hypothesized vortex generators.[35] If embedded vortices are indeed forming in the bristled shark skin microgeometry, the mechanism for separation control closely resembles that of surface patterning consisting of embedded cavities, i.e. dimples on a golf ball. In this case, it is the high-partial-slip velocities over the embedded cavities in the shark skin which appear to maintain high-momentum flow close to the wall. In addition, the riblets appear to serve an additional purpose in promoting the formation of a three-dimensional, yet stable, embedded vortex flow field inside the skin.

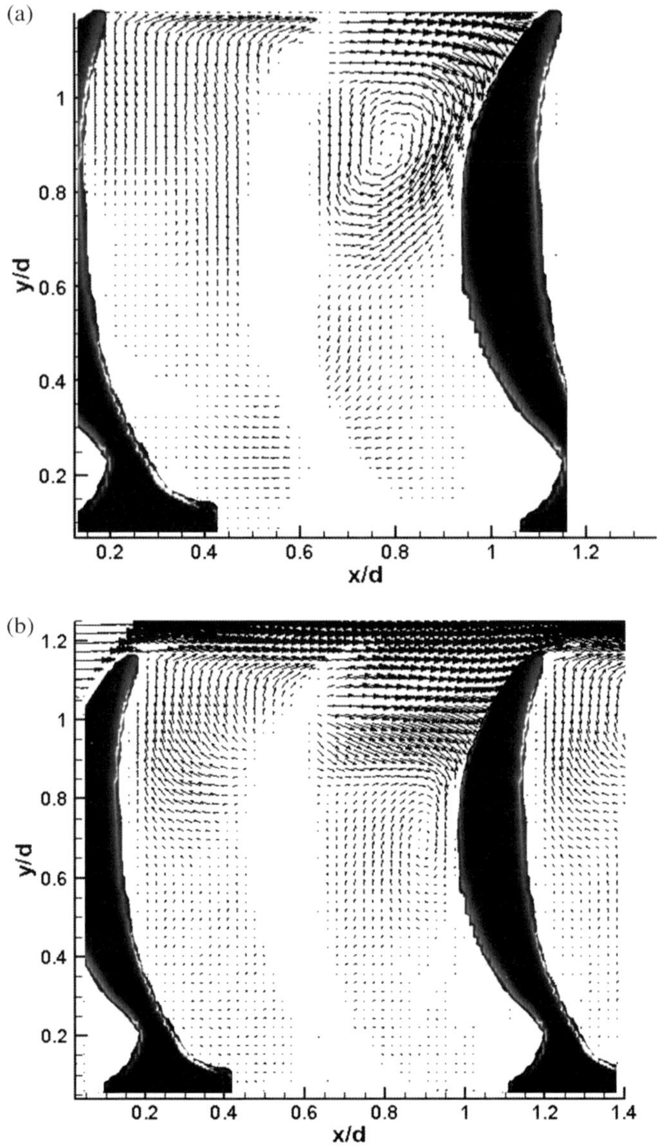

Fig. 10. DPIV vector field forming inside the cavity in a plane passing through a denticle peak for (a) laminar and (b) time-averaged turbulent boundary layer conditions. The formation of the primary vortex is evident for each case. The blank space in the image is due to blockage from the adjacent row of denticles.

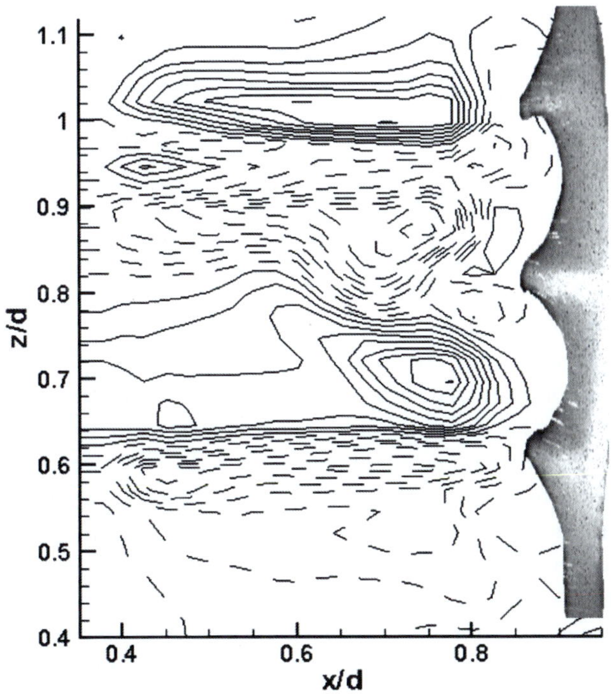

Fig. 11. Vorticity field forming inside the cavity in a plane located at 5.25 mm below the denticle peak (20% cavity depth) under laminar boundary layer conditions. Positive and negative vorticity is indicated by solid and dashed lines, respectively.

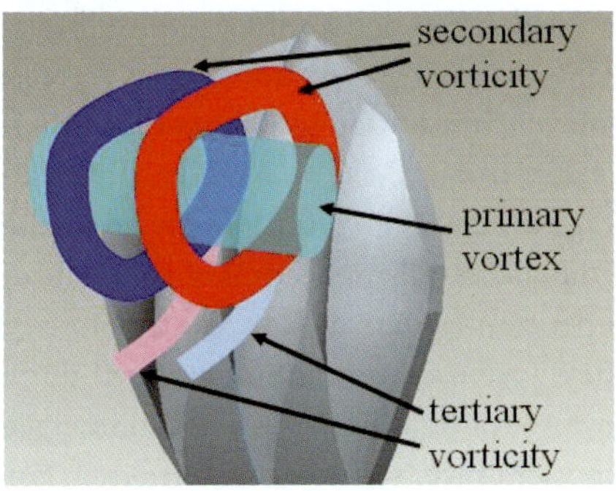

Fig. 12. Schematic of vorticity forming in the cavity in front of each denticle tip. Red and blue coloring designates positive and negative vorticity respectively.

This partial-slip condition, over a proportionately large surface area felt by the flow passing over the shark's skin, inhibits flow reversal which would otherwise lead to separation. Thus, the overall effect for the shark would be a reduction in pressure drag.

5. Summary With Technological Applications

Thus far it has been shown that the shark skin itself has a demonstrated ability to reduce turbulent skin-friction drag on the order of 3% for turbulent boundary layer conditions. In addition, in the case of faster shark species where denticle bristling is feasible, drag reduction may also be achieved through separation control/delay from the development of embedded vortices forming within the bristled shark skin micro-geometry. This may also have the added benefit of increasing contragility, allowing for quicker turning when in pursuit of prey. In the field of biomimetics, engineers are constantly turning to biological observations to provide inspiration to overcome technological obstacles, and this is especially true in the case of drag reduction. Thus far, there have been several attempts to mimic the shark skin pattern whereby a reduction in drag would be achievable for modern modes of transportation.

The first case is that of riblets as applied to sailing boats and aircraft. In the case of aircraft, testing began in the 1980s. One test reported a 1.6% reduction in drag through the use of V-grooved riblets on a 1/11th scale Airbus A320 model, where 2/3 of the wetted model area was covered with the streamwise grooves.[56] As previously discussed, Bechert *et al.*[34] performed exhaustive experiments with optimizing riblet spacing and cavity geometry (Fig. 7). The optimal geometry was deduced to be the slit/blade design, which would not prove optimal for aircraft coatings due to durability. Thus they devised and tested a trapezoidal riblet groove that gave an 8.2% reduction in drag, and working with 3M, a riblet tape for aircraft was manufactured.[57] After all secondary effects are removed, in addition to the fact that only 50% of the drag on long range aircraft is due to skin-friction, the result is an optimal overall drag-reduction potential of about 3% for a typical aircraft.[34] There are additional scattered experimental observations of the successful application of riblets to aircraft.[28] Finally, riblets have been used in competition to date at least twice before being banned by racing officials. Reportedly, they were used on the rowing shell during the 1984 Olympics by the United States rowing team, as well as on the submerged hull of the winner of the 1987 America's Cup yacht race, the *Stars and Stripes*.[28] Finally, more recent work is being performed to explore the feasibility of manufacturing riblet surfaces for turbulent skin-friction reduction over fan, turbine, and compressor blades.[58] However, the biggest obstacles to the sustained use of riblets are their aging and contamination issues, particularly bio-fouling over time in marine applications.[59] It is interesting to note that sharks appear to have overcome both of these problems as their skin appears to stay free of bio-fouling, possibly in part due to the topography and movement of the individual scales during swimming, and they have an inherent biological replacement mechanism whereby lost denticles are quickly re-grown.

This observation points to other interesting research taking place to mimic the shark skin microgeometry for drag-reduction purposes, however in this case to devise a surface that resists marine bio-fouling. With the banned use of apparently hazardous chemical coatings, ocean-going vessels are in need of a new type of coating to prevent bio-fouling of the hull. Carmen *et al.*[60] have designed a biologically inspired textured surface known as Sharklet AF™. This engineered microtopography, which closely resembles the riblet pattern found on shark skin, has been demonstrated to inhibit the settlement of spores of a marine alga (Fig. 13).

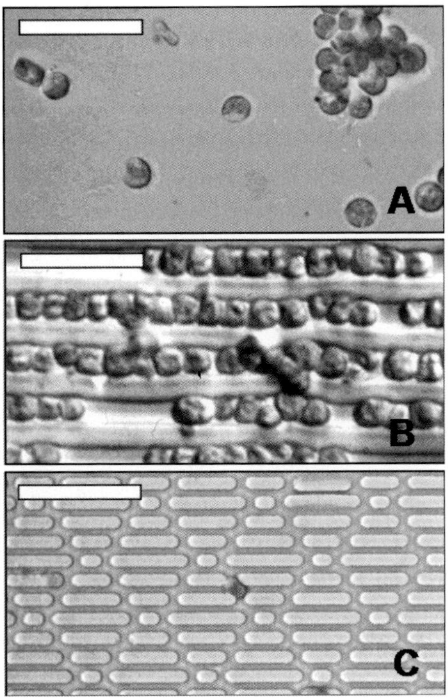

Fig. 13. Images of Ulva settlement on (A) smooth, (B) 5 μm wide, 5 μm spaced, and 5 μm deep channels, and (C) 5 μm deep Sharklet AF™ in PDMSe. Images were taken via light microscopy. Scale bars represent 25 μm. Taken from Carmen *et al.*[60] and reproduced by permission from Taylor & Francis.

Finally, there have been several claims by swimsuit manufacturers that mimicking the shark skin can lead to drag reduction, and thus faster race times, for competitive swimmers. Since the late 1990s, Speedo™ FASTSKIN swimsuits have entered the market; the latest is the FASTSKIN LZR reportedly developed with the help of NASA scientists.[61] The latest suits are made of a specially bonded fabric, where wind tunnel tests were performed to determine which fabric and weave gave the lowest drag. The fabric grooves (mimicking riblets) aligned with the flow could conceivably act to reduce turbulent skin-friction drag if sized correctly. However, the major decrease in drag by the suits also reportedly comes from further streamlining of the body and decreasing flow separation. The LZR Pulse fabric is applied particularly to drag-prone parts of a swimmer's body identified through CFD analysis.[61] Other features, such as body compression thereby controlling skin vibration, may also play a role. However, the overall effect according to the manufacturer is a 10% decrease in drag as compared to their previous model. Since the advent of these engineered swimsuits, scientific studies have been performed, some reporting no advantage to the suits.[62] However, a recent study which analyzed the flow over a swimmer's body observed the flow to be neither laminar nor completely turbulent, and that suits which trip the flow to turbulent may increase skin-friction drag but would favorably decrease pressure drag resulting in an overall decrease in drag for a swimmer wearing such a suit.[63]

Overall, it appears evident that fast-swimming sharks, such as the shortfin mako, have developed a skin for optimal boundary layer control to increase peak swimming speeds and contragility for pursuit of prey. The future will determine if biomimetic adaptations of the shark skin in applications including aircraft and marine vessels, among others, will lead to increased levels of boundary layer control that can decrease drag and increase maneuverability for modern technologies.

Acknowledgments

Support for the research, reported by the author, through an NSF SGER grant CTS-0630489, Lindbergh Foundation Grant and a University of Alabama RAC grant is gratefully acknowledged. Research assistant, Pablo Hidalgo, aided in this research.

References

[1] Raschi W, Tabit C. (1992) Functional aspects of placoid scales: A review and update. *Aust J Mar Freshwater Res* **43**: 123–147.

[2] Reif W, Dinkelacker A. (1982) Hydrodynamics of the squamation in fast-swimming sharks. *N Jahrb Geol Palaontol* **164**: 184–187.

[3] Reif W. (1985) Morphology and hydrodynamic effects of the scales of fast swimming sharks. *Fortschritte der Zoologie* **30**: 483–485.

[4] Bechert DW, Hoppe G, Reif W. (1985) On the drag reduction of the shark skin. AIAA Shear Flow Control Conference, 12–14 March, Boulder, Colorado, AIAA Paper 85-0546.

[5] Raschi W, Musick J. (1986) Hydrodynamic aspects of shark scales. NASA Contractor Rep. 3963.

[6] Bushnell D, Moore K. (1991) Drag reduction in nature. *Ann Rev Fluid Mech* **23**: 65–79.

[7] McCormick HW, Allen T, Young W. (1963). *Shadows in the Sea, Sharks, Skates and Rays.* Weathervane Books, New York.

[8] Robinson BW. (1971) *The Arts of the Japanese Sword.* Faber and Faber, London.

[9] Hertwig O. (1874) Ueber den Bau der Placoidschuppen und der Zahne der Selachier. *Jenaische Zeitschrift fuer Naturwissenschaft 8 (N.F. I)* **33**: 1–404.

[10] Reif W. (1985) *Squamation and Ecology of Sharks.* no. 78, Courier Forschungs-Institut Senckenberg, Frankfurt am Main.

[11] Applegate SP. (1967) A survey of shark hard parts. In: PW Gilbert, RF Mathewson & DP Rall (eds), *Sharks, Skates, and Rays*, pp. 37–67. Johns Hopkins Press, Baltimore.

[12] Reif W. (1978) Protective and hydrodynamic function of the dermal skeleton of elasmobranchs. *Neues Jahrbuch fur Geologie und Palaontologie Abhandlungen* **157**: 133–141.

[13] Bone Q, Howarth JV. (1966) Report to council 1966–67. Marine Biological Association of the United Kingdom 47, 19. [Abstr.]

[14] Walsh MJ, Weinstein LM. (1978) Drag and heat transfer on surfaces with small longitudinal fins. American Institute of Aeronautics and Astronautics, Paper No. 78–1161.

[15] Walsh MJ. (1982) Turbulent boundary layer drag reduction using riblets. American Institute of Aeronautics and Astronautics, Paper No. 82-0169.

[16] Bechert DW, Bruse M, Hage W. (2000) Experiments with three-dimensional riblets as an idealized model of the shark skin. *Exps Fluids* **28**: 403–412.

[17] Videler J. (1995) Body surface adaptions to boundary-layer dynamics. *Symp Soc Exp Biol* **49**: 1–20.

[18] Bruse M, Bechert D, van der Hoeven J, Hage W, Hoppe G. (1993) Experiments with conventional and with novel adjustable drag-reducing surfaces. *Proc Int Cong Near-Wall Turbulent Flows*, Tempe, AZ, 719–738.

[19] Bechert D, Bartenwerfer M, Hoppe G. (1986) Drag reduction mechanisms derived from the shark skin. *ICAS Congress*, 15th, London, 1044–1068.

[20] Wainwright S. (1978) Shark skin: A function in locomotion. *Science* **202**: 747–749.

[21] Fish F. (1998) Imaginative solutions by marine organisms for drag reduction. *Proc Int Symp Seawater Drag Reduction*, 1–8.

[22] Vogel S. (2003) *Comparative Biomechanics: Life's Physical World.* Princeton University Press, Princeton.

[23] Fish F, Lauder G. (2006) Passive and active flow control by swimming fishes and mammals. *Ann Rev Fluid Mech* **38**: 193–224.

[24] Donley JM, Sepulveda CA, Konstantinidis P, Gemballa S, Shadwick RE. (2004) Convergent evolution in mechanical design of lamnid sharks and tunas. *Nature* **429**: 61–65.

[25] Anderson E, McGillis W, Grosenbaugh M. (2001) The boundary layer of swimming fish. *J Exp Biol* **204**: 81–102.

[26] Schulz W, Webb P. (2002) Power requirements of swimming: do new methods resolve old questions? *Integr Comp Biol* **42**: 1018–1025.

[27] Hansen R, Hoyt J. (1984) Laminar-to turbulent transition on a body of revolution with an extended favorable pressure gradient forebody. *ASME J Fluids Eng* **106**: 202–210.

[28] Gad-el-Hak, M. (2000) *Flow Control: Passive, Active and Reactive Flow Management.* Cambridge University Press, Cambridge.

[29] Rohr J, Latz M, Fallon S, Nauen J, Hendricks E. (1998) Experimental approaches towards interpreting dolphin-stimulated bioluminescence. *J Exp Biol* **201**: 1447–1460.

[30] Thorburn, C. (2007) Personal communication. Video evidence available in *Animal Nation: Mako Sharks* available on DVD.

[31] Narasimha R, Sreenivasan K. (1973) Relaminarization in highly accelerated turbulent boundary layers. *J Fluid Mech* **61**: 187–199.

[32] Videler J. (1993) *Fish Swimming.* Chapman and Hall, London.

[33] Blake R. (2004) Fish functional design and swimming performance. *J Fish Biol* **65**: 1193–1222.

[34] Bechert D, Bruse M, Hage W, Van der Hoeven J, Hoppe G. (1997) Experiments on drag-reducing surfaces and their optimization with an adjustable geometry. *J Fluid Mech* **338**: 59–87.

[35] Bechert D, Bruse M, HageW, Meyer R. (2000) Fluid mechanics of biological surfaces and their technological application. *Naturwissenschaften* **80**: 157–171.

[36] Bernard P, Wallace J. (2002) *Turbulent Flow: Analysis, Measurement & Prediction.* John Wiley & Sons, Inc., Hoboken, NJ.

[37] Liu ZC, Adrian RJ. (1999) Evidence for hairpin packet structure in DNS channel flow. *Proc First Int Symp Turbulent Shear Flow Phenomena*, 609–614.

[38] Lee SJ, Lee SH. (2001) Flow field analysis of a turbulent boundary layer over a riblet surface. *Exps Fluids* **30**: 152–166.

[39] Joslin R. (1998) Aircraft laminar flow control. *Ann Rev Fluid Mech* **30**: 1–29.

[40] Davies C, Carpenter P. (1997) Numerical simulation of the evolution of Tollmien-Schlichting waves over finite compliant panels. *J Fluid Mech* **335**: 361–392.

[41] Pavlov V. (2006) Dolphin skin as a natural anisotropic compliant wall. *Bioinsp Biomim* **1**: 31–40.

[42] Grek G, Kozlov V, Titarenko S. (1995) The influence of riblets on a boundary layer with embedded streamwise vortices. *Phys Fluids* 7(10): 2504–2506.

[43] Choi K. (2000) European drag-reduction research — recent developments and current status. *Fluid Dyn Res* **26**: 325–335.

[44] Neumann D, Dinkelacker A. (1989) Drag reduction by longitudinal riblets on the surface of a streamwise aligned body of revolution. In *Drag Reduction in Fluid Flows*, Ellis Horwood.

[45] Doligalski T, Smith C, Walker J. (1994) Vortex interactions with walls. *Ann Rev Fluid Mech* **26**: 573–616.

[46] Tuncer I, Platzer M. (1999) A computational study of flow separation characteristics and wake profiles behind a flapping airfoil. AIAA paper 99-0648.

[47] Lin J. (1999) Control of turbulent boundary-layer separation using micro-vortex generators. AIAA paper 99-3404.

[48] Lin J. (2002) Review of research on low-profile vortex generators to control boundary-layer separation. *Prog Aero Sci* **38**: 389–420.

[49] Choi J, Jeon W, Choi H. (2006) Mechanism of drag reduction by dimples on a sphere. *Phys Fluids* **18** (041702): 1–4.

[50] Gren P. (1987) Structured surfaces and turbulence. Thesis, Lulea University, Sweden.

[51] Bechert D, Hoppe G, Reif W. (1985) On the drag reduction of the shark skin. AIAA paper 85-0546.

[52] Bruse M, Bechert D, van der Hoeven J, Hage W, Hoppe G. (1993) Experiments with conventional and with novel adjustable drag-reducing surfaces. *Proc Int Cong Near-Wall Turbulent Flows*, Tempe, AZ, 719–738.

[53] Bechert D, Bartenwerfer M, Hoppe G. (1989) Turbulent drag reduction by nonplanar surfaces — a survey of the research at TU/DLR Berlin, Structure of Turbulence and Drag Reduction: IUTAM Symposium, Zurich, Switzerland, 525–543.

[54] Bechert D, Bruse M, Hage W. (2000) Experiments with three-dimensional riblets as an idealized model of shark skin. *Exps Fluids* **28**: 403–412.

[55] Lang A, Motta P, Hidalgo P, Westcott M. (2008) Bristled shark skin: A microgeometry for boundary layer control? *Bioinspir Biomim* **3**(4) 046005.

[56] Renaux J. (2004) Overview on drag reduction technologies for civil transport aircraft. *European Congress on Comp Methods in Appl Sci and Eng ECCOMAS.*

[57] Marentic FJ, Morris TL. (1986) Drag reduction article. European Patent Application by the 3M Company, St Paul, Minnesota; Application No. 86304131.5 Int. Cl. B 64C21}10; B 05 D 5}02. Priority 31.05.85 US 740239.

[58] Klocke F, Feldhaus B, Mader S. (2007) Development of an incremental rolling process for the production of defined riblet surface structures. *Prod Eng Res Devel* **1**: 233–237.

[59] Bandyopadhyay P, Henoch C, Hrubes J, Semenov B, Amirov A, Kulil V, Malyuga A, Choi K, Escudler M. (2005) Experiments on the effects of aging on compliant coating drag reduction. *Phys Fluids* **17**(085104): 1–9.

[60] Carman M, Estes T, Feinberg A, Schumacher J, Wilerson W, Wilson L, Callow M, Callow J, Brennan A. (2006) Engineered antifouling microtopographies — correlating wettability with cell attachment. *Biofouling* **22**(1/2): 11–21.

[61] Hutchinson H. (2008) Beyond the sharkskin suit. *Mech Eng* **130**(4): 42–44.

[62] Roberts B, Kamel K, Hedrick C, McLean S, Sharp R. (2003) Effect of a FastSkin^{TM} suit on submaximal freestyle swimming. *Med Sci Sports Exer* **35**(3): 519–524.

[63] Mollendorf J, Termin II A, Oppenheim E, Pendergast D. (2004) Effect of swim suit design on passive drag. *Med Sci Sports Exer* **36**(6): 1029–1035.

Chapter 3

Lotus Effect: Superhydrophobicity and Self-Cleaning

Michael Nosonovsky and Edward Bormashenko

The lotus effect and roughness-induced superhydrophobicity of natural and artificial (biomimetic) surfaces are studied in this chapter. The lotus effect is characterized by extreme water-repellency (superhydrophobicity) as well as self-cleaning resulting from the surface roughness and low surface energy coating. Various natural superhydrophobic surfaces (water-repellent plant leaves, bird feathers, water strider legs) are studied, and the theory of superhydrophobicity is presented. This includes the concepts of the contact angle and contact angle hysteresis, homogeneous (Wenzel) and composite (Cassie) wetting states, the Cassie–Wenzel transitions, and the effect of the multi-scale (hierarchical roughness). Artificial biomimetic superhydrophobic surfaces are discussed, including various ways of manufacturing and design optimization issues.

1. Introduction

Biological and bio-inspired surfaces that utilize the lotus effect are the subject of this chapter. The lotus effect involves two important properties that are typical for many water-repellent plant leaves: the superhydrophobicity and self-cleaning. A surface with the water contact angle greater than 150° and with low contact angle hysteresis (that will be defined and discussed in Sec. 3.2.1) is called superhydrophobic. The large contact angle and low contact angle hysteresis result in a low value of a sliding angle: a water drop rolls along such a surface even when the surface is tilted for a small angle (Fig. 1). Self-cleaning is the ability of many superhydrophobic surfaces to remain clean, since rolling water drops wash out contamination particles such as dust or dirt. The typical example of a natural superhydrophobic self-cleaning surface is the lotus leaf as well as leaves of many similar water-repellent plants, insect and bird wings, etc.

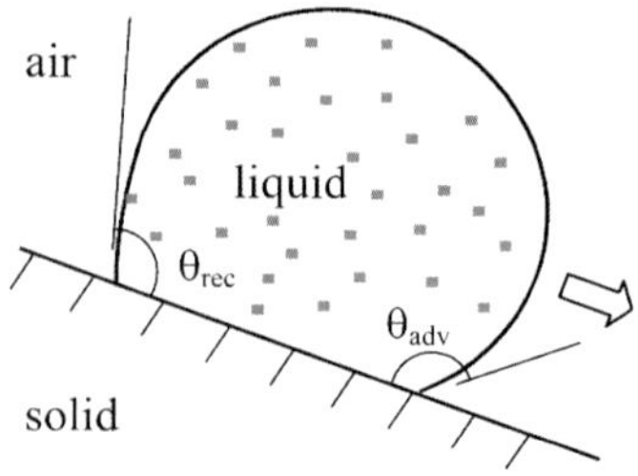

Fig. 1. A water drop on a tilted solid surface. The contact angle at the front is the advancing contact angle, θ_{adv}, whereas that at the back is the receding contact angle, θ_{rec}. The difference, $\theta_{adv} - \theta_{rec}$, constitutes adhesion hysteresis.

Artificial superhydrophobic surfaces have been introduced in the past decade and a variety of production methods (self-assembly, lithography, sol–gel) and materials (polymers, metals, Si, etc.) have been suggested.

Superhydrophobicity is usually a result of hydrophobic (low surface energy) coating (e.g. wax or Teflon) combined with the roughness that magnifies the effect of the coating. However, superhydrophobicity has been reported also for intrinsically hydrophilic (metallic) surfaces. In this case, superhydrophobicity arises from the surface roughness only. The typical size of roughness details in superhydrophobic surfaces is of the micron scale.

Numerous attempts to mimic the lotus effect for various applications have been made in recent years. These attempts are stimulated, on the one hand, by the development of the technology of micropatterned surfaces (such as lithography), and on the other hand, by the increasing need for nonadhesive (nonsticky) surfaces. Since the surface-to-volume ratio grows with miniaturization, the surface effects are crucial for microdevices such as microelectromechanical systems. In addition to microdevices, the lotus effect finds its application for conventional self-cleaning surfaces (paints, glasses, textiles, etc.). Possible applications for underwater devices, microfluidics, and other areas have been discussed.

In the present chapter, we will review natural superhydrophobic surfaces, such as plant leaves, bird feathers, insect wings and legs. There after, we will present the theory of superhydrophobicity and will review biomimetic superhydrophobic surfaces.

2. Superhydrophobic Surfaces in Nature

In this section, superhydrophobic plant leaves, bird feathers, insect wings and legs are discussed.

2.1. *Leaves of Water-Repellent Plants*

Hydrophobic and water-repellent abilities of many plant leaves have been known for a long time. Scanning electron microscopic (SEM) studies of the past 30 years revealed that the hydrophobicity of the leaf surface is related to its microstructure. All primary parts of plants are covered by a cuticle composed of soluble lipids embedded in a polyester matrix, which makes the cuticle hydrophobic in most cases.[1] The hydrophobicity of the leaves is related to self-cleaning, the ability of the hydrophobic leaves to remain clean after being immersed in dirty water. This feature is best known for the Lotus (*Nelumbo nucifera*) leaf that is considered by some oriental cultures as "sacred" due to its purity. The lotus flower is quoted extensively in oriental sacred texts, for example: "*One who performs his duty without attachment, surrendering the results unto the Supreme Lord, is unaffected by sinful action, as the lotus leaf is untouched by water*" (Bhagavad Gita 5.10). Not surprisingly, the ability of lotus-like surfaces for self-cleaning and water repellency was called the "Lotus effect".[1]

The outer single-layered group of cells covering a plant, especially the leaf and young tissues is called epidermis. The epidermal cells of leaves produce the protective waxy covering or the cuticle. The cuticle is composed of an insoluble cuticular membrane covered with epicuticular waxes, which are mixtures of hydrophobic aliphatic components, hydrocarbons with chain lengths typically in the range C16–C36, such as paraffins.[2] The SEM study reveals that the lotus leaf surface is covered by "bumps," more exactly called *papillae* (papillose epidermal cells), which, in turn, are covered by an additional layer of epicuticular waxes.[1] The wax is hydrophobic with water contact angle of about 95°–110°, whereas the experimental values of the static water contact angle with the lotus leaf were reported as about 160°.[1] The increase of the contact angle is a result of the surface roughness due to the papillae (Fig. 2).

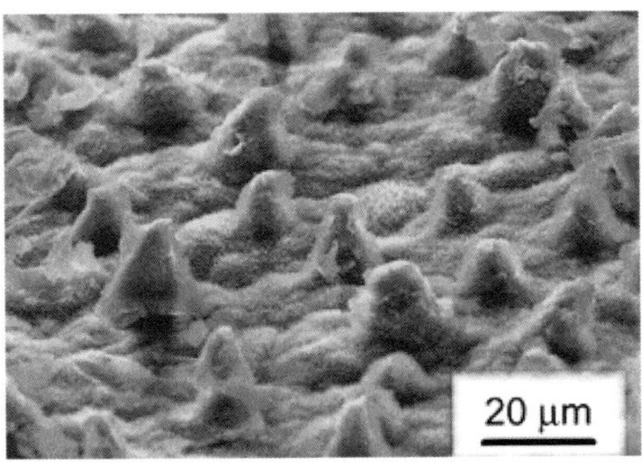

Fig. 2. SEM image of the lotus leaf surface showing papillae.

Neinhuis and Barthlot[3] have studied systematically surfaces and wetting properties of about 200 water-repellent plants. They reported several types of epidermal relieves and epcuticular wax crystals, with the papillose epidermal cells as one type of the relief. In leaves having the papillae, either every epidermal cell forms a single papilla or a cell is divided into papillae. The scale of the epidermal relief ranged from 5 μm in multipapillate cells to 100 μm in large epidermal cells. Some surfaces also were convex (rather than having real papillae) and/or had hairs (*trichomes*). A drop tends to form a heterogeneous (Cassie) interface on the surface of a leaf with air pockets trapped between the papillae. This significantly decreases the solid–liquid contact area.

Wagner *et al.*[4] also found various types and shapes of wax crystals at the surface. Interestingly, the hairy surfaces without wax, being subjected to rain, exhibited water-repellency for short periods (several minutes), after which water penetrated under the hairs, whereas waxy trichomes led to strong water-repellency. The chemical structure of the epicuticular waxes has been studied extensively by plant scientists and lipid chemists in recent decades.[5] The wax crystal surface is very rough, in addition to the roughness created by the papillae. Apparently, roughness plays the dominant role in the lotus effect, since the superhydrophobicity can be achieved independently of the type of wax or other hydrophobic coating.

While it is intuitive that water-repellency and self-cleaning are related to each other, because the ability to repel water is related to the ability to repel contaminants, it is difficult to quantify self-cleaning. As a result, a quantitative relation of the two properties remains to be established. A qualitative explanation of the self-cleaning was proposed by Barthlot and Neinhuis,[1] who suggested that on a smooth surface, contamination particles are mainly redistributed by a water drop, whereas on a rough surface they adhere to the drop and are removed from the leaves when the drop rolls off. A detailed model of this process has not been developed, but, obviously, whether the particle adheres to the drop depends upon the interactions at the triple line (liquid–solid–air contact line) and on whether the magnitude and direction of the surface force allows the particle to be detached from the surface and flown away (Fig. 3).

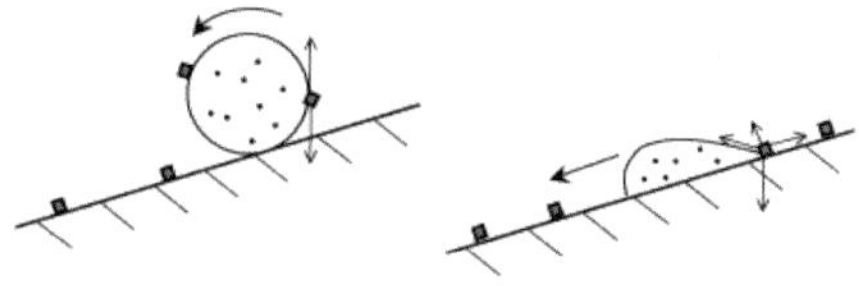

Fig. 3. Self-cleaning of a superhydrophobic surface. A drop with a very high contact angle (left) washes out contamination particles, while a drop with low contact angle (right) does not clean the surface. Forces acting upon a particle are shown.

As far as the biological implications of the lotus effect, the self-cleaning plays an important role in the defense against pathogens, such as spores and conidia of pathogenic microorganisms.[1]

2.2. *Insect and Birds Wings*

One of the recently studied biological objects demonstrating pronounced superhydrophobic properties is the dove's feather. The Biblical patriarch Noah sent a dove in order to discover whether the earth had become dry after the Flood (Genesis, 8:8). The bird returned safely to the Ark. It does mean that the dove did not suffer from surrounding water due to unusual properties of its feather (*pennae*). Feathers of birds perform several important functions: they form an insulating layer around the body, they create wing and tail surfaces which are essential for a flight, their coloration can provide a bird with a camouflage and, last but not least, they keep a body waterproofed. It is generally agreed that water-repelling properties of the outer feathers (pennae) are due to the preen oil (preen wax) spread by the bird under the preening procedure.[6] This is true for water-fowls. However for doves, the waterproof properties are at least partially stipulated by the complicated make-up of the feather.[7]

The details of the dove's feather morphology, namely network formed by keratin built *barbs* and *barbules* are depicted in Fig. 4. The drops deposited on the thoroughly cleansed feather vanes demonstrated obtuse apparent contact angles (APCAs) as high as 140°–150°. Experimentally established local contact angle measured on the *rachis* (or vertebral column) indicates that large apparent angles on feather's vane cannot be explained by the increase of the surface underlying the liquid since this can only amplify the wetting (the Wenzel effect). Moreover, measured APCAs follow the prediction of Cassie model discussed below in detail.[7] Actually, the water drop, supported by network formed by barbs and barbules, sits partially on air pockets. Small static apparent angle hysteresis also justifies Cassie–Baxter wetting hypothesis. The hierarchical two-fold structure of the dove's feather is noteworthy. An importance of hierarchical roughness, promoting water-repellency will be discussed below.[8–11]

Extremely high contact angles (177°) were also observed recently on some insect (dragonfly) wings. Superhydrophobicity on these objects was also attributed to Cassie wetting regime.[12]

2.3. *Insect Legs*

The hydrophobic abilities of rough surfaces are also utilized by insects that live close to water. A water strider (*Gerris remigis*) has the ability to walk upon water surface without getting wet. Even the impact of rain drops with a size greater than the strider's size does

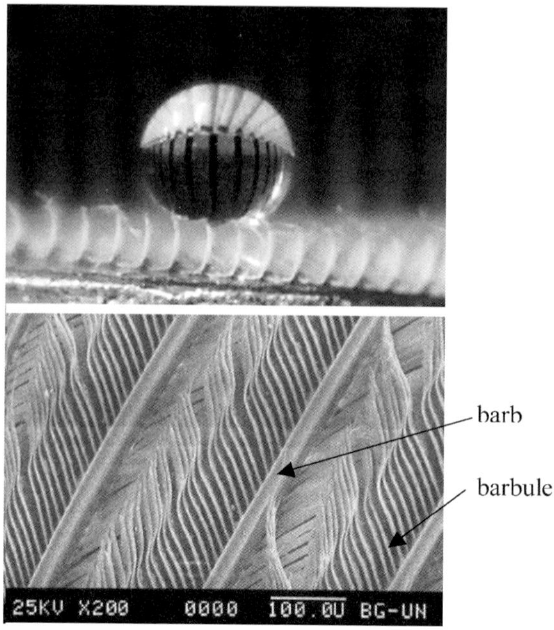

Fig. 4. Water drop upon pigeon feather. Microstructure of the feather.

not make it immerse into water. Gao and Jiang[13] showed that the special hierarchical structure of strider legs, which are covered by large numbers of oriented tiny hairs (microsetae) with fine nanogrooves, might be responsible for the water resistance. According to their measurements, a leg does not pierce the water surface until a dimple of 4.38 mm depth is formed. They found that the maximal supporting force of a single leg is 1.52 mN, or about 15 times the total body weight of the insect. The corresponding volume of water ejected is roughly 300 times that of the leg itself. Gao and Jiang[13] suggested that superhydrophobicity of the water strider leg is responsible for these abilities. They measured the contact angle of the insect's legs with water and found it to be 167.6°. Scanning electronic micrographs revealed numerous oriented setae on the legs. The setae are needle-shaped hairs, with diameters ranging from 3 μm down to several hundred nanometers. Most setae are roughly 50 μm in length and arranged at an inclined angle of about 20° from the surface of leg. Nanoscale grooves were found on each microseta, and these form a unique hierarchical structure. This hierarchical micro- and nanostructuring on the leg's surface seems to be responsible for its water resistance and the strong supporting force.

3. Modeling Superhydrophobicity

In this section, theoretical models of superhydrophobicity will be discussed. First, the Young, Wenzel, and Cassie equations of wetting of a rough surface will be introduced. Next, we will discuss the contact angle hysteresis as a measure of energy dissipation during liquid flow, transition between the wetting regimes, and hierarchical surfaces.

3.1. *Wetting of Flat and Rough Surfaces: The Governing Equations*

From the thermodynamic point of view, wetting of a rough solid surface is governed by Young, Wenzel, and Cassie equations that relate the contact angle between liquid and solid to interface free energies and to surface roughness.

3.1.1. *The Young equation*

Let us consider liquid in contact with a solid in presence of vapor, with the free surface energies of the solid–liquid, solid–vapor, and liquid–vapor interfaces equal to γ_{SL}, γ_{SV}, and γ_{LV}, respectively. When a liquid front comes in contact with a flat solid surface under the contact angle θ, the propagation of the liquid front for a small distance *ds* results in a net energy change of $ds(\gamma_{SL} - \gamma_{SV} + \gamma_{LV}\cos\theta)$. Therefore, for the liquid front to be at equilibrium, the Young equation should be satisfied[14]:

$$\gamma_{LV}\cos\theta = \gamma_{SV} - \gamma_{SL}. \tag{1}$$

The Young equation allows the calculation of the contact angle for given values of the interface energies. It is obvious from Eq. (1) that three situations are possible. If $(\gamma_{SV} - \gamma_{SL})/\gamma_{LV} > 1$, complete wetting takes place with the liquid fully adsorbed by the solid surface ($\theta = 0$). If $(\gamma_{SV} - \gamma_{SL})/\gamma_{LV} < -1$, complete rejection of the liquid by solid surface takes place ($\theta = 180°$). The most common situation is the intermediate situation of partial wetting ($-1 < (\gamma_{SV} - \gamma_{SL})/\gamma_{LV} < 1$, $0 < \theta < 180°$). A liquid that has the contact angle $\theta < 90°$ is often referred to as a "wetting liquid," while that with $\theta > 90°$ is "a nonwetting liquid." Corresponding surfaces are called, in the case of the contact with water, "hydrophilic" and "hydrophobic," respectively.

The free interface energies can also be viewed as surface tension forces. These forces are applied to the three-phase contact line (the triple line) and directed toward the corresponding interface. The surface tensions are measured in force units per length of the contact line, N/m, the same units as the interface energy, J/m^2. Historically, the Young equation was formulated in early 1800s in terms of forces, after the thermodynamic concept of free energy in general and free surface energy in particular was introduced since the middle and in the second half of the 19th century by Helmholtz, Gibbs, and other scientists.[14]

Note that Young's formulation implies that the solid is nondeformable and insoluble, so that only the horizontal projection of the tensions is considered. Assuming three tension forces act upon the triple line from the directions of the three phases, the balance of the horizontal projection of these forces leads to Eq. (1). As far as the vertical component of the force, it is balanced by the elastic response of the solid surface. This is a valid assumption for the thermodynamic equilibrium (the quasi-thermodynamic approximation). If the system is not at equilibrium, the tensile stress, caused by the vertical component, would locally increase the chemical potential of the system, so that the material can dissolve and change its shape.[15,16]

Young's equation does not take into account a number of factors, which can significantly affect the contact angle at the micro- and nanoscale. It is emphasized that the contact angle provided by the Young equation is a macroscale parameter, so it is sometimes called "the apparent contact angle" (APCA). The actual angle under which the liquid–vapor interface comes in contact with the solid surface at the micro- and nanoscale can be different. There are several reasons for that. First, water molecules tend to form a thin layer upon the surfaces of many materials. This is because of a long-distance van der Waals adhesion force that creates the so-called disjoining pressure.[17] This pressure is dependent upon the liquid layer thickness and may lead to the formation of stable thin films or precursors. In this case, the shape of the drop near the triple line gradually transforms from a spherical surface into a precursor layer, and thus the nanoscale contact angle is much smaller than the apparent contact angle. In addition, adsorbed water monolayers and multilayers are common for many materials.

Second, even carefully prepared atomically smooth surfaces exhibit certain roughness and chemical heterogeneity. Water tends to cover at first the hydrophilic spots with high surface energy and low contact angle.[18] The tilt angle of the surface due to roughness can also contribute to the APCA.

Third, Young's equation provides the value of the so-called static contact angle, that is, it ignores any dynamic effects related to the change of the drop's shape. The very concept of the static contact angle is not well defined. For practical purposes, the contact angle, which is formed after a drop is gently placed upon a surface and stops propagating, is considered the static contact angle. However, depositing the drop involves adding liquid while leaving, which may involve evaporation, so it is difficult to avoid dynamic effects.

Fourth, for a small drop and curved triple lines, the effect of the contact line tension may be significant. Molecules at the surface of a liquid or solid phase have higher energy because they are bonded to fewer molecules than those in the bulk. This leads to surface tension and surface energy. In a similar manner, molecules at the edge have fewer bonds than those at the surface, which leads to line tension and the curvature dependence of the surface energy. This effect becomes important when the radius of curvature is comparable with the Tolman's length.[19] However, the triple line at the nanoscale can be bending, and the radius of curvature can be very small, so that the line tension effects become important.[20] Thus while the contact angle is a convenient macroscale parameter, wetting is governed by interactions at the micro- and nanoscale, which determine the contact angle hysteresis and other wetting properties.

3.1.2. *The Wenzel and Cassie equations*

The Wenzel[21] equation relates the contact angle of a water drop upon a rough solid surface (Fig. 5), θ, with that upon a smooth surface, θ_0, through the nondimensional surface roughness factor, $R_f \geq 1$, equal to the ratio of the surface area to its flat projection:

$$\cos\theta = R_f \cos\theta_0. \tag{2}$$

The equation was originally derived for the homogeneous solid–liquid interface (no air pockets) (Fig. 6a) using the surface force balance and empirical considerations; however, it was later put in a proper thermodynamic framework. It is important that according to Wenzel model the inherently hydrophilic flat surface will be more hydrophilic when rough, and inherently hydrophobic surface will become more hydrophobic.

For a surface composed of two fractions, one is with the fractional area f_1 and the contact angle θ_1 and the other with f_2 and θ_2, respectively ($f_1 + f_2 = 1$), the contact angle is given by the Cassie equation:

$$\cos\theta = f_1\cos\theta_1 + f_2\cos\theta_2. \tag{3}$$

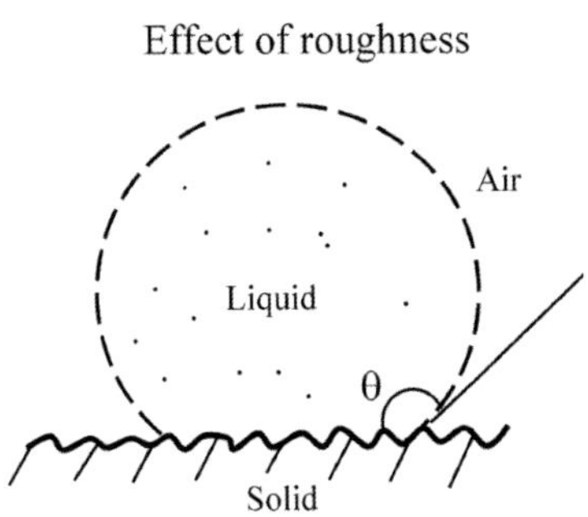

Fig. 5. Surface roughness enhances the contact angle.

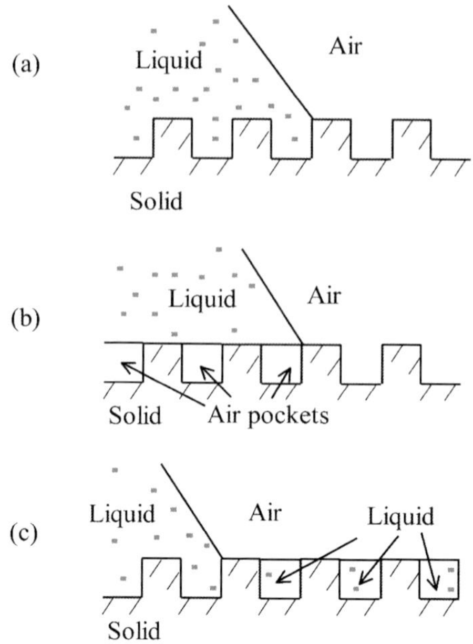

Fig. 6. Solid–liquid–air interface (a) homogeneous (Wenzel), (b) composite (Cassie), and (c) with filled holes (Cassie with water penetration).

For the case of a composite interface (Fig. 6b), consisting of the solid–liquid fraction ($f_1 = f_{SL}, \theta_1 = \theta_0$) and liquid–vapor fraction ($f_2 = 1 - f_{SL}, \cos\theta_2 = -1$), combining the Eqs. (2) and (3) yields the Cassie–Baxter[22] equation:

$$\cos\theta = f_{SL} R_f \cos\theta_0 - 1 + f_{SL}. \tag{4}$$

It is seen from Eq. (4) that an inherently hydrophobic surface ($\cos\theta_0 > 0$) can turn into a hydrophobic one ($\cos\theta < 0$) in the Cassie regime.

The opposite limiting case of $\cos\theta_2 = 1$ ($\theta_2 = 0°$ corresponds to the water-on-water contact) yields:

$$\cos\theta = 1 + f_{SL}(\cos\theta_0 - 1). \tag{5}$$

Equation (5) is sometimes used for the homogeneous interface instead of Eq. (1), if the rough surface is covered by holes filled with water (Fig. 6c).

Gao and McCarthy[23] showed experimentally that the contact angle of a drop is defined by the triple line and does not depend upon the roughness under the bulk of the drop. A similar result for chemically heterogeneous surfaces was obtained by Extrand.[24]

Gao and McCarthy[23] concluded that the Wenzel and Cassie equations "should be used with the knowledge of their fault." The question remained, however, under what circumstances the Wenzel and Cassie equations can be safely used and under what circumstances do they become irrelevant. In their paper, entitled "How Wenzel and Cassie were wrong," Gao and McCarthy[23] did not provide any suggestion when these equations can be used; but their paper initiated a discussion. Both Wenzel and Cassie equations were derived recently in the framework of the general thermodynamic approach,[25] Nosonovsky[26] suggested generalization of these equations for the case of non uniform roughness. Independently and simultaneously, a similar generalization for the axisymmetric case was proposed by McHale[27], Panchagnula and Vedantam.[28]

The most convenient way to derive the Wenzel and Cassie equations is to use the principle of virtual work and consider a liquid advancing for a small distance δx. For a liquid front propagating along a rough two-dimensional profile, the derivative of the surface free energy (per liquid front length), W, with respect to the profile length, s, yields the surface tension force $\sigma = dW/ds = \gamma_{SL} - \gamma_{SV}$. The quantity of practical interest is the component of the tension force that corresponds to the advancing of the liquid front in the horizontal direction for dx. This component is given by $dW/dx = (dW/ds)(ds/dx) = (\gamma_{SL} - \gamma_{SV})ds/dx$. The local value of dW/dx depends upon x, however, if the virtual distance δx is large enough, the average value of $\overline{dW/dx}$ can be considered (where the bar indicates the average value). It is noted that the derivative $R_f = \overline{ds/dx}$ is equal to Wenzel's roughness factor. Therefore, the Young equation, which relates the contact angle with solid, liquid, and vapor interface tensions, $\gamma_{LV} \cos\theta = \gamma_{SV} - \gamma_{SL}$, is modified as:

$$\gamma_{LV} \cos\theta = R_f\,(\gamma_{SV} - \gamma_{SL})\,. \tag{6}$$

The empirical Wenzel equation (Eq. (2)) is a consequence of Eq. (6) combined with the Young equation.

3.2. *Contact Angle Hysteresis*

In addition to high contact angle, superhydrophobic surfaces should have low contact angle hysteresis, which serves as a measure of energy dissipation during the flow of a drop along the solid.

3.2.1. *Definition of contact angle hysteresis*

The Young, Wenzel, and Cassie equations were deduced on the basis of simple semi-empirical considerations. Later investigations put these equations into a thermodynamic framework, however, they also showed that there is no one single value of the contact angle for a rough or heterogeneous surface. The contact angle can be in a range of values between the so-called receding contact angle, θ_{rec}, and the advancing contact angle, θ_{adv}. The system tends to achieve the advancing contact angle when liquid is added (for example, at the front end of a moving or inflated drop), whereas the receding contact angle is achieved when the liquid is removed (for example, at the read end of a moving or deflated drop), Fig. 1. When no liquid is added or removed, the system tends to have a static or "most stable" contact angle, which can be approximated by Eqs. (1)–(6). The difference between θ_{adv} and θ_{rec} is known as "contact angle hysteresis" and it reflects a fundamental asymmetry of wetting and dewetting and the irreversibility of the wetting/dewetting cycle. Although for surfaces with carefully controlled roughness on the molecular scale, it is possible to achieve contact angle hysteresis as low as $< 1^\circ$,[29] it cannot be eliminated completely, since even atomically smooth surfaces have a certain roughness and heterogeneity. The contact angle hysteresis is present even in the quasi-static case of very slow liquid propagation and it can serve as a measure of energy dissipation. A water-repellent surface should have a low contact angle hysteresis to allow water to flow easily along the surface. It is generally agreed that the Wenzel wetting regime provides much larger contact angle hysteresis than the Cassie–Baxter one.

3.2.2. *Empirical models of contact angle hysteresis*

Three types of dissipative processes contribute to contact angle hysteresis: interactions in the bulk of the drop, at the solid–liquid contact surface, and at the triple line. While the bulk interactions are the result of liquid viscosity and they can be excluded by considering the quasi-static case (slow motion), the contact surface and triple line interactions remain significant even in the quasi-static limit. The contribution of the contact surface is similar to that of the adhesion hysteresis — the difference in energy required to separate two media (the solid and liquid in this case) and the energy gained by bringing them together. The contribution of the triple line interaction is due to the roughness and pinning of the triple line at rough surface edges. It has been suggested that these two mechanisms are similar to the mechanisms of dry sliding and rolling friction.[15,30]

For a micropatterned surface built of flat-top columns ($R_f = 1$), the contact angle hysteresis involves a term inherent to the nominally smooth surface and the term dependent upon the surface roughness, H_r:

$$\cos\theta_{adv} - \cos\theta_{rec} = f_{SL}\,(\cos\theta_{adv0} - \cos\theta_{rec0}) + H_r, \qquad (7)$$

where θ_{adv0} and θ_{rec0} are the advancing and receding contact angles for the smooth surface.[31] The first term in the right-hand part of Eq. (7), which corresponds to the inherent contact angle hysteresis of a smooth surface is proportional to the fraction of the solid–liquid contact area, f_{SL}. The second term, H_r, may be assumed to be proportional to the length density of the pillar edges, or, in other words, to the length density of the triple line. Thus, Eq. (7) involves both the term proportional to the solid–liquid interface area and to the triple line length.

3.2.3. *Simulation and semi-empirical models*

The task of predicting contact angle hysteresis for a rough surface with a given topography remains a difficult one. A number of simulations have been conducted; however, in most cases they are limited to 2D topography. Kusumaatmaja and Yeomans[32] showed that the contact angle hysteresis is sensitive to the details of the surface patterning.

Vedantam and Panchagunula[33] suggested to use a semi-empirical phase field method, which had been used successfully to study various wetting transitions. In this method, the order-parameter $\eta(x, y)$ is selected in such a manner that $\eta = 0$ for the nonwetted regions of the surface and $\eta = 1$ for wetted regions, whereas $0 < \eta < 1$ for partially wetted regions. The energy function $f(\eta)$ is then constructed in such a manner that the free surface energy can be expressed as:

$$E = \int_A f(\eta)dA = \int_{A_{SV}} f(0)dA + \int_{A_{SL}} f(1)dA + \int_{A(0<\eta<1)} f(\eta)dA, \qquad (8)$$

where integration is performed by the area. The function $f(\eta)$ can be built in a somewhat arbitrary manner, however, its minima should correspond to the equilibrium states of the system (e.g. the Wenzel and Cassie states). The energy functional is then written as:

$$L = \int_A \left\{ f(\eta) + \frac{\lambda}{2} \left| \nabla f(\eta) \right|^2 \right\} dA, \qquad (9)$$

where λ is the gradient coefficient. The functional that should be minimized involves the free energy and the gradient of the free energy. The later term is needed to account for the fact that creating an interface between two phases is energetically unprofitable. The kinetic equation is given in the form:

$$\beta\dot{\eta} = -\frac{dL}{d\eta} = \lambda\nabla^2\eta - \frac{\partial f}{\partial \eta}, \qquad (10)$$

where $\beta > 0$ is the kinetic coefficient. In the simplest case it may be assumed that $\beta = \text{const}$, whereas in the general case a more complicated functional dependence upon η and its derivatives may exist. Vedantam and Panchagunula[33] showed that in the case of $\beta = \text{const}$ for an axisymmetric drop flowing with the velocity V, Eq. (10) leads to:

$$\cos\theta_{\text{adv}} - \cos\theta_{\text{rec}} = 2\alpha\beta V. \tag{11}$$

In other words, assuming that the kinetic coefficient is constant, the contact angle hysteresis is expected to be proportional to the flow velocity. A more complicated form of the kinetic coefficient may lead to a more realistic dependence of the contact angle hysteresis on the velocity.

The phase field method also accounts in a natural way for the contact line tension term and leads to the Boruvka and Neumann[34] equation:

$$\cos\theta = \frac{\gamma_{\text{SV}} - \gamma_{\text{SL}}}{\gamma_{\text{LV}}} - \frac{\tau K}{\gamma_{\text{LV}}}, \tag{12}$$

where K is the curvature and τ is the contact line tension.[33] And understandable so because the gradient term in Eq. (9) implies that the transition between the wetted and unwetted parts of the surface is associated with excess energy, or, in other words, bending of the phase boundary line at the phase field plane is energetically unprofitable. The contact line term in Eq. (12) corresponds to this additional contour line energy and it is important for micrometrically scaled drops only.[20] The phase field method also seems promising for the analysis of the Cassie–Wenzel transition.

A semi-phenomenological model of the contact angle hysteresis has been proposed recently by Whyman, Bormashenko and Stein. According to their model, the contact angle hysteresis is given by the equation: $\theta_{\text{adv}} - \theta_{\text{rec}} = \left(\frac{8U}{\gamma R_0}\right)^{1/2} h(\theta^*)$, where U is the height of the potential barrier connected with the motion of the triple line along a substrate, $\gamma = \gamma_{\text{LV}}$, R_0 is the initial radius of the spherical drop before deposition on the substrate, $h(\theta^*)$ is the dimensionless function of APCA θ^*.[25]

3.3. *Stability and the Cassie–Wenzel Transition*

Stability of the Cassie wetting regime and the Cassie–Wenzel transition have been extensively studied in recent years.

3.3.1. *Vibration-induced transition*

Application of various external stimuli, such as pressure or vibration, could convert Cassie wetting state into the Wenzel one.[35–40] It has been accepted for a long time that the Cassie wetting regime corresponds to a higher energetic state, and the Cassie–Wenzel transition is irreversible. Actually the situation is more complicated (the reversible Cassie–Wenzel transitions will be discussed below) and it calls for additional theoretical and experimental insights.

First of all it has to be emphasized, that the kinetics of the transition remains unclear. Ishino and Okumura[41] suggested that the Cassie–Wenzel transition occurs via a nucleation mechanism starting from the drop center. On the other hand, recent investigations performed with polymer honeycomb surfaces demonstrated that the Cassie–Wenzel transition is more likely due to the displacement of a triple line under external stimuli.[40]

The most comprehensive experimental method supplying the qualitative information about wetting transitions is based on vibration of drops deposited on the rough surfaces. First attempts to characterize the rough surfaces with vibrated droplets were by Della Volpe and Marmur.[42,43] A qualitative interpretation of experiments performed with vibrated drops was proposed recently by Bormashenko *et al.*[39,40] When a drop of volume V is vibrated with frequency ω and amplitude A, the additional pressure p_i due to inertial force exerted on the substrate is given by:

$$p_i = \frac{\rho V A \omega^2}{\pi R(\theta)^2 \sin^2 \theta}, \tag{13}$$

where ρ is the density of liquid, $R(\theta)$ is the radius of the drop depending on APCA θ. The excess over the atmospheric pressure inside the drop is the sum of Eq. (13) and the Laplace and hydrostatic pressure:

$$p = p_i + \frac{2\gamma}{R(\theta)} + \rho g h, \tag{14}$$

where h is the drop height. Thus, if the wetting transition originates from the nucleation mechanism introduced by Ishino and Okumura,[41] it could be expected that it occurs under the constant pressure $p = const$ (the idea that the critical pressure governs the transition on the rough surfaces was supported by investigations performed by Zhang *et al.*[44] Recent experiments carried out with drops of various radii demonstrated that actually the wetting transition occurs when the condition $pR = \text{const}$ is fulfilled, where p is the total pressure in the drop and R is the radius of the drop surface.[40] Thus it was supposed that wetting regime changes when the certain threshold value of the force $\frac{dF}{dl}$ acting on the unit length of a triple line, given by Eq. (14), is exceeded. $\frac{dF}{dl}$ is given by:

$$\frac{dF}{dl} = \frac{R(\theta)p}{4\sin\theta}(2\theta - \sin 2\theta), \tag{15}$$

where p is given by Eq. (14). The experiments previously performed on various polymer and metallic rough surfaces demonstrated that wetting transition occurs when the condition $\frac{dF}{dl}$ = const is satisfied.[45] Thus it looks like the wetting transition is 1D, stipulated by the displacement of the triple line exposed to external factors, such as pressure, vibrations, etc.

It is also noteworthy that Cassie–Wenzel transition is not the only possible scenario of change in the wetting regime. It should be remembered that Cassie and Wenzel states do not exhaust all possible wetting regimes. There also exists a Cassie impregnating wetting regime when the drop finds itself on a wet substrate, viewed as a patchwork of solid and liquid.[46] The Gibbs free energy of a drop deposited on a rough surface demonstrates multiple minima (Fig. 7), and the lowest energy state corresponds to a Cassie impregnating wetting regime; thus alternative pathways of wetting transitions could be expected.[47] Indeed, single and two-stage pathways of wetting transitions were already observed, including Cassie (air-trapping)–Wenzel–Cassie (impregnating), Wenzel–Cassie (impregnating) and Cassie (air-trapping)–Cassie (impregnating) transitions.[48]

It has to be emphasized that there exist surfaces where vibration-induced wetting was not observed at all, in particular dove's pennae. The energy supplied by vibrations in this case is not sufficient for surmounting the potential barriers separating wetting states (Fig. 6). Perhaps, this is due to the hierarchical structure of a feather, which enhances the stability of the Cassie (air trapping) state.

When the drop is vibrated horizontally, the wetting transition is of much more complicated nature due to excitation of surface waves.[49] The resonance wetting transition has been recently reported in this experimental situation, however, even in this case the wetting transition is due to the displacement of the triple line and does not occur according to the nucleation mechanism.[49]

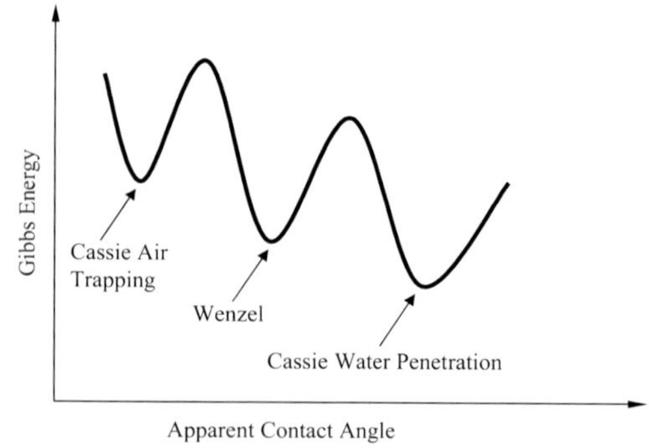

Fig. 7. Free surface energy (Gibbs Energy) as a function of contact angle showing the three stable states.

3.3.2. *Transition during evaporation*

Evaporation of the drop also could promote the wetting transition. Evaporation induced Cassie–Wenzel transitions were first reported by McHale *et al.*[37] Then evaporation induced wetting transition on the polymer honeycomb and hierarchical superhydrophobic surfaces was studied by Bormashenko *et al.*[48,49] The strong dependence of APCA on the drops volume illustrated with Fig. 8 evidences the wetting transition.[49] The quantitative explanation of the phenomenon is based on the simple assumption that takes into account the growth in the Laplace pressure $P = 2\gamma/R$ when a drop turns small under evaporation.[49] The Laplace pressure exerted on a substrate promotes water penetration in the pores of the relief. The water intrusions in the microstructure of the relief promote the inhomogeneous wetting regime.[50] The water penetration is governed to a much greater extent by the local contact angle.[11,46] However, the rigorous theory of the phenomenon considering the displacement of the triple line is not developed still.

3.3.3. *Reversible superhydrophobicity*

The important area of application of superhydrophobic surfaces is reversible superhydrophobicity, that is, the ability of a surface to switch between the hydrophobic and hydrophilic properties under the influence of the electric potential, ultraviolet or light irradiation, or temperature.[51–54] This area has emerged in recent years, since 2004, and a number of important findings have been made, including the ability to switch between the Cassie and Wenzel states. Krupenkin *et al.*[55] reported that

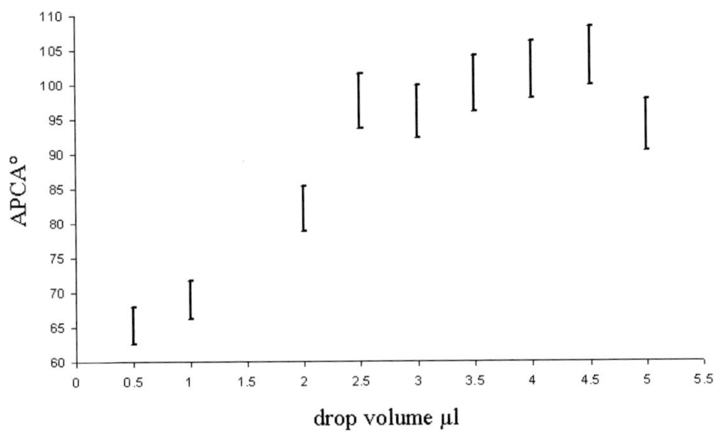

Fig. 8. APCA vs. drop volume for the polymer honeycomb surfaces.

drop behavior can be reversibly switched between the superhydrophobic Cassie state and the hydrophilic Wenzel state by the application of electrical voltage and current (electrowetting). Wang *et al.*[54] created a surface that can switch between stable superhydrophilic, metastable superhydrophobic, and stable superhydrophobic states. Interestingly, their switchible surface was driven by DNA nanodevices. A number of other approaches to reversible superhydrophobicity have been suggested as well.[56]

3.4. *Role of Hierarchical Roughness*

The role of surface hierarchy in the lotus effect is also not completely clear, although a number of explanations of why most natural surfaces are hierarchical has been suggested.[8,9,48,57–59] Nosonovsky and Bhushan[30,57,59] suggested that the mechanisms involved with superhydrophobicity are scale-dependent and thus the roughness must be hierarchical in order to respond to these mechanisms. Nosonovsky[8] showed also that small-scale roughness could enhance the stability of the composite interface. It is also noteworthy that the dependence of APCA on the Young angle for the hierarchical surfaces is not given by the simple Cassie equation (Eq. (3)), but it is much more complicated.[48,61] Usually this dependence is much weaker then the one predicted by the Cassie equation.

The hierarchy may have to do also with the simple fact that the surface must be able to repel both macroscopic and microscopic drops. Experiments with artificial fog (microdrops) and artificial rain (large drops) show that surfaces with only one scale of roughness repel rain drops well. However, they cannot repel small fog drops which are trapped in the valleys between the bumps.[60]

3.5. *Dynamic Effects: Bouncing Drops*

Another important phenomenon related to hydrophobicity is bouncing drops (Fig. 9). When a drop impacts a superhydrophobic surface with a certain velocity, in some cases it can be bounced of in an almost elastic manner.[61,62] The kinetic energy of the drop is stored in the surface deformation during the impact. A deformed drop has a higher surface area and therefore higher free surface energy. Therefore, during the impact when

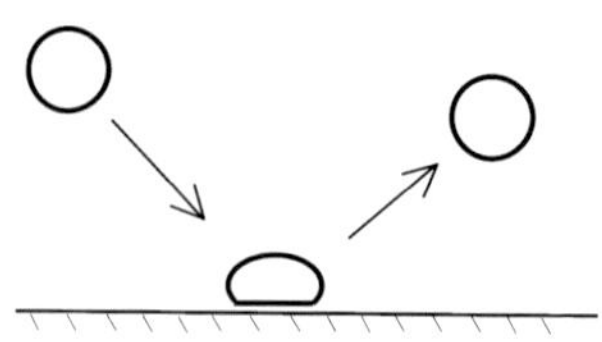

Fig. 9. A bouncing "elastic" droplet hits a hydrophobic surface.

the drop is deformed, it can accommodate the kinetic energy. It was suggested that the effect might be of practical interest for agriculture, in particular, for treating leaves with pesticides.[62] Since the leaf is likely to repel the drop, it will not be treated; in addition, the scattering of the drops contaminates the soil. If small amounts of a polymer soluble in water (such as polyoxyethylene) are added to water,[63] the bulk viscosity of water remains the same, however, the so-called elongation viscosity (resistance of the liquid to large extensions) is high. Drops of such a mixture do not bounce, and has been used as an agricultural spray.[62] On the other hand, the bouncing drops are useful for waterproof fabrics or concrete that should preserve their dryness under a rain.

The dissipation which prevents drop bouncing is due to viscosity and the moving triple line. For the bouncing drops, the triple lines are small or absent due to the presence of a thin film of compressed air during the impact. It was reported that full rebounding takes place for the surfaces with contact angles larger than 150°–160° with the ratio of the velocities before and after the impact of 0.9 and higher.[63] To estimate the maximum velocities of impact, V, for which the effect takes place is obtained by comparison of the kinetic and free surface energy. The kinetic energy of a drop of radius R_0 and density ρ is given by:

$$E_{\mathrm{kin}} = \frac{(4/3)\pi\rho R_0^3 V^2}{2}, \tag{16}$$

while the free surface energy is given by:

$$E_{\mathrm{surf}} = 4\pi R_0^2 \gamma_{\mathrm{LV}}. \tag{17}$$

In order for the surface deformation to accommodate the kinetic energy, the latter should be of the same order of magnitude as the former, which yields:

$$V^2 = \frac{6\gamma_{\mathrm{LV}}}{\rho R_0}. \tag{18}$$

For a water ($\gamma_{\mathrm{LV}} = 0.072\,\mathrm{N/m}$, $\rho = 1000\,\mathrm{kg/m^3}$) drop of $R_0 = 0.001$, the corresponding impact velocity is $V = 0.66\,\mathrm{m/s}$. Indeed, with increasing velocity, the elastic behavior decreases.[64] The ratio of the kinetic and surface energy (or inertial to the capillary forces) is characterized by the nondimensional Weber number:

$$We = \frac{\rho R_0 V^2}{\gamma_{\mathrm{LV}}}. \tag{19}$$

Small *We* corresponds to low kinetic energy in comparison with the free surface energy. Two other important nondimensional numbers are the capillary number:

$$Ca = \frac{\mu_L V}{\gamma_{LV}} \tag{20}$$

that characterizes the ratio of the dynamic viscosity (μ_L) to capillary forces, and the Reynolds number:

$$Re = \frac{\rho V R_0}{\mu_L} \tag{21}$$

that characterizes the ratio of the inertial to viscous forces (note that $Re = We/Ca$). Bouncing usually requires small *We*, *Ca*, and *Re*. Big velocity (high *We*) would result in the rupture of the drop into parts, while big viscosity (high *Ca* and *Re*) would result in dissipation and eventual wetting of the surface (Fig. 10).

3.6. *A Drop on an Inclined Surface*

For a drop of radius R moving along the inclined plane with tilt angle α, the energy gain corresponding to the inclined distance l is given by $lmg \sin\alpha$, where mg is the weight of the drop. The contact angle hysteresis is related to the tilt angle by:

$$t\gamma(\cos\theta_{adv} - \cos\theta_{rec}) = gm \sin\alpha, \tag{22}$$

where $\gamma = \gamma_{LV}$ is the surface tension and t is the length of the triple line.[65] Equation (22) was formulated by Macdougall and Ockrent[66] based on their experimental observations of drops on inclined surfaces and independently by Frenkel[67] in his theoretical study of the contact angle hysteresis. Equation (22) is based on several assumptions, in particular, that the front edge of the drop exhibits the advancing

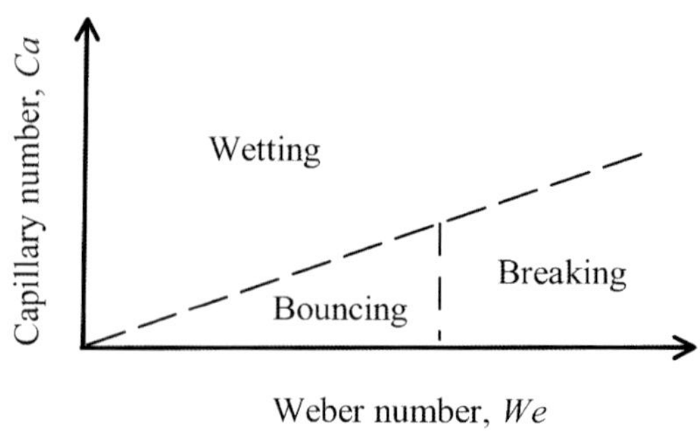

Fig. 10. Various regimes for bouncing droplets.

and the back end the receding contact angle and that the drop foundation is well approximated by the cylindrical or rectangular shape. Marmur[47] and Krasovitski and Marmur[68] showed that the drop on an inclined plane does not necessarily exhibit the receding or advancing contact angle, since the former is based on the geometry of the drop, whereas the latter is based on the nature of the solid surface.

4. Self-Cleaning

Self-cleaning is another important feature associated with the lotus effect. At this point, no quantitative theory has been proposed on self-cleaning that would relate, for example, the size and contact angle of a drop with the size of a contaminating particle being washed away. There is a qualitative understanding that water-repellent surfaces do also repel other contaminants and that dust can easily be washed from them by flowing water. A number of experimental studies have been conducted. The self-cleaning abilities of patterned surfaces were investigated by Fürstner *et al.*[60] They studied Si wafer specimens with regular patterns of spikes that were manufactured by X-ray lithography. The specimens were hydrophobized with Au thiol. For comparison, they also studied replicates of plant surfaces, made by a two-component silicon molding mass applied onto the leaf surface. The negative replica is flexible and rubber-like. A melted hydrophobic polyether was applied onto this mold. They also studied several metal foil specimens, hydrophobized by means of a fluorinated agent. In order to investigate the self-cleaning effect, a luminescent and hydrophobic powder was used as a contaminant. Following contamination, the specimens were subjected to an artificial fog and rain.[60]

Drops of water rolled off easily from Si samples with a microstructure consisting of rather slender and sufficiently high spikes; this is attributed to the occurence of the Cassie wetting state. These samples could be cleaned after artificial contamination by means of the fog treatment almost completely. The behavior of water drops was different upon surfaces with low spikes and a rather high pitch. The researchers found that a considerable decrease of the contact angles and a distinct rise in the sliding angles apparently corresponding to the Wenzel state. Some metal foils and some replicates had two levels of roughness. These specimens did not show a total removal of all contaminating particles when they were subjected to artificial fog, but water drops impinging with sufficient kinetic energy could clean them perfectly. A substrate without structures smaller than 5 μm could not be cleaned by means of fog consisting of water drops with diameter 8–20 μm because this treatment resulted in a continuous water film on the samples. However, artificial rain removed all the contamination. On the other hand, smooth specimens made of the same material could not be cleaned completely by impinging drops. This is a clear indication of the different contact phenomena on smooth hydrophobic in contrast to self-cleaning microstructured surfaces. Another interesting observation of this group was that despite the missing structure of the wax crystals, the water contact angle of the Lotus replica was the highest of all the replicates, indicating that the microstructure formed by the papillae alone is already optimized with regard to water repellency.[60]

5. Biomimetics: Artificial Superhydrophobic Surfaces

Two general requirements for a superhydrophobic surface are that the surface should be rough and that it should have a hydrophobic (low surface energy) coating. These two requirements lead to two methods of producing a superhydrophobic surface: first, it is possible to make a rough surface from an initially hydrophobic material and, second, to modify an initially rough surface by changing the surface chemistry or applying a hydrophobic material upon it. Note that roughness is usually a more critical property than low surface energy, since both moderately hydrophobic and very hydrophobic materials exhibit similar wetting behavior when roughened.[56] And understandably so, based on the simple Wenzel model, the cosine of the contact angle is given by $R_f \cos\theta_0$, so even small (negative) $\cos\theta_0$ will result in a high contact angle when combined with a big roughness factor.

A proper control of roughness constitutes the main challenge in producing a reliable superhydrophobic surface, while if the initial material is hydrophilic, a surface treatment or coating is required which will decrease the surface energy. While the two factors — roughness and low surface energy — are required for superhydrophobicty, the role of roughness clearly dominates. For example, it is not really important whether the low energy surface is built of typical paraffin $-CH_2-$ groups or $-CH_3-$ groups of fluorocarbons having much lower energy. Furthermore, many rough surfaces without any lower energy characteristics still exhibit superhydrophobicity. The role of hierarchical roughness still remains to be investigated. While many suggestions have been made regarding why superhydrophobic surfaces in nature are hierarchical, experiments with nonhierarchical patterned surfaces demonstrate superhydrophobicity as well.

5.1. *Micropatterned Surfaces Produced by Lithography and Other Methods*

Fabrication of superhydrophobic surfaces has been an area of active research since the mid 1990s. In general, the same techniques that are used for micro- and nanostructure fabrication, such as lithography, etching, and deposition, have been utilized for producing superhydrophobic surfaces. Among especially interesting developments is the creation of switchable surfaces that can be turned from hydrophobic to hydrophilic by applying electric potential, heat or ultraviolet (UV) irradiation.[52,53,56,69,70] Another important requirement for potential applications for optics and self-cleaning glasses is the creation of transparent superhydrophobic surfaces. In order for the surface to be transparent, roughness details should be smaller than the wavelength of the visible light (about 400–700 nm).[71]

Lithography is a well-established technique applied for creating large area of periodic micro/nanopatterns. It includes photo, E-beam, X-ray, and soft lithography. Martines *et al.*[72] fabricated ordered arrays of nanopits and nanopillars by using electron beam lithography. They obtained a superhydrophobic surface with a contact angle of 164° and hysteresis of 1° for a surface consisting of tall pillars with cusped tops after a hydrophobization with octadecyltrichlorosilane (OTS). Furstner *et al.*[60] created silicon wafers with regular patterns of spikes by X-ray lithography. The wafer was hydrophobized by sputtering a layer of gold and subsequent immersion in a hexadecanethiol solution. Bhushan and Jung[73] investigated patterned Si produced with photolithography. To obtain a sample that is hydrophobic, an SAM of 1,1,2,2-tetrahydroperfluorodecyltrichlorosilane (PF_3) was deposited on the sample surfaces using the vapor phase deposition technique. They obtained a superhydrophobic surface with a contact angle up to 170°.[74]

Another well-known and effective way to make rough surfaces is etching using either plasma, laser, chemical or electrochemical techniques.[56] Jansen *et al.*[75] etched a silicon wafer using a fluorine-based plasma by utilizing the black silicon method to obtain isotropic, positively and negatively tapered as well as vertical walls with smooth surfaces. Coulson *et al.*[76] described an approach in plasma chemical roughening of poly(tetrafluoroethylene) (PTFE) substrates followed by the deposition of low surface energy plasma polymer layers, which give rise to high repellency towards polar and non-polar probe liquids. A different approach was taken by Shi *et al.*,[77] who treated a Teflon film with oxygen plasma and obtained a superhydrophobic surface with a contact angle of 168°. Fluorinated materials have a limited solubility, which makes it difficult to roughen them. However, they may be linked or blended with other materials, which are often easier to roughen, in order to make superhydrophobic surfaces. Teshima *et al.*[78] obtained a transparent superhydrophobic surface from a poly(ethylene terephthalate) (PET) substrate via selective oxygen plasma etching followed by plasma-enhanced chemical-vapor deposition (PECVD) using tetramethylsilane (TMS) as the precursor. Khorasani *et al.*[79] produced porous PDMS surfaces with a contact angle of 175° using a CO_2-pulsed laser etching method as an excitation source for the surface. Qian and Shen[80] described a simple surface roughening method by dislocation selective chemical etching on polycrystalline metals such as aluminum. After treatment with fluoroalkylsilane, the etched metallic surfaces exhibited superhydrophobicity. Xu *et al.*[52] fabricated a reversible superhydrophobic surface with a double-roughened perfluorooctanesulfonate (PFOS) doped conducting polypyrrole (PPy) film by a combination of electropolymerization and chemical polymerization. The reversibility was achieved by switching between superhydrophobic doped or oxidized states and superhydrophilicity dedoped or neutral states with changing the applied electrochemical potential.

Stretching is another method to produce a superhydrophobic surface. Zhang *et al.*[81] stretched a Teflon film and converted it into fibrous crystals with a large fraction of void space in the surface, leading to high roughness and superhydrophobicity.

There are several ways to make a rough surface including adsorption, dip coating, electrospinning, anodization, electrochemical, evaporation, chemical-vapor deposition (CVD), and plasma. Solidification of wax can be used to produce a superhydrophobic surface. Shibuichi *et al.*[82] used alkylketene dimer (AKD) wax on a glass plate to spontaneously form a fractal structure in its surfaces. They obtained a surface with a contact angle larger than 170° without any fluorination treatments. Klein *et al.*[83] obtained superhydrophobic surfaces by simply dip coating a substrate with slurry containing nano-silica spheres, which adhered to a substrate after a low temperature heat treatment. After reaction of the surface with a fluoroalkyltrichlorosilane, the hydrophobicity increased with decreasing area fraction of spheres. Abdelsalam *et al.*[84] studied the wetting of structured gold surfaces formed by electrodeposition through a template of submicrometer spheres and discussed the role of the pore size and shape in controlling wetting. Lau *et al.*[85] created superhydrophobic carbon nanotube forests by modifying the surface of vertically aligned nanotubes with PECVD. Superhydrophobicity was achieved down to the microscopic level where essentially spherical, micrometer-sized water drops can be suspended on top of the nanotube forest. Zhu *et al.*[86] and Huang *et al.*[87] prepared surfaces with two-scale roughness by controlled growth of carbon nanotube (CNT) arrays by CVD. Zhao *et al.*[88] also synthesized the vertically aligned multiwalled carbon nanotube (MWCNT) arrays by CVD on Si substrates using thin film of iron (Fe) as catalyst layer and aluminum (Al) film.[74]

Attempts to create a superhydrophobic surface by casting and nanoimprint methods have been successful. Sun *et al.*[70] reported a nanocasting method to make a superhydrophobic PDMS surface. They first made a negative PDMS template using the lotus leaf as an original template and then used the negative template to make a positive PDMS template — a replica of the original lotus leaf. Zhao *et al.*[88] prepared a superhydrophobic surface by casting a micellar solution of a copolymer poly(styrene-*b*-dimethylsiloxane) (PS-PDMS) in humid air based on the cooperation of vapor-induced phase separation and surface enrichment of PDMS block. Lee *et al.*[89] produced vertically aligned PS nanofibers by using nanoporous anodic aluminum oxide as a replication template in a heat- and pressure-driven nanoimprint pattern transfer process. As the aspect ratio of the polystyrene (PS) nanofibers increased, the nanofibers could not stand upright but formed twisted bundles resulting in a three-dimensionally rough surface with advancing and receding contact angles of 155.8° and 147.6°, respectively.[74]

Modifying the surface chemistry with a hydrophobic coating widens the potential applications of superhydrophobic surfaces. There are several ways to modify the chemistry of a surface including sol–gel, dip coating, self-assembly, electrochemical and chemical/physical vapor deposition. Shirtcliffe *et al.*[53] prepared porous sol–gel foams from organo-triethoxysilanes which exhibited switching between superhydrophobicity and superhydrophilicity when exposed to different temperatures. The critical switching temperature was between 275 and 550°C for different materials, and when the foam was heated above the critical temperature, complete rejection of water by the cavities to complete filling of the pores. Hikita *et al.*[90] used colloidal silica particles and fluoroalkylsilane as the starting materials and prepared a sol–gel film with superliquid-repellency by hydrolysis and condensation of alkoxysilane compounds. Feng *et al.*[51] produced superhydrophobic surfaces using ZnO nanorods by sol–gel method. They showed that superhydrophobic surfaces can be switched into hydrophilic surfaces by alternation of ultraviolet (UV) irradiation. Shang *et al.*[91] did not blend low surface energy materials in the sols, but described a procedure to make transparent superhydrophobic surfaces by modifying silica based gel films with a fluorinated silane. In a similar way, Wu *et al.*[92] made a microstructured ZnO-based surface via a wet chemical process and obtained superhydrophobicity after coating the surface with long-chain alkanoic acids.

Zhai *et al.*[93] used a layer-by-layer (LBL) self-assembly technique to create a poly(allylamine hydrochloride)/poly(acrylic acid) (PAH/PAA) multilayer which formed a honeycomb-like structure on the surface after an appropriate combination of acidic treatments. After cross-linking the structure, they deposited silica nanoparticles on the surface via alternating dipping of the substrates into an aqueous suspension of the negatively charged nanoparticles and an aqueous PAH solution, followed by a final dipping into the nanoparticle suspension. Superhydrophobicity was obtained after the surface was modified by a CVD of (tridecafluoro-1,1,2,2-tetrahydrooctyl)-1-trichlorosilane followed by a thermal annealing.

Zhang *et al.*[94] showed that the surface covered with dendritic gold clusters, which was formed by electrochemical deposition onto indium tin oxide (ITO) electrode modified with a polyelectrolyte multilayer, showed superhydrophobic properties after further deposition of an *n*-dodecanethiol monolayer. The chemical bath deposition (CBD) has also been used to make nanostructured surfaces, thus, Hosono *et al.*[95] fabricated a nanopin film of brucite-type cobalt hydroxide (BCH) and achieved a contact angle of 178° after further modification of lauric acid (LA). Shi *et al.*[77] described the use of galvanic cell reaction as a facile method to chemically deposit Ag nanostructures on the *p*-silicon wafer on a large scale. When the Ag covered silicon wafer was further modified with a self-assembled monolayer of *n*-dodecanethiol, a superhydrophobic surface was obtained with a contact angle of about 154° and a tilt angle lower than 5°.

5.2. *Evaporation Induced Honeycomb Polymer Surfaces*

Micrometrically scaled honeycomb patterns obtained under rapid evaporation of polymer solutions were first reported by Widawski *et al.*[96] and Pitois and François,[97] and then they were studied by various groups including Srinivasarao *et al.*,[98] Bormashenko *et al.*,[99,100] and Kabuto *et al.*[101] The typical honeycomb relief obtained with evaporated solution of PS in a mixture of chlorinated solvents is presented in Fig. 11. Various polymers were already used for manufacturing porous relieves, including PC, PMMA, and PS. It was shown that the make-up of the relief (diameter of pores and their depth) could be controlled effectively by variation of the concentration of polymer solution, molecular mass of a polymer, etc.[99,100] The formation of the honeycomb pattern is related usually to condensation of water drops on the surface of rapidly evaporated polymer solution, thus the process is also called water-assisted self-organization.[98] In spite of the fact that the physical mechanism of the patterning is not profoundly understood,[102] the ease of fabrication of large-area close-packed micrometrically scaled patterns and the flexibility of the process make it attractive for investigators.

Yabu and Shimomura[103] prepared a porous superhydrophobic transparent membrane by casting a fluorinated block polymer solution under humid environment. The transparency was achieved because the honeycomb-patterned films had sub-wavelength pore size. Yabu and Shimomura also recently used fluorinated polymer honeycomb relief for manufacturing a pincushion-like structure demonstrating APCA as high as 170°.[104] Polymer honeycomb patterns could be also used as a matrix for manufacturing metallic highly developed relieves, demonstrating pronounced hydrophobic properties.[11]

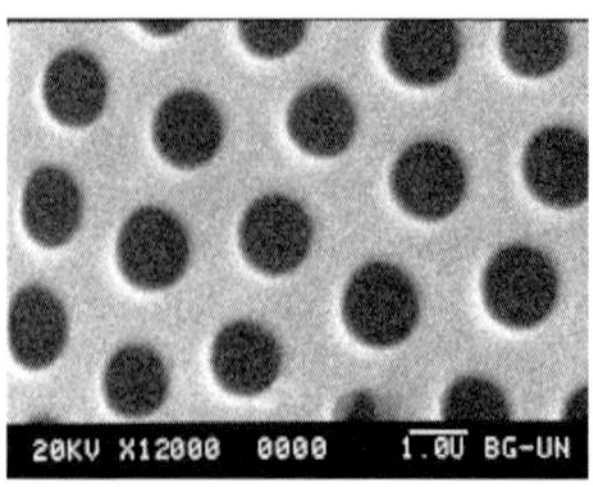

Fig. 11. SEM image of the honeycomb structure. (the scale bar is 1 μm)

The methods used for the production of micropatterned superhydrophobic surfaces are summarized in Table 1.[74]

5.3. *Commercially Available Lotus-Effect Products*

A number of products that utilize the Lotus effect are already commercially available or being developed. In addition, many patents have been granted for various possible applications of the self-cleaning surfaces.[104] Most of these applications utilize the self-cleaning effect, especially in the case of glasses (for architecture, automotive, optical sensor, and other applications), roof tiles, and other architectural applications. Besides that, sprays and paints which create clean surfaces (e.g. graffiti-resistant) have been suggested, as well as water-repellent textiles. Some agricultural applications are also discussed (e.g. pesticide additives that can decrease bouncing off plant surfaces or increase penetration into the soil).

From a commercial point of view, cleaning of windows is expensive and cumbersome, especially if the windows are on a skyscraper. The hydrophilic route to self-cleaning windows has meanwhile been realized by several companies and has been released to the market. How far these windows will be a commercial success remains to be seen.[108] The Germany-based website Lotus-Effekt.de, dedicated to the commercial application of patented self-cleaning superhydrophobic micro-to-nano structured surfaces, states the following:

"*Lotusan*®, an exterior paint from the firm *Sto* is marketed already with the greatest success since 1999. It is used by professional firms of house painters and is, not yet, available for the general building trade. Up to the present, *Lotusan*® has been used to paint c. 300,000 buildings. In 2004, *Degussa* (daughter company of *Goldschmidt*) has introduced the first spray: *Tegotop*® 105 which can be used to impregnate surfaces. Self-cleaning textiles are being tested at present and will be commercially available from summer 2005. Marquees will probably be the first to receive such treatment. Optical sensors in public high impact areas (for instance, toll bridge sensors on highways) are already furnished throughout Germany with Lotus-*Effect*® glass manufactured by *Ferro AG*. A series of further products is being tested, among these *Aeroxide*® LE of the *Degussa* for plastics. For years self-cleaning glasses have appeared in advertisements, featuring the so-called photo catalytic stratification. The firm *Ferro* keeps prototypes of architectural glass with Lotus-*Effect*® in permanent test conditions. In the region of optical sensors (toll bridges), Lotus-*Effect*® glasses are already used successfully. For architectural glass and rear windows of cars, applications will probably follow soon. With *Erlus-Lotus*®, the first self-cleaning roof in the world came on the market. For demonstrations we employ a spoon with a perfect Lotus-*Effect*® surface. Honey and many other substances roll off without a hint of residue. The

Table 1. Typical materials and corresponding techniques to produce micro/nanoroughness (based on Ref. [74]).

Material	Technique	Contact angle, °	Notes	Source
Teflon	Plasma	168		Zhang *et al.*[94]; Shiu *et al.*[105]
PFOS	Electro- and chemical polymerization	152	Reversible (electric potential)	Xu *et al.*[52]
PDMS	Laser treatment	166		Khorasani *et al.*[79]
PS-PDMS block copolymer	Electrospining	>150		Ma *et al.*[106]
PVDF copolymer	Embossing	>150		Bormashenko *et al.*[11]
Copolymer comprising fluorinated and methyl methacrylate repeat units	Evaporation	160–170		Yabu and Shimomura[103]
PS nanofiber	Nanoimprint	156		Lee *et al.*[89]
PET	Oxygen plasma etching	>150		Teshima *et al.*[78]
Organo-triethoxysilanes	Sol–gel	155	Reversible (temperature)	Shirtcliffe *et al.*[53]
Al	Chemical etching	>150		Qian and Shen[80]
Copper	Electrodeposition	160	Hierarchical	Shirtcliffe *et al.*[107]
Si	Photolithography	170		Bhushan and Jung[73]
Si	E-beam lithography	164		Martines *et al.*[72]
Si	X-ray lithography	>166		Fürstner *et al.*[60]

spoon is a prototype that is, unfortunately, not for sale. Firms can already order the first spray Tegotop® 105 for testing from the *Degussa-Goldschmidt AG.* The properties of the new-fangled intelligent textiles are astonishing. Not only does water roll off, but ketchup and red wine do likewise. The area of use will hardly be in suits, ties or shirts, but rather in outdoor clothing, marquees, tents, and in tarpaulins for lorries.[109]

In addition to household and "conventional" products, possible use of roughness- or heterogeneity-induced superhydrophobicity in nano- and biotechnology applications is often discussed. This includes, for example, nonsticky surfaces for the components of micro/nanoelectromechanical systems (NEMS/MEMS). Since adhesion plays an important role for small devices, the so-called "stiction" of two component surfaces is a significant problem in that industry, which may lead to device failure. Making a surface hydrophobic can reduce meniscus force and stiction.[57]

Controlling drops containing biologically relevant molecules (DNA and proteins) is important in biotechnology. Superhydrophobicity is useful for these applications: the almost fully spherical drops on a superhydrophobic surface can shrink exactly like a drop in free air. Furthermore, the positioning and shape of water drops can be controlled by a pattern that combines hydrophilic and hydrophobic elements. Interestingly, some desert beetles capture their drinking water by a hydrophobic/philic structured back.[110] At a patterned heterogeneous substrate, hydrophilic regions can help to contain small liquid volumes of DNA, which may improve spotting and analyzing DNA and proteins by avoiding wall contact.[105,111]

In micro/nanofluidics, a guided motion of drops on heterogeneous hydrophobic/philic surfaces gives the opportunity to develop drop-based microfluidics systems, as opposed to the classical concept based on microfluidic channels. Drops moving freely on open surfaces and bulk liquids flowing in channels constitute two extremes, with the patterned heterogeneous hydrophobic/philic surfaces being the intermediate between these two.[112] Driving the liquids along the channels and making them merge at predefined locations offers a novel way to mix reactants or steer biochemical reactions, defining the concept of a "liquid microchip"[113] or "surface-tension confined microfluidics".[114] These open structures have advantages over capillaries, because blocking of the capillary by unforeseen chemical reactions cannot occur. Drops have very low contact areas with the substrate, and they are easy to move by external fields, for example, electrostatic forces or surface capillary waves. Systems that make use of a drop-based actuation mechanism are also being developed, and their aim is to control drop positioning and motion on the substrates with as little surface contact as possible, and to turn the drop-based system into a programmable reactor, by which the liquid positions are prescribed and tuned.[105]

6. Closure

In this chapter, we have discussed the superhydrophobic and self-cleaning surfaces in nature, the theory of superhydrophobicity, artificial superhydrophobic surfaces, and the prospects of their application in commercial products. We have demonstrated that a high contact angle does not necessarily provide self-cleaning properties of the surface, whereas a combination of a high contact angle and low contact angle hysteresis yields a true superhydrophobicity. The crucial role of hierarchical scaling of the relief in constituting water-repelling properties of the surface was discussed. The superhydrophobicity and lotus effect constitute a rapidly developing area with a number of applications commercially available and growing potential for future applications. With the technology of surface micropatterning becoming more mature, new opportunities emerge. This includes combining the superhydrophobicity with optical properties and producing transparent and nonreflective (using the moth-eye effect) surfaces, reversible superhydrophobicity with electrowetting, and other ways to control surface energy, dynamic effects (such as bouncing drops), underwater superhydrophobicity. Superhydrophobic surface is quite diverse including various materials and types of roughness. Therefore, the superhydrophobicity should be treated as a universal physical phenomenon. Interesting theoretical questions that emerged during these studies (multiscale nature of wetting, proper selection of roughness parameters, approaches to wetting regime transition) are being solved successfully by scientists.

References

[1] Barthlott W, Neinhuis C. (1997) Purity of the sacred lotus, or escape from contamination in biological surfaces. *Planta* **202**: 1–8.

[2] Baker EA. (1982) Chemistry and morphology of plant epicuticular waxes. In DF Cutler, KL Alvin & CE Price (eds), *The Plant Cuticle*, pp. 139–165. Academic Press, New York.

[3] Neinhuis C, Barthlott W. (1997) Characterization and distribution of water-repellent, self-cleaning plant surfaces. *Ann Bot* **79**: 667–677.

[4] Wagner P, Furstner R, Barthlott W, Neinhuis, C. (2003) Quantitative assessment to the structural basis of water repellency in natural and technical surfaces. *J Exp Bot* **54**: 1295–1303.

[5] Jetter R, Kunst L, Samuels AL. (2006) Composition of plant cuticular waxes. In M Riederer & C Müller (eds), *Biology of the Plant Cuticle*, pp. 145–181. Blackwell Publishing, Oxford.

[6] Damste JSS, Dekker M, van Dongen BE, Schouten S, Piersma T. (2000) Structural identification of the diester preen-gland waxes of the red knot (*Calidris canutus*). *J Nat Prod* **63**: 381–384.

[7] Bormashenko E, Bormashenko Y, Stein T, Whyman G, Bormashenko E. (2007) Why do pigeon feathers repel water? Hydrophobicity of Pennae, Cassie–Baxter wetting hypothesis and Cassie–Wenzel capillarity-induced wetting transition. *J Colloid Interf Sci* **311**: 212–216.

[8] Nosonovsky M. (2007) Multiscale roughness and stability of superhydrophobic biomimetic interfaces. *Langmuir* **23**: 3157–3161.

[9] Nosonovsky M, Bhushan B. (2007) Hierarchical roughness optimization for biomimetic superhydrophobic surfaces. *Ultramicroscopy* **107**: 969–979.

[10] Nosonovsky M, Bhushan B. (2007) Biomimetic superhydrophobic surfaces: multiscale approach. *Nano Lett* 7: 2633–2637.

[11] Bormashenko E, Bormashenko Y, Whyman G, Pogreb R, Stanevsky O. (2006) Micrometrically scaled textured metallic hydrophobic interfaces validate the Cassie–Baxter wetting hypothesis. *J Colloid Interf Sci* **302**: 308–311.

[12] Song F, Xiao KW, Bai K, Bai YL. (2007) Microstructure and nanomechanical properties of the wing membrane of dragonfly. *Mater Sci Eng A* **457**: 254–260.

[13] Gao XF, Jiang L. (2004) Water-repellent legs of water striders. *Nature* **432**: 36.

[14] Rowlinson JS, Widom B. (1982) *Molecular Theory of Capillarity.* Clarendon, Oxford.

[15] Yaminsky V. (1995) Molecular mechanisms of hydrophobic transitions. In J Drelich, JS Laskowski & KL Mittal (eds), *Apparent and Microscopic Contact Angles*, pp. 47–95. VSP, Utrecht.

[16] Shanahan MER, Carré A. (1995) Viscoelastic dissipation in wetting and adhesion phenomena. *Langmuir* **11**: 1396–1402.

[17] Derjaguin BV, Churaev NV. (1974) Structural component of disjoining pressure. *J Colloid Interf Sci* **49**: 249–255.

[18] Checco A, Guenoun P, Daillant J. (2003) Nonlinear dependence of the contact angle of nanodroplets on contact line curvatures. *Phys Rev Lett* **91**: 186101.

[19] Anisimov MA. (2007) Divergence of Tolman's length for a droplet near the critical point. *Phys Rev Lett* **98**: 035702.

[20] Pompe T, Fery A, Herminghaus S. (2000) Measurement of contact line tension by analysis of the three-phase boundary with nanometer resolution. In J Drelich, JS Laskowski & KL Mittal (eds), *Apparent and Microscopic Contact Angles*, pp. 3–12. VSP, Utrecht.

[21] Wenzel RN. (1936) Resistance of solid surfaces to wetting by water. *Indust Eng Chem* **28**: 988–994.

[22] Cassie A, Baxter S. (1944) Wettability of porous surfaces. *Trans Faraday Soc* **40**: 546–551.

[23] Gao L, McCarthy TJ. (2007) How Wenzel and Cassie were wrong. *Langmuir* **23**: 3762–3765.

[24] Extrand CW. (2003) Contact angle hysteresis on surfaces with chemically heterogeneous islands. *Langmuir* **19**: 3793–3796.

[25] Whyman G, Bormashenko E, Stein T. (2008) The rigorous derivation of Young, Cassie–Baxter and Wenzel equations and the analysis of the contact angle hysteresis phenomenon. *Chem Phys Lett* **450**: 355–359.

[26] Nosonovsky M. (2007) On the range of applicability of the Wenzel and Cassie equations. *Langmuir* **23**: 9919–9920.
[27] McHale G. (2007) Cassie and Wenzel: were they really so wrong? *Langmuir* **23**: 8200–8205.
[28] Panchagnula MV, Vedantam S. (2007) Comment on how Wenzel and Cassie were wrong by Gao and McCarthy. *Langmuir* **23**: 13242.
[29] Gupta P, Ulman A, Fanfan F, Korniakov A, Loos K. (2005) Mixed self-assembled monolayer of alkanethiolates on ultrasmooth gold do not exhibit contact angle hysteresis. *J Am Chem Soc* **127**: 4–5.
[30] Nosonovsky M. (2007) Model for solid–liquid and solid–solid friction for rough surfaces with adhesion hysteresis. *J Chem Phys* **126**: 224701.
[31] Nosonovsky M, Bhushan B. (2007) Multiscale friction mechanisms and hierarchical surfaces in nano- and bio-tribology. *Mater Sci Eng R* **58**: 162–193.
[32] Kusumaatmaja H, Yeomans JM. (2007) Modeling contact angle hysteresis on chemically patterned and superhydrophobic surfaces. *Langmuir* **23**: 6019–6032.
[33] Vedantam S, Panchagnula MV. (2007) Phase field modeling of hysteresis in sessile drops. *Phys Rev Lett* **99**: 176102.
[34] Boruvka L, Neumann AW. (1977) Generalization of the classical theory of capillarity. *J Chem Phys* **66**: 5464–5476.
[35] Lafuma A, Quéré D. (2003) Superhydrophobic states. *Nat Mater* **2**: 457–460.
[36] Liu B, Lange FF. (2006) Pressure induced transition between superhydrophobic states: Configuration diagrams and effect of surface feature size. *J Colloid Interf Sci* **298**: 899–909.
[37] McHale G, Aqil S, Shirtcliffe NJ, Newton MI, Erbil HY. (2005) Analysis of droplet evaporation on a superhydrophobic surface, *Langmuir* **21**: 11053–11060.
[38] Yoshimitsu Z, Nakajima A, Watanabe T, Hashimoto K. (2002) Effects of surface structure on the hydrophobicity and sliding behavior of water droplets. *Langmuir* **18**: 5818–5822.
[39] Bormashenko E, Pogreb R, Whyman G, Erlich M. (2007) Cassie–Wenzel wetting transition in vibrated drops deposited on the rough surfaces: Is dynamic Cassie–Wenzel transition 2D or 1D affair? *Langmuir* **23**: 6501–6503.
[40] Bormashenko E, Pogreb R, Whyman G, Erlich M. (2007) Resonance Cassie–Wenzel wetting transition for horizontally vibrated drops deposited on a rough surface. *Langmuir* **23**: 12217–12221.
[41] Ishino C, Okumura K. (2006) Nucleation scenarios for wetting transition on textured surfaces: The effect of contact angle hysteresis. *Europhys Lett* **76**: 464–470.
[42] Della Volpe C, Maniglio D, Morra M, Siboni S. (2002) The determination of a 'stable-equilibrium' contact angle on heterogeneous and rough surfaces. *Colloids Surf A*, **206**: 47–67.
[43] Meiron TS, Marmur A, Saguy IS. (2004) Contact angle measurement on rough surfaces. *J Colloid Interf Sci* **274**: 637–644.
[44] Zhang J, Gao X, Jiang L. (2007) Application of superhydrophobic edge effects in solving the liquid outflow phenomena. *Langmuir* **23**: 3230–3235.

[45] Bormashenko E, Pogreb R, Stein T, Whyman G, Erlich M, Musin A, Machavariani V, Aurbach D. (2008) Characterization of rough surfaces with vibrated drops. *Phys Chem Chem Phys* **10**: 4056–4061.

[46] de Gennes PG, Brochard-Wyart F, Quéré D. (2003) *Capillarity and Wetting Phenomena*. Springer, Berlin.

[47] Marmur A. (2006) Soft contact: Measurement and interpretation of contact angles. *Soft Matter* **2**: 12–17.

[48] Bormashenko E, Stein T, Whyman G, Bormashenko Y, Pogreb E. (2006) Wetting properties of the multiscaled nanostructured polymer and metallic superhydrophobic surfaces. *Langmuir* **22**: 9982–9985.

[49] Bormashenko E, Bormashenko Y, Stein T, Whyman G, Pogreb R, Barkay Z. (2007) Environmental scanning electron microscope study of the fine structure of the triple line and Cassie–Wenzel wetting transition for sessile drops deposited on rough polymer substrates. *Langmuir* **23**: 4378–4382.

[50] Marmur A. (2003) Wetting on hydrophobic rough surfaces: To be heterogeneous or not to be? *Langmuir* **19**: 8343–8348.

[51] Feng XJ, Feng L, Jin MH, Zhai J, Jiang L, Zhu, DB. (2004) Reversible superhydrophobicity to super-hydrophilicity transition of aligned ZnO nanorod films. *J Am Chem Soc* **126**: 62–63.

[52] Xu L, Chen W, Mulchandani A, Yan Y. (2005) Reversible conversion of conducting polymer films from superhydrophobic to superhydrophilic. *Angew Chem Int Ed* **44**: 6009–6012.

[53] Shirtcliffe NJ, McHale G, Newton MI, Perry CC, Roach P. (2005) Porous materials show superhydrophobic to superhydrophilic switching. *Chem Commun* **31**: 3135–3137.

[54] Wang S, Liu H, Liu D, Ma X, Fang X, Jiang L. (2007) Enthalpy driven three state switching of a superhydrophilic/superhydrophobic surface. *Angew Chem Int Ed* **46**: 3915–3917.

[55] Krupenkin TN, Taylor JA, Wang EN, Kolodner P, Hodes M, Salamon TR. (2007) Reversible wetting-dewetting transitions on electrically tunable superhydrophobic nanostructured surfaces. *Langmuir* **23**: 9128–9133.

[56] Ma M, Hill RM. (2006) Superhydrophobic surfaces. *Curr Opin Colloid Interf Sci* **11**: 193–202.

[57] Nosonovsky M, Bhushan B. (2008) Patterned non-adhesive surfaces: Superhydrophobicity and wetting regime transitions. *Langmuir* **24**: 1525–1533.

[58] Patankar NA. (2004) Mimicking the lotus effect: Influence of double roughness structures and slender pillars. *Langmuir* **20**: 8209–8213.

[59] Nosonovsky M, Bhushan B. (2008) Biologically-inspired surfaces: Broadening the scope of roughness. *Adv Func Mater* **18**: 843–855.

[60] Fürstner R, Barthlott W, Neinhuis C, Walzel P. (2005) Wetting and self-cleaning properties of artificial superhydrophobic surfaces. *Langmuir* **21**: 956–961.

[61] Hartley GS, Brunskill RT. (1958) Reflection of water drops from surfaces. In JF Danielli (ed), *Surface Phenomena in Chemistry and Biology*, pp. 214–223. Pergamon, London.

[62] Quéré D. (2005) Non-sticking drops. *Rep Prog Phys* **68**: 2495–2532.

[63] Bergeron V, Bonn D, Martin JY, Vovelle L. (2000) Controlling droplet deposition with polymer additives. *Nature* **405**: 772–775.

[64] Richard D, Quéré D. (2000) Bouncing water drops. *Europhys Lett* **50**: 769–775.

[65] Furmidge CGL. (1962) Studies at phase interfaces. I. The sliding of liquid drops on solid surfaces and a theory for spray retention. *J Colloid Sci* **17**: 309–324.

[66] Macdougall G, Ockrent C. (1942) Surface energy relations in liquid solid systems. *Proc R Soc London* **180**: 0151–0173.

[67] Frenkel YI. (1948) On the behavior of liquid drops on a solid surface. 1. The sliding of drops on an inclined surface. *J Exp Theoret Phys (USSR)* **18**: 659.

[68] Krasovitski B, Marmur A. (2004) Drops down the hill: Theoretical study of limiting contact angles and the hysteresis range on a tilted plane. *Langmuir* **21**: 3881–3885.

[69] Lahann J, Mitragotri S, Tran T, Kaido H, Sundaram J, Choi IS, Hoffer S, Somorjai GA, Langer R. (2003) A reversibly switching surface. *Science* **299**: 371–374.

[70] Sun M, Luo C, Xu L, Ji H, Ouyang Q, Yu D, Chen Y. (2005) Artificial lotus leaf by nanocasting. *Langmuir* **21**: 8978–8981.

[71] Nakajima A, Fujishima A, Hashimoto K, Watanabe T. (1999) Preparation of transparent superhydrophobic boehmite and silica films by sublimation of aluminum acetylacetonate. *Adv Mater* **11**: 1365–1368.

[72] Martines E, Seunarine K, Morgan H, Gadegaard N, Wilkinson CDW, Riehle MO. (2005) Superhydrophobicity and superhydrophilicity of regular nanopatterns. *Nano Lett* **5**: 2097–2103.

[73] Bhushan B, Jung YC. (2007) Wetting study of patterned surfaces for superhydrophobicity. *Ultramicroscopy* **107**: 1033–1041.

[74] Bhushan B, Nosonovsky M, Jung YC. (2008) Lotus effect: Roughness-induced superhydrophobic surfaces. In B Bhushan (ed). *Nanotribology and Nanomechanics: An Introduction*, pp. 995–1072. Springer Verlag, Berlin.

[75] Jansen H, de Boer M, Legtenberg R, Elwenspoek M. (1995) The black silicon method: A universal method for determining the parameter setting of a fluorine-based reactive ion etcher in deep silicon trench etching with profile control. *J Micromech Microeng* **5**: 115–120.

[76] Coulson SR, Woodward I, Badyal JPS, Brewer SA, Willis C. (2000) Super-repellent composite fluoropolymer surfaces. *J Phys Chem B* **104**: 8836–8840.

[77] Shi F, Song Y, Niu J, Xia X, Wang Z, Zhang X. (2006) Facile method to fabricate a large-scale superhydrophobic surface by Galvanic cell reaction. *Chem Mater* **18**: 1365–1368.

[78] Teshima K, Sugimura H, Inoue Y, Takai O, Takano A. (2005) Transparent ultra water-repellent poly(ethylene terephthalate) substrates fabricated by oxygen plasma treatment and subsequent hydrophobic coating. *Appl Surf Sci* **244**: 619–622.

[79] Khorasani MT, Mirzadeh H, Kermani Z. (2005) Wettability of porous polydimethylsiloxane surface: Morphology study. *Appl Surf Sci* **242**: 339–345.

[80] Qian BT, Shen ZQ. (2005) Fabrication of superhydrophobic surfaces by dislocation-selective chemical etching on aluminum, copper, and zinc substrates. *Langmuir* **21**: 9007–9009.

[81] Zhang JL, Li JA, Han YC. (2004) Superhydrophobic PTFE surfaces by extension. *Macromol Rapid Commun* **25**: 1105–1108.

[82] Shibuichi S, Onda T, Satoh N, Tsujii K. (1996) Super-water-repellent surfaces resulting from fractal structure. *J Phys Chem* **100**: 19512–19517.

[83] Klein RJ, Biesheuvel PM, Yu BC, Meinhart CD, Lange FF. (2003) Producing superhydrophobic surfaces with nano-silica spheres. *Z Metallkd* **94**: 377–380.

[84] Abdelsalam ME, Bartlett PN, Kelf T, Baumberg J. (2005) Wetting of regularly structured gold surfaces. *Langmuir* **21**: 1753–1757.

[85] Lau KKS, Bico J, Teo KBK, Chhowalla M, Amaratunga GAJ, Milne WI, McKinley GH, Gleason KK. (2003) Superhydrophobic carbon nanotube forests. *Nano Lett* **3**: 1701–1705.

[86] Zhu L, Xiu Y, Xu J, Tamirisa PA, Hess DW, Wong C. (2005) Superhydrophobicity on two-tier rough surfaces fabricated by controlled growth of aligned carbon nanotube arrays coated with fluorocarbon. *Langmuir* **21**: 11208–11212.

[87] Huang L, Lau SP, Yang HY, Leong ESP, Yu SF, Prawer S. (2005) Stable superhydrophobic surface via carbon nanotubes coated with a ZnO thin film. *J Phys Chem B* **109**: 7746–7748.

[88] Zhao N, Xie QD, Weng LH, Wang SQ, Zhang XY, Xu J. (2005) Superhydrophobic surface from vapor-induced phase separation of copolymer micellar solution. *Macromolecules* **38**: 8996–8999.

[89] Lee W, Jin M, Yoo W, Lee J. (2004) Nanostructuring of a polymeric substrate with well-defined nanometer-scale topography and tailored surface wettability. *Langmuir* **20**: 7665–7669.

[90] Hikita M, Tanaka K, Nakamura T, Kajiyama T, Takahara A. (2005) Superliquid-repellent surfaces prepared by colloidal silica nanoparticles covered with fluoroalkyl groups. *Langmuir* **21**: 7299–7302.

[91] Shang HM, Wang Y, Limmer SJ, Chou TP, Takahashi K, Cao GZ. (2005) Optically transparent superhydrophobic silica-based films. *Thin Solid Films* **472**: 37–43.

[92] Wu XD, Zheng LJ, Wu D. (2005) Fabrication of superhydrophobic surfaces from microstructured ZnO-based surfaces via a wet-chemical route. *Langmuir* **21**: 2665–2667.

[93] Zhai L, Cebeci FC, Cohen RE, Rubner MF. (2004) Stable superhydrophobic coatings from polyelectrolyte multilayers. *Nano Lett* **4**: 1349–1353.

[94] Zhang X, Shi F, Yu X, Liu H, Fu Y, Wang ZQ, Jiang L, and Li X. (2004) Polyelectrolyte multilayer as matrix for electrochemical deposition of gold clusters: Toward superhydrophobic surface. *J Am Chem Soc* **126**: 3064–3065.

[95] Hosono E, Fujihara S, Honma I, Zhou HS. (2005) Superhydrophobic perpendicular nanopin film by the bottom-up process. *J Am Chem Soc* **127**: 13458–13459.

[96] Widawski G, Rawiso M, François B. (1994) Self-organized honeycomb morphology of star-polymer polystyrene films. *Nature* **369**: 387–389.

[97] Pitois O, François B. (1999) Formation of ordered micro-porous membranes. *Eur Phys J B* **8**: 225–231.

[98] Srinivasarao M, Collings D, Philips A, Patel S. (2001) Three-dimensionally ordered array of air bubbles in a polymer film. *Science* **292**: 79–83.

[99] Bormashenko E, Pogreb R, Stanevsky O, Bormashenko Y, Stein T, Cohen R, Nunberg M, Gaisin VZ, Gorelik M, Gendelman O. (2005) Mesoscopic and submicroscopic patterning in thin polymer films: Impact of the solvent. *Mater Lett* **59**: 2461–2464.

[100] Bormashenko E, Pogreb R, Stanevsky O, Bormashenko Y, Gendelman O. (2005) Formation of honeycomb patterns in evaporated polymer solutions: Influence of the molecular weight. *Mater Lett* **59**: 3553–3557.

[101] Kabuto T, Hashimoto Y, Karthaus O. (2007) Thermally stable and solvent resistant mesoporous honeycomb films from a crosslinkable polymer. *Adv Func Mater* **17**: 3569–3573.

[102] Bormashenko E, Musin A, Bormashenko Y, Whyman G, Pogreb R, Gendelman O. (2007) Formation of films on water droplets floating on a polymer solution surface. *Macromol Phys Chem* **208**: 702–709.

[103] Yabu H, Shimomura M. (2005) Single-step fabrication of transparent superhydrophobic porous polymer films. *Chem Mater* **17**: 5231–5234.

[104] Yabu H, Takebayashi M, Tanaka M, Shimomura M. (2005) Superhydrophobic and lipophobic properties of self-organized honeycomb and pincushion structures. *Langmuir* **21**: 3235–3237.

[105] Shiu J, Kuo C, Chen P, Mou C. (2004) Fabrication of tunable superhydrophobic surfaces by nanosphere lithography. *Chem Mater* **16**: 561–564.

[106] Ma M, Hill RM, Lowery JL, Fridrikh SV, Rutledge GC. (2005) Electrospun poly(styrene-block-dimethylsiloxane) block copolymer fibers exhibiting superhydrophobicity. *Langmuir* **21**: 5549–5554.

[107] Shirtcliffe NJ, McHale G, Newton MI, Chabrol G, Perry CC. (2004) Dual-scale roughness produces unusually water-repellent surfaces. *Adv Mater* **16**: 1929–1932.

[108] Blossey R. (2003) Self-cleaning surfaces — virtual realities. *Nat Mater* **2**: 301–306.

[109] "Lotus Effect FAQ" web-site www.lotus-effekt.de (accessed on 10 August 2007).

[110] Parker AR, Lawrence CR. (2001) Water capture by a desert beetle. *Nature* **414**: 33–34.

[111] Gillmor SD, Thiel AJ, Strother TC, Smith LM, Lagally MG. (2000) Hydrophilic/hydrophobic patterned surfaces as templates for DNA arrays. *Langmuir* **16**: 7223–7228.

[112] Drelich J, Wilbur JL, Miller JD, Whitesides GM. (1996) Contact angles for liquid drops at a model heterogeneous surface consisting of alternating and parallel hydrophobic/hydrophilic strips. *Langmuir* **12**: 1913–1922.

[113] Gau H, Herminghaus S, Lenz P, Lipowsky R. (1999) Liquid morphologies on structured surfaces: From microchannels to microchips. *Science* **283**: 46–49.

[114] Lam P, Wynne KJ, Wnek GE. (2002) Surface-tension-confined microfluidics. *Langmuir* **18**: 948–951.

Chapter 4

The Moth-Eye Effect — From Fundamentals to Commercial Exploitation

Andreas Gombert and Benedikt Bläsi

The historical developments of artificial "moth-eye" structures as anti-reflective surfaces are described. In theory, "moth-eye" structures are regarded as zero-order diffraction gratings as the relation of the grating period to the wavelength is such that only zero-order diffracted waves can propagate. The optical properties of the grating can be described by effective medium theories. It is shown that propagation of zero-order diffracted waves alone is very difficult to achieve for visible wavelengths in general due to limitations concerning the manufacturability on large areas. Therefore, nonideal anti-reflective moth-eye structures had to be optimized with respect to the grating periods depending on the grating type and of course on the specific application. The results of the optimization methods are presented. Then the manufacturing methods especially suited for the origination and replication on large areas are described in detail. Finally, the applications of "moth-eye" structures and developments beyond the mimicry of nature are presented.

1. Introduction

The anti-reflective effect of surface-relief structures with lateral dimensions smaller than the wavelengths of light was first observed by Bernhard.[1] In 1967, he discovered that the cornea of many night-flying insects (Lepidoptera, Neuroptera, Trichoptera) has a nanostructure for the purpose of camouflage (see Fig. 1). The structure can be described as a two-dimensional (hexagonal) array of conical nipples. The distance between two nipples is in the range of 200–250 nm and the height is up to 250 nm. Based on Bernhard's publication, the anti-reflective effect of this type of surface-relief structures is often called the "moth-eye principle".[2–4] He showed by up-scaled experiments with radar waves and by theoretical calculations that these surface-relief

Fig. 1. SEM picture of natural moth-eye structures (flour moth, *Ephestia kuehniella*). © Fraunhofer ISE.

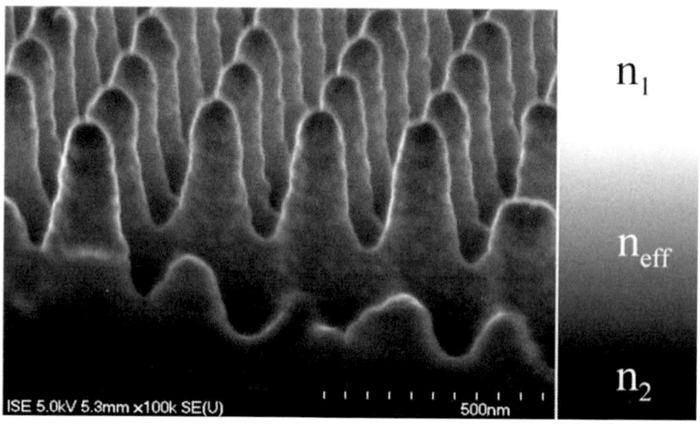

Fig. 2. SEM picture of an artificial moth-eye structure made at Fraunhofer ISE and illustration of the graded refractive index. The periods of the crossed grating are 250 nm each. © Fraunhofer ISE.

structures are anti-reflective (AR) over a broad spectral range. The AR effect is caused by the fact that incident light cannot resolve the fine array of protuberances. At any point in the cross-section profile of the structure, the optical properties are described by effective refractive indices. They are an average of those of the material of the cornea and air depending upon the relative proportions of the two at that point. Surface-relief structures with a continuous cross-section profile are similar to graded index interference coatings. This is illustrated in Fig. 2, in which an artificially made moth-eye structure is shown together with a bar indicating the smooth transition between the refractive indices of the homogeneous media. Moth-eye structures were always regarded as commercially very attractive because a sophisticated optical component, a very broadband anti-reflective surface, can be mass produced via relatively simple replication processes like embossing, for example.

In the 1970s, scientists were interested in the moth-eye effect for the creation of solar selective absorbers, i.e. surfaces which absorb the solar radiation very well but do not have too much thermal radiation losses after being heated by the absorption of the solar radiation.[5–8] At the beginning of the 1980s, subwavelength gratings in photoresist were studied as anti-reflective and birefringent artificial dielectrics.[4,9–11] Later, theoretical work was published on anti-reflective rectangular groove subwavelength gratings on dielectric and lossy materials using effective medium theory (EMT).[12–14]

In the 1990s, the interest in subwavelength gratings was increasing, presumedly because the possibilities of numerical simulation and of micro and nanofabrication improved. The homogenous properties of subwavelength gratings have been addressed and studied using different EMT's which have been proposed for one- and two-dimensional gratings.[15–20] The validity of the EMT's has been investigated mainly for binary subwavelength gratings by comparison with results obtained by rigorous diffraction theory. Multilevel stairstep gratings have been studied by Glytsis and Gaylord[21] and continuous one-dimensional surface profiles by Raguin and Morris[22] for large wavelength-to-period ratios (i.e. the wavelength is much larger than the grating period). New structure profiles like the quintic[23] or the surface-relief equivalent of the Klopfenstein taper[24] have been proposed.

It is worth noting that there are only a few publications on experimental work in general. Some of them published in the 1990s are on structures on semiconductors which are anti-reflective in the infrared.[25,26] Others including the work of the authors were motivated by solar energy applications for which the AR effect has to be effective over a very broad spectral range, a requirement which can best be fulfilled by mixing the refractive indices of air and the transparent material on a subwavelength scale.[27–29]

In the late 1990s, several commercial applications started. Large-area applications in particular, were based on the work at Fraunhofer ISE. From late 1997 on, the company Fresnel Optics[30] in Germany produced many thousands of Fresnel lenses for LC projectors made anti-reflective by moth-eye structures. In 2001, the company Holotools[31] was founded which originates moth-eye and other micro and nanostructures by interference lithography homogeneously on areas of up to $960 \times 720\,\text{mm}^2$. The company MacDermid Autotype[32] sells various optical films with moth-eye structures (see Fig. 3). In the first years of this century, most of the work related to AR surface-relief structures is on the fabrication of aperiodic structures. Manufacturing methods like phase separation in polymer blends,[33] self-organized surface-relief structures in sputtered coatings[34] or plasma etching processes[35] are proposed. For some of these structures, the wording "moth-eye" is used as a label.

Fig. 3. Image of a flat-panel display with standard anti-glare surface (left side) and combined moth-eye anti-reflective and anti-glare (MARAG™) surface (right side). On the flat-panel display, the head of a flour moth (SEM image) is shown. © MacDermid Autotype.

2. Theory

Moth-eye structures are so-called subwavelength surface-relief gratings, i.e. diffractive optical elements that do not bend light but produce zero-order diffraction only. All higher-order diffracted fields are evanescent and only the zero-order propagates as in a homogeneous medium. Anti-reflective surface-relief gratings can be periodic in one (1D) or two directions (2D). Figure 4 is a sketch of a 1D surface-relief diffraction grating. One generally distinguishes between the two homogeneous regions I and III, and the modulated region II.

An incident electromagnetic wave with a wave vector **k** lying in the x–z plane is diffracted into a certain number of diffraction orders according to the so-called grating equation[39]:

$$n_{1,3} \sin \theta_{1,3} - n_1 \sin \theta_i = \frac{m\lambda_0}{\Lambda} \tag{1}$$

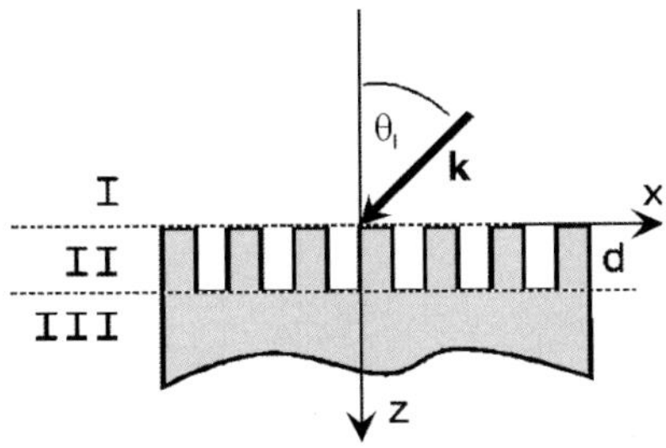

Fig. 4. Sketch of a 1D surface-relief diffraction grating. © Fraunhofer ISE.

where $n_{1,3}$ is the appropriate refractive index, $\theta_{1,3m}$ the angle of diffraction of the mth order in regions I or III, θ_i the angle of incidence, λ_0 the incident vacuum wavelength, and Λ the grating period.

The situation described in Fig. 4 corresponds to the so-called classical diffraction, where the plane of incidence contains the grating vector **K** with the components $K_x = 2\pi/\Lambda$, $K_y = 0$, and $K_z = 0$. If this is not the case, then the situation is called conical diffraction because the wave vectors of the diffracted waves lie on the surface of a cone.

As seen from the grating equation, the actual ratio λ_0/Λ required for zero-order diffraction only depends on the refractive indices and the angle of incidence. If only zero-order diffraction occurs, the inhomogeneous region II can be regarded as an effective medium with homogeneous optical properties.

In other words, if the spatial variation of electromagnetic fields over the boundaries of the two materials in region II is very small, then the wave cannot resolve the structure and the inhomogeneous medium behaves like a homogeneous one. The optical properties of these homogeneous effective media can then be described by an effective dielectric tensor. Of course, this effective dielectric tensor depends on the dielectric tensors of the single components, on the volume fraction of the components and on the form of the components.

2.1. *Effective Medium Theories (EMTs) for Subwavelength Gratings*

The major developments in effective medium theories for periodically stratified media shall be addressed here. One has generally to distinguish between one and two-dimensional subwavelength gratings. The earliest work which can now be related to one-dimensional subwavelength gratings is the work of Wiener on form birefringence.[36] The work of Wiener assumes that the wavelength λ_0 is much larger

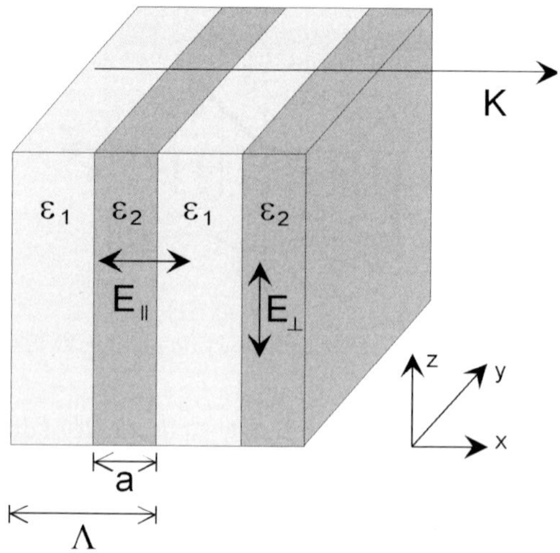

Fig. 5. Periodically stratified medium. © Fraunhofer ISE.

than the period Λ of the structure and uses static field expressions. The theory of form birefringence is based on the consideration of field continuity. According to Fig. 5, we see immediately that only two different decoupled situations for the field exist: the electric field is parallel or perpendicular to the grating vector **K**. Thus, the one-dimensional subwavelength grating behaves like a uniaxial birefringent medium with its optical axis parallel to the grating vector. In a z- and y-invariant medium, an electromagnetic field is said to be $E_\perp$ $(H_\perp)$ if the x-component of its electric (magnetic) field vanishes.

Let us first consider the polarization for which the electrical field is perpendicular to the grating vector $(E_\perp)$. Due to the continuity of the tangential component of the electrical field, the electrical field is the same in regions 1 and 2. If we define the filling factor f by:

$$f = \frac{a}{\Lambda}, \tag{2}$$

then we can write the effective electric displacement as:

$$D_{\text{eff}}^{(\perp)} = \varepsilon_{\text{eff}}^{(\perp)} E_\perp = (1-f)\varepsilon_1 E_\perp + f\varepsilon_2 E_\perp, \tag{3}$$

and thus:

$$\varepsilon_{\text{eff}}^{(\perp)} = (1-f)\varepsilon_1 + f\varepsilon_2. \tag{4}$$

In analogy, we find the expression for the polarization for which the electrical field is parallel to the grating vector ($E_{\|}$) by using the continuity of the electric displacement:

$$E_{\mathrm{eff}}^{(\|)} = \frac{D_{\|}}{\varepsilon_{\mathrm{eff}}^{(\|)}} = \frac{(1-f)D_{\|}}{\varepsilon_1} + \frac{fD_{\|}}{\varepsilon_2}, \tag{5}$$

$$\frac{1}{\varepsilon_{\mathrm{eff}}^{(\|)}} = \frac{(1-f)}{\varepsilon_1} + \frac{f}{\varepsilon_2}. \tag{6}$$

Equations (4) and (6) can also be written by using an ordinary (O) and an extraordinary (E) effective refractive index:

$$n_O^{(0)} = \sqrt{(1-f)n_1^2 + fn_2^2}, \tag{7}$$

$$n_E^{(0)} = \frac{1}{\sqrt{\frac{(1-f)}{n_1^2} + \frac{f}{n_2^2}}}. \tag{8}$$

The effective refractive indices in (7) and (8) are labeled with a zero because in the literature, effective refractive indices based on static field considerations are called zero-order effective refractive indices as they have no correction term taking the ratio Λ/λ_0 into account and this will be described below.

The lamellar surface-relief grating is not z-invariant. This in general, results in a depth dependence of the effective dielectric tensor, as shown by Lalanne.[37] This depth dependence is negligible for low refractive indices and depths. A more severe limitation of the model described so far is the fact that the fields are not static in the reality of optics. With increasing period of the grating, more energy from the propagating zero-order diffracted waves is coupled into the first-order evanescent fields. This results in a stronger modulation of the near field at a small distance from the modulated region (region II in Fig. 4) although only the zeroth order diffracted waves propagate. In EMT, this effect has to be taken into account by a change of the effective refractive index which is not possible by using the EMT with the static field approximation, of course. All the different attempts to develop an EMT without a static field approximation for one-dimensional gratings are based on the same characteristic equation. This transcendental eigenvalue equation was found by Rytov[38] and Yariv and Yeh,[39] who worked on periodically stratified media. It was also found by Botten *et al.* in their modal approach of the lamellar grating as the eigenvalue equation of the lamellar diffraction grating.[40] Li has shown that the characteristic equation holds also in the conical case of incidence, if the fields are decomposed.[41] For simplicity, we will only show the result of Rytov's work. By using the Taylor series expansion of the tangent terms of the transcendental equations and truncating, he found closed-form expressions for the effective refractive indices which take the ratio Λ/λ_0 with the power of 2 into account. Therefore, they are called second-order effective refractive indices:

$$n_O^{(2)} = \left[(n_O^{(0)})^2 + \frac{\pi^2}{3} \left(\frac{\Lambda}{\lambda_0} \right)^2 f^2(1-f)^2(n_2^2 - n_1^2)^2 \right]^{\frac{1}{2}}, \quad (9)$$

$$n_E^{(2)} = \left[(n_E^{(0)})^2 + \frac{\pi^2}{3} \left(\frac{\Lambda}{\lambda_0} \right)^2 f^2(1-f)^2 \left(\frac{1}{n_2^2} - \frac{1}{n_1^2} \right)^2 (n_E^{(0)})^6 (n_O^{(0)})^2 \right]^{\frac{1}{2}}. \quad (10)$$

For media with relatively low refractive indices in the range of 1.5 (a typical value for glass and transparent polymers), the difference between zero- and second-order effective refractive indices is moderate even for the largest Λ/λ_0 ratios for which only the zeroth order diffracted waves propagate under normal incidence as seen in Fig. 6.

In the case of a two-dimensional grating, the description of subwavelength gratings by effective medium theories is much more complicated due to the fact that the parallel and perpendicular components of the electromagnetic fields are always coupled at boundaries. Therefore, no closed-form solution of the problem exists. The first attempt to achieve effective refractive indices was made by Wilson and Hutley, who averaged the refractive indices of the components weighted by the fill factor.[4] Bräuer proposed a procedure which is based on EMT for one-dimensional gratings.[42] He replaces rows and columns of two-dimensional gratings by effective grating lines

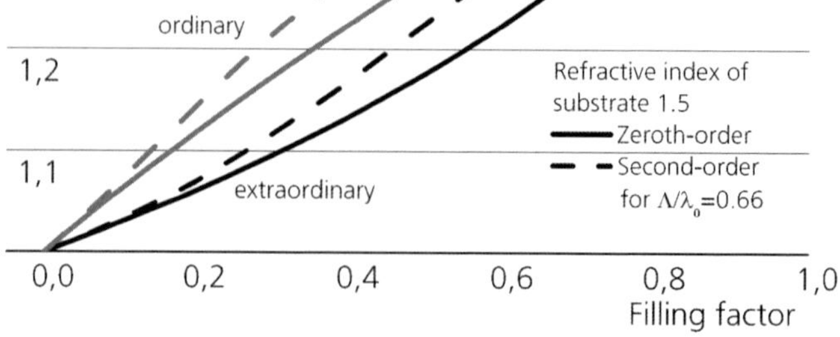

Fig. 6. Zero- and second-order effective refractive indices of periodically stratified media with refractive indices $n_1 = 1.0$ and $n_2 = 1.5$. © Fraunhofer ISE.

for a TE and a TM mode and subsequently calculates an effective refractive index by averaging the "effective" one-dimensional gratings. A totally different attempt to achieve effective refractive indices of two-dimensional gratings was made by Grann *et al.*[43] He calculated the reflected power using rigorous coupled wave analysis and determined the refractive index of a homogeneous film with the same depth as the grating which would provide the same reflected power. His attempt is therefore not a theory in its original sense. Lalanne and Lemercier-Lalanne proposed a method based on a Fourier decomposition of the propagating wave for the determination of the effective refractive indices of two-dimensional periodic structures.[44] The determination of the effective refractive indices according to their method requires the inversion of an infinite matrix which has to be truncated and solved numerically. We did not use their method, because we were satisfied for our application with the method proposed by Bräuer, which was in a good agreement with the data calculated by Grann.

3. Design Considerations

When designing artificial moth-eye structures, the goal is to achieve a good AR effect for the wavelength spectrum of interest with a reasonable effort. It is made base on the fact that a graded index between the two adjacent media can lead to a spectrally very broadband AR effect. A graded index layer system made out of homogenous layers would be a high pass filter but subwavelength gratings with continuous profiles show a band pass characteristic as AR surfaces. The short wave limit is given by the propagation of higher-order diffracted waves as defined by the period, the long wave limit by a minimum depth of the grating. Thus, it is necessary to have a closer look at the period-to-wavelength ratio and the depth-to-wavelength ratio which define the lower and upper bounds of the band pass and are coupled by the aspect ratio of the surface-relief structures. The aspect ratio shall, in this context, be defined as grating depth divided by the grating period. The period-to-wavelength ratio should ideally be such that no higher-order diffracted waves propagate in every possible case of incidence. This condition can be derived from the grating equation:

$$\frac{\Lambda}{\lambda_0} < \frac{1}{n_i + n_s}. \tag{11}$$

Thus, for a substrate material with a refractive index of $n_s = 1.5$ and air as medium from which the radiation is incident, the period of the grating should be 2.5 times smaller than the smallest wavelength under consideration.

For the required depth-to-wavelength ratio, the reflectance values resulting from different effective refractive index profiles at different depths were calculated by using

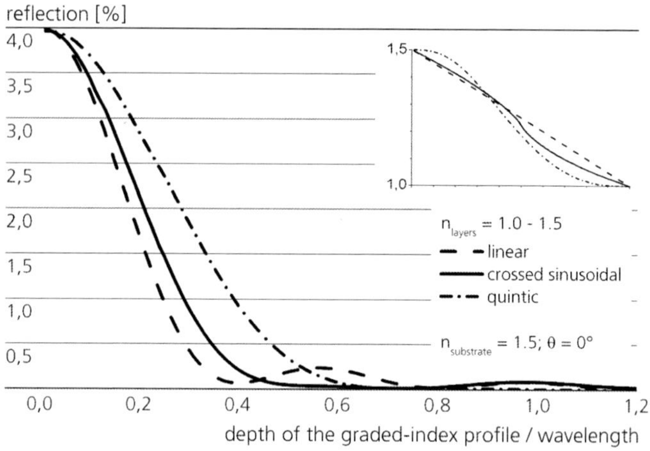

Fig. 7. Calculated reflectance at normal incidence of graded index surfaces made by corrugations with different effective refractive index profiles as a function of the depth-to-wavelength ratio. The inset shows the values of the refractive indices over the normalized depth which were taken for the calculation of the reflectance values. © Fraunhofer ISE.

thin film theory. In Fig. 7, the results are shown for a linear graded index profile, for the graded index profile of a crossed grating with sinusoidal shape in each of the modulation directions, and for the so-called quintic profile proposed by Southwell[45] which is very efficient at larger depth-to-wavelength ratios but almost impossible to manufacture. The inset shows the refractive index profile over the normalized depth.

Assuming that for an ideal AR surface the reflectance should be below 0.5% for all the wavelengths, the minimum of the grating depth should be in the range of 0.4 of the longest wavelength.

If we now consider applications in which the anti-reflective surfaces are designed for visible wavelengths or for solar radiation by using the above mentioned design rules, we get the results in Table 1 for a substrate material with refractive index $n_s = 1.5$:

Neither the required grating periods nor the required aspect ratios can easily be originated and replicated on large areas. The origination technique on large areas is interference lithography with positive photoresists and this will be described later. The minimum period which could be obtained experimentally with good quality was ~ 200 nm. Therefore, in practice, one has to deal with subwavelength AR surfaces which are not ideal in the following sense: higher-order diffracted waves cannot be avoided for every case of incidence and at longer wavelengths, the reflectance exceeds 0.5% per surface.

Table 1. Grating periods, depths, and aspect ratios for AR gratings not exceeding 0.5% reflectance per surface at the interface between air and a material with $n_s = 1.5$ for two different applications.

Wavelength range	Maximum grating period	Minimum grating depth	Minimum aspect ratio
Visible wavelengths 380–780 nm	~ 150 nm	~ 310 nm	~ 2.1
Solar radiation 300–2500 nm	~ 120 nm	~ 1000 nm	~ 8.3

For nonideal AR subwavelength gratings where the propagation of higher-order diffracted waves can occur for some of the wavelengths, the validity of the different EMT's had to be proven by comparison with results obtained by rigorous diffraction theory. We found that for substrate materials with a refractive index $n_s = 1.5$, EMT holds very well even if the wavelength is just slightly larger than the cut-off wavelength for the propagation of higher-order diffracted waves.[46]

When higher-order diffracted waves propagate in the substrate, a dip in transmittance was found when measuring samples which had AR subwavelength gratings on both sides.[28,29] This effect was studied theoretically for a one-dimensional grating by rigorous calculation of the diffraction efficiencies and summing up all the waves which propagate due to multiple reflection and diffraction (see Fig. 8).[46] It was found that a reduction of the hemispherical transmittance occurs especially when the 1st diffracted order propagates in the substrate but not in air. As soon as the $(-1, 0)$ transmitted

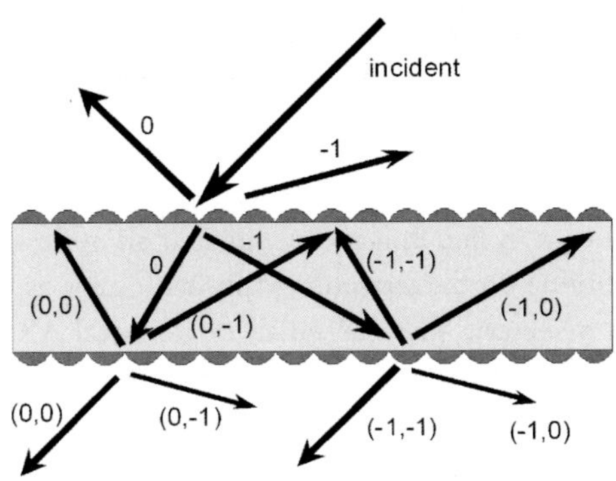

Fig. 8. Sketch of a transparent sheet with a linear AR grating on both sides and higher-order diffracted waves propagating. The indices of the diffracted waves indicate the order of diffraction of the respective wave after interaction with the first and the second diffraction grating. Only the first two interactions are shown in this sketch. © Fraunhofer ISE.

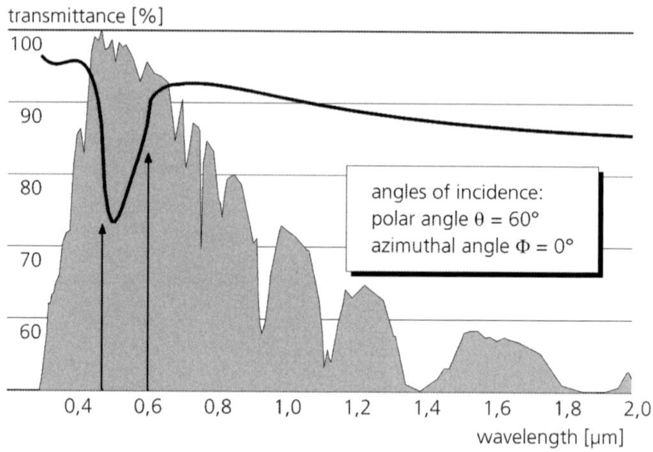

Fig. 9. Calculated hemispherical transmittance of a transparent sheet with refractive index $n = 1.5$ and sinusoidal linear AR gratings on both surfaces. The grating period and depth is 260 nm. Also shown is the global solar spectrum AM 1.5 g (in arbitrary units). The arrows indicate the position of the cut-off wavelengths of the 1st diffracted orders propagating in the substrate at a wavelength of 615 nm and in air at a wavelength of 485 nm. © Fraunhofer ISE.

order propagates, the hemispherical transmittance of such an AR structured sheet increases due to the high diffraction efficiency of the $(-1, 0)$ transmitted order. The hemispherical transmittance of a sinusoidal linear grating with a period and a depth of 260 nm at a polar angle of incidence of 60° is shown in Fig. 9. The arrows indicate the position of the cut-off wavelengths of the 1st diffracted orders propagating in the substrate at a wavelength of 615 nm and in air at a wavelength of 485 nm. It is clear that such a grating is not a good choice for visual applications at such a large angle of incidence. However, the average solar transmittance (AM 1.5) is 88.5% which is still above the solar transmittance of 84.8% of a sheet with no AR surfaces.

With the result shown in Fig. 9, it is obvious that an optimization of the grating type and period depending on the technologically achievable aspect ratio and on the application is always necessary. The AR effect of nonideal AR gratings is strongly influenced by the propagation of higher-order diffracted waves which depends on the period-to-wavelength ratio, the grating type, the refractive index of the substrate, and on the polar and azimuthal angles of incidence of the incident wave. Therefore, the cut-off wavelengths for higher-order diffracted waves were investigated for a range of angles of incidence[29,46,47] and compared for different gratings types (linear, crossed, and hexagonal). The technologically achievable aspect ratios were assessed by manufacture of different grating types at different periods.

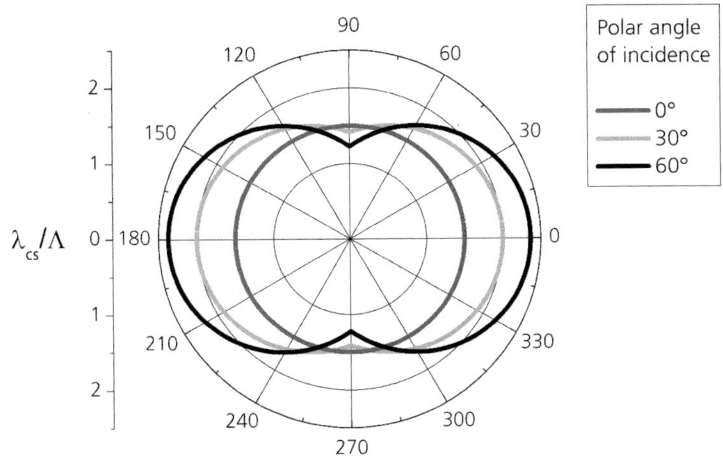

Fig. 10. Polar plot of the ratio of the cut-off wavelength of the 1st order diffracted wave in the substrate to the grating period λ_{cs}/Λ for a linear grating and air in region I and glass in region III ($n_i = 1.0$ and $n_s = 1.5$). © Fraunhofer ISE.

In Fig. 10, the cut-off wavelengths for the propagation of the 1st diffracted order in the substrate which is the last higher diffracted order to become cut-off are shown for a linear grating. The refractive indices of the two media are $n_i = 1.0$ and $n_s = 1.5$. The grating vector of the linear grating points in the 0° direction. One can see very well that the cut-off wavelength increases with increasing polar angle along the axis of the grating vector but not perpendicular to it. This means that the linear grating seems to be advantageous when the polar angles of incidence are relatively large and the azimuthal angles of incidence not equally distributed. In the case of a crossed grating, the pattern shown in Fig. 10 is repeated additionally along the 90°–270° axis due to a second grating vector pointing in the 90° direction. This makes the optical properties of the crossed grating more independent of the azimuthal angle of incidence in comparison to the linear grating. This holds even more for hexagonal gratings where the pattern is repeated along axes every 60°. This type of grating shows the interesting effect that all cut-off wavelengths are smaller than those for crossed gratings due to the effect that the base vectors of the reciprocal grating have the norm $2\pi/\Lambda \sin 60°$ instead of $2\pi/\Lambda$.

With the developed optimization methods,[46,47] it was found that for solar applications linear gratings with periods in the range of 200–210 nm were best suited at polar angles of incidence larger than 30°, and square or hexagonal gratings with periods of 340–360 nm at polar angles of incidence ranging from 0° to 30°.

For visual applications, the light reflected into directions where observers are located shall be reduced as far as possible. As deduced from Fig. 9 which shows the transmittance for a nonattenuating sheet, the reflectance is also increased as soon as higher-order diffracted waves propagate. By defining angular ranges of observation, i.e. the polar and the azimuthal angles with respect to the AR surface under which observers typically see it, an optimization of the grating parameters can be done.[47] The following angular ranges of observation have been studied: symmetrical ones where the maximum polar angle is 30° or 60° and asymmetrical ones where the angles of observation reach different maximum values horizontally (*h*) and vertically (*v*), i.e. the combinations $\pm 30°$ (*h*) $\pm 10°$ (*v*) and $\pm 60°$ (*h*) $\pm 20°$ (*v*). The optimization results are very similar to those for solar applications. Linear gratings with periods of 170–200 nm were best suited for angular ranges of observation in which the maximum angle is 60°. Hexagonal gratings with periods of 320–330 nm are best suited for angular ranges of observation in which the maximum angle is 30°. In both cases, the result is almost independent of the symmetry of the angular range. Of course, in the case of linear gratings it makes sense to orient the grating vector of the grating always in the direction of the plane of incidence in which the smallest polar angles of incidence are expected (see also Fig. 10). The optimized nonideal AR gratings have grating parameters which are technologically feasible on large areas with the restriction that gratings periods below 200 nm have not been manufactured on large areas yet and that gratings with periods of 250 nm are still made more easily than those with periods of 200 nm. The required aspect ratios are all in a range which can be originated and replicated throughout the whole process chain.

4. Manufacturing

For a successful commercial exploitation of the moth-eye effect, the sophisticated optical effect which is given by the surface-relief gratings must be replicated in high volume mass production. Replication of nano and microstructured surfaces normally comprises three steps: the origination of the master structure, a first replication of the master structure to produce a number of sufficiently durable production tools and finally the high volume, high speed replication into the material of the end product. Often, the first step is also the most difficult one. We have chosen interference lithography as an origination technology which is suitable for structures with periods $<1\,\mu$m on large areas because it is a parallel process, i.e. the whole area is exposed at the same time. All other known "top–down" techniques to originate engineered microstructures of those dimensions are writing (serial) techniques, e.g. e-beam writing. Although parallelization efforts are made by using masks in the beam, e-beam writing is not suitable to originate structures on half a square meter, for example.

4.1. *Origination by Interference Lithography*

Interference lithography makes use of the interference pattern which is formed when at least two coherent waves are superimposed. A photoresist layer on a substrate is exposed to this interference pattern. After development of the photoresist, a surface-relief profile is formed which nonlinearly depends on the interference pattern. This process differs in two main points from the technology described in the chapter *Laser Interference Metallurgy* by Frank Mücklich and Andreás Fabián Lasagni: (1) In our process, the structure formation is not a direct effect of the laser radiation. The interference pattern causes a chemical modification in the photoresist which then leads to the formation of a height profile in the development step. (2) We use microreplication processes to manufacture structured products. So the exposure and development step does not need to be applied to every product, but only to a master structure.

The extensive use of interference lithography started in the early 1970s after the development of high power lasers that emitted sufficiently short wavelengths for the manufacture of diffraction gratings.[48] A very high stability of the whole optical set-up is required during the exposure. The exposure times can be relatively long (even hours) due to the limited laser power and depend on the area to be homogeneously structured, on the structure depth, and the photoresist used. A variation of the interferogram of more than a tenth of the period during exposure affects the illumination contrast significantly and thus should be avoided. Unstable conditions during exposure will result in microstructures with low aspect ratio. Over years, we took numerous different measures to optimize the mechanical and thermal stability of the optical set-ups. It turned out that the temperature variation is the parameter which is the most difficult to control. Air conditioning was not possible within the required range of temperature variation of 1/100 K. Thus, after mounting of a photoresist plate in the sample holder, the laboratory was left untouched so that the room could reach a sufficient thermal stability before the shutter was opened. As this can take several hours, the climatic data were acquired and the laser, the shutter, and the rotating sample holder were controlled by computer.

In Fig. 11, a sketch of the optical set-up for interference lithography and in Fig. 12, a picture of the largest interference lithography laboratory at Fraunhofer ISE are shown. A laser beam is split, directed with mirrors and then spatially filtered and expanded. A sample holder with the photoresist plate is positioned where the expanded beams are superimposed. A shutter defines the exposure time. Behind the spatial filters, no optical components are in the optical path in order to avoid parasitic interference effects such as Airy patterns from dust particles. The nonplanarity of the interfering beams results in a small variation of the grating periods which is tolerated for the above mentioned applications. If one assumes symmetrical angles of incidence α, then the

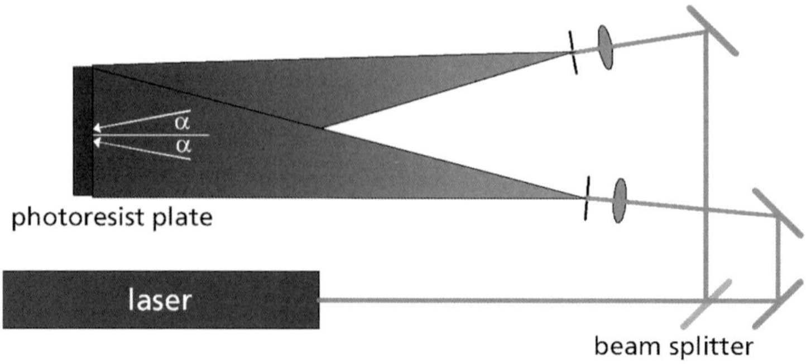

Fig. 11. Sketch of an optical set-up used for interference lithography showing the most important optical components. © Fraunhofer ISE.

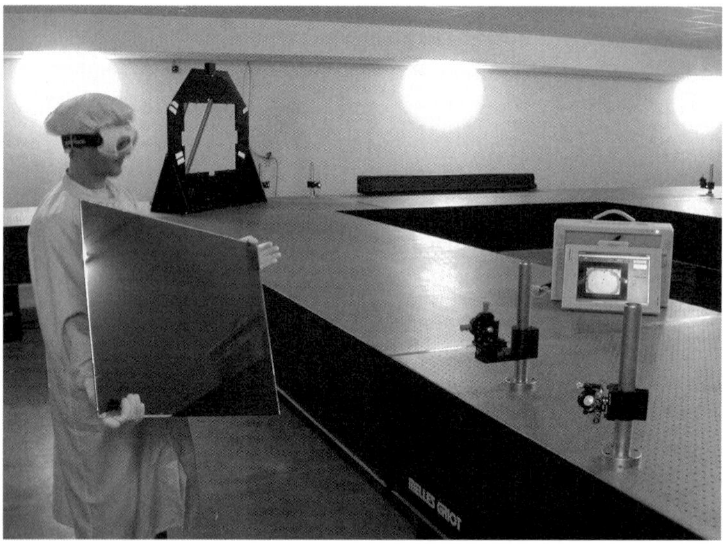

Fig. 12. Photo of the largest interference lithography laboratory at Fraunhofer ISE. © Fraunhofer ISE.

grating period is given by:

$$\Lambda = \frac{\lambda_L}{2 \sin \alpha} \tag{12}$$

where λ_L is the wavelength of the laser.

Structuring large areas requires large distances from the spatial filters to the photoresist plate in order to maintain the period variation of the photoresist structure small. Another reason is the maximum power density at the spatial filter. If the distance from the spatial filters to the photoresist plate is small, the beams have to be expanded at a large divergence angle in order to irradiate the same area which causes a small focal spot in the spatial filter. The risks to exceed the damage threshold or to obtain thermal instabilities in the spatial filter are high. As a result, large optical tables are recommended (Fig. 12) to provide stability. The variable optical set-ups all follow the same principle: The beams are guided almost parallel to the optical table and the photoresist plate stands upright. With the interference of two plane waves, only intensity profiles can be generated which are sinusoidal in one direction and invariant in the orthogonal direction. Thus, the resulting grating type is a linear grating. In order to originate gratings which are periodic in two directions (so-called two-dimensional gratings), two subsequent exposures were found to be the most suitable method on large areas. By two exposures and an intermediate rotation of the sample holder by 60° or 90°, hexagonal and crossed 2D gratings can be manufactured. With interference lithography, the team at Fraunhofer ISE was able to originate microstructures with lower aspect ratio and larger period than moth-eye structures homogeneously on areas of up to $1.2 \times 1.2\,m^2$. The largest area on which moth-eye structures were originated is $960 \times 720\,mm^2$ to date. This great success was achieved at the company Holotools in Germany.

4.2. *Choice of Laser and Photoresist*

Lasers which are suitable for interference lithography have to emit blue or UV wavelengths because all currently available photoresists are sensitive in this spectral region. The laser wavelength also determines the smallest structure period. Theoretically, the period is half of the laser wavelength if the interfering beams are incident from air under polar angles of + and −90° (see Eq. (12)). Experimentally, polar angles larger than 65° should be avoided because then a major part of the incident light is reflected at the air–photoresist interface and may cause parasitic interference patterns due to stray light. Thus, the smallest period is approx 0.55 times the laser wavelength. At Fraunhofer ISE, argon ion lasers emitting up to 1.3 W single frequency at a wavelength of 364 nm are used. Choosing this wavelength instead of the wavelength of 458 nm which was formerly used in most cases,[4,49] was a big step in the direction of the origination of microstructures on large areas because the photoresist is much more sensitive at the shorter wavelength. The manufacture of gratings with smaller periods is also allowed.

Different types of diazonaphtochinone (DNQ)-based photoresists[50] were used. These photoresists are sensitive from 300 to 460 nm. They are relatively highly absorbing and show a moderate contrast. These properties are normally regarded as disadvantages, but for the origination of structures with continuous surface-relief profiles, as described in this work, they are essential. With low absorbing high contrast photoresists, only binary resist profiles can be realized which do not form a graded effective refractive index and are more difficult to replicate.

4.3. *Replication*

After origination of the photoresist structures, metal coatings with thicknesses of less than 100 nm were applied by sputtering to serve as a plating base on the photoresist surface-relief profile for subsequent electroforming. The materials Ni–V (93%/7%) and gold were used for the plating base. Nickel films (so-called nickel shims) with thicknesses of up to 3 mm (typically $<400\ \mu m$) were electroformed in a nickel sulfamate solution by four external partners of Fraunhofer ISE. In electroforming, a whole tree of embossing tools can be made by further copying the first nickel shim. In order to be able to separate nickel from nickel, the first nickel shim is passivated in a 0.1%–2.5% solution of potassium or sodium dichromate.[51] The aspect ratio (depth/period of the structure) of the structures described here was in general much higher and the periods smaller compared to what the partners typically electroform. This resulted in difficulties in separating the nickel copy from the nickel shim. It was found empirically that nickel copies from nickel shims can be made without damage up to an aspect ratio of approximately 1.5.

The nickel shims were used for replication of the microstructures in different polymer materials and sol–gel materials by different processes: hot embossing (also called hot compression molding) and casting UV curable systems like acrylics or inorganic–organic sol–gel hybrids called ORMOCERs™,[28,52] injection molding, and injection compression molding. Each of the processes is suitable for replication of the described microstructures but the processing parameters have to be adjusted carefully in order to completely fill the cavities of the microstructure and to avoid elastic deformation. Limitations are also given for polymers which show significant shrinkage. For optical applications, the transparent polymers poly(methyl methacrylate) (PMMA) and poly(carbonate) (PC) are well suited.

Curing polymers by UV or electron beam radiation offers several advantages. Firstly, a continuous production is possible by coating carrier films like poly(ethylene terephthalate) with a curable lacquer and cross-linking the lacquer while being in contact with nickel shims which are mounted on a rotating drum. Secondly, the wear of the nickel shims is low because no high pressure is applied. Thirdly, microstructures in cross-linked polymers possess a much higher mechanical and chemical stability than thermoplastically formed polymers. For applications in which a certain scratch resistance is required, only hard coatings like cross-linked polymers can be used. The high mechanical stability of UV replicated microstructures on films makes them also suitable for in-mould decoration processes.[53] Here, a microstructured film is inserted in the mould of an injection molding machine with the structured side facing towards the mold and the polymer is injected into the rear side. Despite the high pressure and temperatures, the cross-linked structure was not destroyed and good optical values were achieved.

The UV-replicated films were also used as flexible tools for embossing sol–gel films on glass. These films are purely inorganic after a temperature treatment. The sol–gel process starts from hydrolyzable molecular compounds such as tetraalkoxysilane. The formation of networks is induced by hydrolysis and condensation as soon as water is added.[54] Sols can be used in almost every wet coating process. After applying the sol, the solvent evaporates and a gel film is formed. In the case of purely inorganic films, the gel film is solidified by tempering at temperatures higher than 300°C. In order to enable the replication of microstructures into gel films which are inorganic at the end, sols containing an organic additive for embossing were developed. On small areas, it was possible to use a classical hot embossing process. On larger areas, dip coating an unstructured sol–gel film on a glass substrate, casting the same sol onto the flexible tool and then using hot compression molding during which the cast film is transferred to the substrate was found to be much more reproducible. After embossing, the organic additive is removed by tempering at 600°C. The shrinkage of the film could be reduced to approximately 30%.

Generally, the replication fidelity depends largely on the process used. A good replication fidelity could be obtained with the UV curing processes which are most promising for large-area mass fabrication also. The current status is that several companies offer moth-eye AR surfaces. The largest homogeneously structured surfaces are produced by the company MacDermid Autotype by using UV roller embossing. They use nickel shims of the size $850 \times 650\ mm^2$ on their embossing roller.

5. Applications

Our initial goal was to develop AR surfaces for solar applications. This could not be reached so far as AR subwavelength gratings are not used for components of solar systems yet. The main reason is that the AR surfaces can be replicated very well in polymers which are not sufficiently durable outdoors. As mentioned, the replication of AR gratings in an inorganic sol–gel material was successful but the process has only been tested on small areas so far.

For applications in which an AR solution is required only for visible wavelengths, "moth-eye" structures compete with interference coatings. In most existing application cases, the lower cost of "moth-eye" structures is the reason to use them although they have several disadvantages like lower mechanical stability and the fact that fingerprints cannot be removed dryly. The fat from the fingerprints fills the grooves of the "moth-eye" structure and can only be removed by solution of the fat, e.g. by using water with a drop of dish cleaner.

As a result, one can say that "moth-eye" structures certainly will not replace AR interference coatings. Modern highly prized AR coatings must have a variety of properties: anti-reflective, anti-soiling, anti-fingerprint, and anti-static. "Moth-eye" structured AR surfaces cannot fulfill all the requirements. Their advantages are lower cost and the variety of functionalities which can be obtained by surface-relief microstructures. It is definitely worth to think about using "moth-eye" structures in those cases in which the AR surface is not exposed. Therefore, master structures have been used to produce replicated polymer films and sheets with AR surfaces including optical components like Fresnel lenses on a commercial basis. Mainly in the flat-panel display market, the request for new and cost-effective AR surfaces is large.

On information displays, the readability of the information is significantly reduced by reflection of ambient light on the display screen. Even indoors, the mirror images of lamps or windows on the screen are very disturbing. Due to the high luminance of the respective light sources, it is not sufficient in most cases to use AR surfaces only because the reflectance cannot be suppressed sufficiently. Anti-glare surfaces which scatter the reflected light into small angles are a well-known solution.[55] Our idea was to combine anti-glare structures with anti-reflective subwavelength gratings in order to obtain an optimum readability by one surface.[56] Based on this idea, replicated films with this microstructure combination are now produced by the company MacDermid Autotype in Wantage, UK, and are available on the market under the trademark MARAG™ (Moth-eye Anti-Reflection and Anti-Glare).

Technically, many problems had to be solved in order to realize such combined structures. It turned out that it would be best to manufacture both structure types by interference lithography. It is important that the anti-glare surface has no high-frequency structural components in order to maintain the readability of the display and to avoid a "milky" appearance. With interference lithography, it is principally possible to manufacture aperiodic surface-relief structures with an exactly adjusted spectrum of the spatial frequencies of the resulting surface.[57] By inserting a masked diffuser in the expanded laser beam, well-defined speckle patterns can be produced. The difficulty was to find exposure and development parameters which result in a very smooth ("poor contrast") anti-glare structure and subwavelength gratings with high aspect ratio ("high contrast") taking the nonlinear response of the photoresist into account.

By optimizing the process and incorporating five different exposures, it was possible to manufacture MARAG™ structures with very good optical properties.[58] In Fig. 13, SEM pictures of a good MARAG™ structure are shown in two different

Fig. 13. SEM pictures of an MARAG™ structure in photoresist at two different magnifications. The smooth undulation necessary for a good anti-glare effect can be seen in the left picture in which the white rectangle indicates the location of the right picture. At the 22 times larger magnification of the right picture, the moth-eye structures can be made visible. © Fraunhofer ISE.

magnifications. It is very difficult to show both structure types in one SEM picture because of the large mismatch of structure dimensions. It was also possible to originate MARAG™ structures on areas of $720 \times 960\,\text{mm}^2$ which indicates a good reproducibility of the process taking the number of exposures into account.

Apart from anti-reflective applications, interference lithography has also been used to originate structures for light management (diffusers, microlenses, microprism arrays) in buildings and display applications[59] for light trapping in solar cells[60] and for artificial "Lotus Effect" structures.[61,62]

6. Summary

"Moth-eye" structures are now commercially available as AR surfaces and are mainly used for applications in information displays. Although the working principle of "moth-eye" structures was already rather well explained by Bernhard in 1967, it took almost 30 years until the first commercial exploitation took place. One of the major reasons was that most of the applications require the "moth-eye" structures to be homogeneously distributed over relatively large areas, e.g. $>0.1\,\text{m}^2$. Therefore, the technological challenge was to develop a process chain in which submicron structures with a relatively large aspect ratio can be reproducibly manufactured on large areas. It can be regarded as a very big success that it was even possible to enlarge the homogeneously structured area to $0.7\,\text{m}^2$. A very interesting approach is to combine "moth-eye" structures with other microstructures with different optical or nonoptical functions. A first commercial realization is the combination of "moth-eye" structures with anti-glare structures in order to provide an optimum readability of information displays.

References

[1] Bernhard CG. (1967) Structural and functional adaption in a visual system. *Endeavour* **26**: 79–84.
[2] Clapham PB, Hutley MC. (1973) Reduction of lens reflection by the 'moth eye' principle. *Nature (London)* **244**: 281–282.
[3] Thornton BS. (1975) Limit of the moth's eye principle and other impedance-matching corrugations for solar-absorber design. *J Opt Soc Am* **65**: 267–270.
[4] Wilson SJ, Hutley MC. (1982) The optical properties of 'moth eye' AR surfaces. *Opt Acta* **29**(7): 993–1009.
[5] Horwitz CM. (1974) A new solar selective surface. *Opt Comm* **11**(2): 210–212.
[6] Mc Phedran RC, Maystre D. (1977) On the theory and solar application of inductive grids. *Appl Phys* **14**: 1–20.
[7] Vincent P. (1978) A finite-difference method for dielectric and conducting crossed gratings. *Opt Comm* **26**(3): 293–296.
[8] Derrick GH, *et al.* (1979) Crossed gratings: A theory and its applications. *Appl Phys* **18**: 39–52.
[9] Ono Y, *et al.* (1987) AR effect in ultrahigh spatial-frequency holographic relief gratings. *Appl Opt* **26**(6): 1142–1146.
[10] Enger RC, Case SK. (1983) Optical elements with ultrahigh spatial-frequency surface corrugations. *Appl Opt* **22**(20): 3220–3228.
[11] Flanders DC. (1983) Submicrometer periodicity gratings as artificial anisotropic dielectrics. *Appl Phys Lett* **58**: 492–494.
[12] Gaylord TK, Baird WE, Moharam MG. (1986) Zero-reflectivity high spatial-frequency rectangular-groove dielectric surface-relief gratings. *Appl Opt* **25**(24): 4562–4567.
[13] Gaylord TK, Glytsis EN, Moharam MG. (1987) Zero-reflectivity homogeneous layers and high spatial-frequency surface-relief gratings on lossy materials. *Appl Opt* **26**(15): 3123–3135.
[14] Glytsis EN, Gaylord TK. (1988) Antireflection surface structure: dielectric layer(s) over a high spatial-frequency surface-relief grating on a lossy substrate. *Appl Opt* **27**(20): 4288–4304.
[15] Brundrett DL, Glytsis EN, Gaylord TK. (1994) Homogeneous layer models for high-spatial frequency dielectric surface-relief gratings: Conical diffraction and AR designs. *Appl Opt* **33**: 2695–2706.
[16] Haggans CW, Li L, Kostuk RK. (1993) Effective-medium theory of zeroth-order lamellar gratings in conical mountings. *J Opt Soc Am A* **10**(10): 2217–2225.
[17] Bräuer R, Bryngdahl A. (1994) Design of antireflection gratings with approximate and rigorous methods. *Appl Opt* **33**: 7875–7882.

[18] Grann EB, Moharam MG, Pommet DA. (1994) Artificial uniaxial and biaxial dielectrics with use of two-dimensional subwavelength binary gratings. *J Opt Soc Am A* **11**(10): 2695–2703.

[19] Lalanne P, Lemercier-Lalanne D. (1996) On the effective medium theory of subwavelength periodic structures. *J Mod Opt* **43**(10): 2063–2085.

[20] Lalanne P, Lemercier-Lalanne D. (1997) Depth dependence of the effective properties of subwavelength gratings. *J Opt Soc Am A* **14**(2): 450–458.

[21] Glytsis EN, Gaylord TK. (1992) High-spatial-frequency binary and multilevel stairstep gratings: Polarization-selective mirrors and broadband antireflection surfaces. *Appl Opt* **31**(22): 4459–4470.

[22] Raguin DH, Morris GM. (1993) Analysis of antireflection-structured surfaces with continuous one-dimensional surface profiles. *Appl Opt* **32**(14): 2582–2598.

[23] Southwell WH. (1991) Pyramid-array surface-relief structures producing antireflection index matching on optical surfaces. *J Opt Soc Am A* **8**(3): 549–553.

[24] Grann EB, Moharam MG, Pommet DA. (1995) Optimal design for antireflective tapered two-dimensional subwavelength grating structures. *J Opt Soc Am A* **12**(2): 333–339.

[25] Motamedi ME, Southwell WH, Gunning WJ. (1992) Antireflection surfaces in silicon using binary optics technology. *Appl Opt* **31**(22): 4371–4376.

[26] DeNatale JF, *et al.* (1992) Fabrication and characterization of diamond moth eye antireflective surfaces on Ge. *J Appl Phys* **71**: 1388–1393.

[27] Heine C, Morf RH. (1995) Submicrometer gratings for solar energy applications. *Appl Opt* **34**(14): 2476–2482.

[28] Gombert A, *et al.* (1998) Antireflective submicrometer surface-relief gratings for solar applications. *Solar Energy Mater Solar Cells* **54**: 333–342.

[29] Gombert A, *et al.* (1998) Glazing with very high solar transmittance. *Solar Energy* **62**(3): 177–188.

[30] www.fresnel-optics.de.

[31] www.holotools.de.

[32] www.macdermidautotype.com.

[33] Walheim S. (2000) Dissertation. University of Konstanz, Germany.

[34] Burmeister F, *et al.* (2005) Investigation of structured TiAlN- and TiO_2-coatings with moth-eye like surface morphologies. *Surf Coat Technol* **200**: 1088–1092.

[35] Schulz U, *et al.* (2007) Antireflection of transparent polymers by advanced plasma etching procedures. *Opt Exp* **15**(20): 13108–13113.

[36] Wiener O. (1912) Die Theorie des Mischkörpers für das Feld der stationären Strömung. *Abh Math Phys Kl Saechs Akad Wiss* **32**: 505–604.

[37] Lalanne P, Lemercier-Lalanne D. (1997) Depth dependence of the effective properties of subwavelength gratings. *J Opt Soc Am A* **14**(2): 450–458.

[38] Rytov SM. (1956) Electromagnetic properties of a finely stratified medium. *Soviet Phys JETP* **2**: 466–474.

[39] Yariv A, Yeh P. (1977) Electromagnetic propagation in periodic stratified media. II. Birefringence, phase matching, and X-ray lasers. *J Opt Soc Am* **67**(4): 438–448.

[40] Botten IC, *et al.* (1981) The dielectric lamellar diffraction grating. *Opt Acta* **28**: 413–428.
[41] Li L. (1993) A modal analysis of lamellar diffraction gratings in conical mountings. *J Mod Opt* **40**(4): 553–573.
[42] Bräuer R. (1994) Dissertation, University of Essen GHS, Germany.
[43] Grann EB, Moharam MG, Pommet DA. (1994) Artificial uniaxial and biaxial dielectrics with use of two-dimensional subwavelength binary gratings. *J Opt Soc Am A* **11**(10): 2695–2703.
[44] Lalanne P, Lemercier-Lalanne D. (1996) On the effective medium theory of subwavelength periodic structures. *J Mod Opt* **43**(10): 2063–2085.
[45] Southwell WH. (1991) Pyramid-array surface-relief structures producing antireflection index matching on optical surfaces. *J Opt Soc Am A* **8**: 549.
[46] Gombert A. (1998) Dissertation, University of Strasbourg, France.
[47] Bläsi B. (2000) Dissertation, University of Freiburg, Germany.
[48] Labeyrie A. (1971) Holo-gratings — the new diffractive breed. *Elect Opt Syst Des* **2**: 32–38.
[49] Mello BA, *et al.* (1995) Developed profile of holographically exposed photoresist gratings. *Appl Opt* **34**: 597.
[50] Dammel R. (1993) Diazonaphtochinone-based resists. Tutorial Texts in Optical Engineering, Vol. 11. SPIE Optical Engineering Press, Bellingham, Wa.
[51] Ahlhorn T. (1994) Elektrolytische Abformung mikrostrukturierter Oberflächen. Fortschr.-Ber. VDI Reihe 2 Nr. 312, VDI-Verlag, Düsseldorf.
[52] Gombert A, *et al.* (1999) Subwavelength-structured antireflective surfaces on glass. *Thin Solid Films* **351**: 73.
[53] Bläsi B, Aufderheide K, Abbott S. (2002) Antireflection with moth-eye structures. *KU Kunststoffe plast europe* **92**(5): 15–18.
[54] Brinker C, Scherer G. (1990) Sol–gel science: The physics and chemistry of sol–gel processing. Academic Press, San Diego.
[55] Nuijs AM, Horikxs JJL. (1994) Diffraction and scattering at antiglare structures for display devices. *Appl Opt* **33**: 4058.
[56] Gombert A, Lerchenmüller H. Antireflective coating and method of manufacturing same. Patent US 6,359,735 B1.
[57] Wadle S, *et al.* (1994) Holographic diffusers. *Opt Eng* **31**: 213.
[58] Boerner V, *et al.* (2003) Holographic antiglare and antireflection films for flat panel displays. *Proc Int Symp Society for Information Display SID*, Baltimore, Md, 19.-22.5.2003: 68–71.
[59] Gombert A, *et al.* (2004) Some application cases and related manufacturing techniques for optically functional micro structures on large areas. *Opt Eng* **43**(11): 2525–2533.
[60] Bläsi B, *et al.* (2007) Innovative concepts for texturing multicrystalline silicon solar cells. Annual Report 2007 of the Fraunhofer-Institut für Solare Energiesysteme, 38–39.
[61] Gombert A. (2000) Surface patterns to serve a purpose. *Mater World* **8**(9): 25–27.
[62] Bläsi B, *et al.* (2001) Functional surface-relief structures manufactured by holography and subsequent microreplication. *Proc 2nd Euspen Int Conf*, Torino, 12–15.

Chapter 5

The Gecko Effect: Design Principles of the Gekkotan Adhesive System Across Scales of Organization

Anthony P. Russell and Megan K. Johnson

In this chapter, we consider the functional basis of adhesion in gekkotan lizards (geckos) and examine this across a broad spectrum of scales, integrating the processes and mechanisms of attachment with features of the configuration of the adhesive system, and attributes of real-world surfaces that are exploited by these lizards. The setae-locomotor surface is the direct site of organism–environment interaction, and the adhesive bond is created and broken at this point. In order to employ such a mechanism, which is based upon structural design principles of outgrowths of the skin, geckos must not only precisely attach and detach the setae, but also control them on unpredictable surfaces and must retain them as a functioning mechanism between sequential sheddings of the skin surface. Examining both the nature of the bond and also how the lizards themselves deploy the minute, hair-like outgrowths on their digits that are responsible for attachment provides insights into not only the wherewithal of the gecko effect, but also the potential for developing biomimetic simulations of this system.

1. Introduction

The last decade has seen an explosion of interest and research in the area of gecko-inspired adhesion (Fig. 1). The catalyst for this was the publication of information about attachment forces measured from a single isolated seta taken from the adhesive array of the tokay gecko.[1] For gekkotans (the higher taxon that includes gekkonids, carphodactylids, pygopodids, and eublepharids), this was a crucial step that bridged the gap between assessing the adhesive capabilities of the entire adhesive apparatus and obtaining empirical evidence of the adhesive capabilities of the actual agents of

Fig. 1. *Chondrodactylus turneri* scaling a vertical rock surface in Namibia, southern Africa. The rock surface is macroscopically rough and the lizard is able to scale it with the use of setal fields alone. The lack of claws ably demonstrates the astounding capabilities of the gecko effect.

adhesion, the setae. Although there has been long-standing interest in the phenomenon of gecko adhesion, the technological ability to probe this phenomenon at the level of the seta has only been available recently. Effectively, the natural nanotechnology of the gekkotan adhesive system, that evolved at least 54 million years ago,[2] has become the subject of technological investigation and adoption as the basis of gecko-inspired synthetic adhesive nanostructures.

So, what is the gecko effect? Essentially it is the ability of relatively large animals (in the range of body masses of 1 g to in excess of 50 g) to temporarily and reversibly bond with substrata ranging from the molecularly smooth[1] to the macroscopically rough and undulant[3] by way of integumentary modifications of the ventral surface of the digits, and occasionally the tail tip.[4] Attachment is able to be established and broken repetitively within milliseconds[5,6] and is effected by microfibrillar arrays of keratin outgrowths, the subdigital setae, that are arranged in orderly structured and intricately patterned fields (Figs. 2(A) and 2(B)). Attachment occurs without the aid of secretions or the formation of vaccua, and the setae are able to establish and

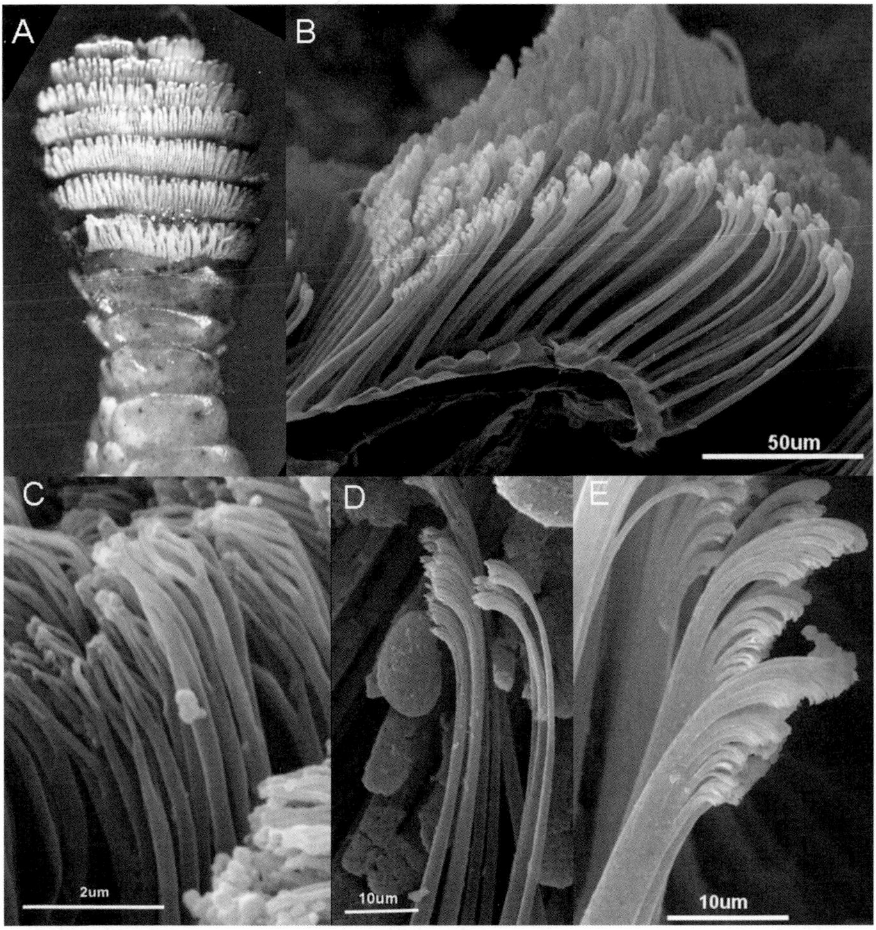

Fig. 2. Setal fields and setal structure in *Rhoptropus* cf. *biporosus*. (A) A setal field of an entire digit, composed of a series of scansors bearing setae. (B) Longitudinal section of a scansor illustrating setal field arrangement from proximal (left) to distal (right). Note the variation in setal length and form even within the confines of a single scansor. (C) Setal tip configuration showing the pattern of branching. (D) Long slender setae located at the distal ends of the scansors. (E) Shorter, thicker setae located at the proximal ends of the scansors.

break bonds repeatedly, apparently without ostensible wear, over protracted periods of time (renewal of the outer epidermal generation of the skin typically occurring at monthly or longer intervals[7]). This dry adhesion creates strong bonds and the adhesive agent (the setae) is directional, self-cleaning,[8] and controllable by a hierarchy of linked components that permit seemingly instantaneous adjustment to local circumstance.

The direct contact between the organism and the substratum is via the tips of the setae that are finely subdivided (Fig. 2(C)) to create the potential for extremely close and really extensive contact in localized regions. Setae themselves exhibit a wide array of structural variation of expression, not only between different species of gecko (and other lizards such as anoline polychrotids that also exhibit similar modifications) but also along the digits of a single individual.[9] Even for a single species there is no such entity as a typical seta, and their form and structure changes in subtle ways from region to region (Figs. 2(D) and 2(E)). Investigations into the gecko effect have largely focused on extrapolations from generalizations about the form and function of "typical" individual setae, and a great deal has been deduced from this. For practical and dynamic applications of the gecko effect in terms of biomimetic devices, it is likely however, that a deeper understanding of the local regional variation of structure of the adhesive (the setae), their geometric patterning into fields, and their underlying control mechanisms will lead to more effective models being developed.

In order to present an overview of the gecko effect, we first focus on the ways in which the setae interact with surfaces, and then broaden our scope to consider the hierarchy of components beyond the individual setae that geckos themselves employ to use this remarkable process at the whole organism level. This entails a consideration of the organization of setal fields and the integrative anatomy of the entirety of the adhesive apparatus. We conclude with a brief consideration of the biomimetic applications that endeavor to exploit knowledge gained about the gecko effect.

2. The Gecko Effect — How is Attachment Achieved?

Pioneering work attempting to address how attachment is brought about in geckos stretches back well over a century but the ability to empirically test the ideas being advocated has often lagged behind explanatory theory, because of a lack of appropriate technology that would allow investigation at the necessary scale.

The existence of subdigital setae was revealed via macroscopic and light-microscopic observation,[10–12] and setal branching was initially observed using light microscopy.[13] Electron microscopy uncovered more detail of the structure, geometry, and interspecific variation of setae.[14–16] Coincident with observations on gekkonid (and polychrotid) setal structure came various hypotheses about the phenomenon of attachment itself.

One of the earliest theories hypothesized was that attachment occurs by way of a sticky secretion, but this was quickly refuted because no glands on the feet of geckos capable of producing such secretions were found.[11]

Suction was advanced as a mechanism,[17–20] but was soon refuted experimentally,[21,22] although this mode of attachment is still frequently, although erroneously, advocated as the means by which gekkonid attachment is brought about. Electrostatic attraction was advocated,[13] but was criticized,[23] and then experimentally refuted,[22] at least as a major mechanism of attachment, even though it may possibly be supplementary.[24,25] The suggestion that setae have a very high coefficient of friction[23] led to the hypothesis that geckos attach by way of frictional forces between the setae and the substrate. This idea remained dominant for many years and was further advanced[22,26] following an elaborate series of experiments.

The so-called "climbers boot" theory was advanced,[27] implicating the setae as structures capable of microinterlocking with substrate asperities. This was rejected[24] on structural grounds, based upon physical properties of the epidermal outgrowths, and later by the demonstration that gecko adhesion could occur on molecularly smooth surfaces.[1]

A major limitation of many of the early theories of attachment was that, although the presence of setae had been established, their structure, morphology, and composition were not yet understood. The advent of scanning electron microscopic studies of setal morphology[15,25,28] and histological structure[24,29,30] led to new theories that accounted for setal form and composition. Additionally, these studies were able to refute previously advanced theories by demonstrating that the shape of setae is not consistent with the "climber's boot" theory,[15] and that the flexible, β-keratin composition of setae is inconsistent with interlocking, as this mechanism would require much more rigid structures.[24]

These new findings led Hiller[25] to re-examine the possibility that intermolecular forces, or true "adhesive" forces, are responsible for attachment in geckos, as originally suggested by Haase.[31] Hiller's[25,32] experiments revealed that a gecko's ability to adhere depends upon the surface energy of the substrate, as measured by its water contact angle. His experiments revealed that setae could stick to hydrophilic surfaces (those with a low water contact angle), but were less able to adhere to hydrophobic surfaces (those with a high water contact angle), indicating that the adhesive force generated by setae seemingly increases with increasing surface energy. Based on this, Hiller[25,32,33] concluded that attachment in geckos occurs by way of intermolecular forces, although he could not conclusively determine which type of intermolecular forces dominate. The re-investigation of adhesion by intermolecular forces by Hiller[25,32,33] led to a renewed interest in gekkotan attachment and ultimately to an explosion of recent studies focusing on the mechanism by which attachment is achieved.

Such investigations implicated intermolecular forces but did not allow for capillary adhesion and van der Waals forces to be distinguished from each other. Van der Waals interactions are the most universal and also the weakest of the intermolecular forces.

Nevertheless, they are capable of producing substantial forces between molecules and objects composed of molecules,[34] although the strength of these bonds depends upon a variety of factors. Primarily, van der Waals interactions are highly dependent on the amount of contact area and the separation distance between the contacting surfaces, and thus require extremely close contact for the generation of substantial adhesive forces.[34] The strength of these forces also depends directly upon the polarizability of the surfaces and is not directly related to surface polarity.[34] Since van der Waals forces are universal and do not require the presence of a permanent dipole, an adhesive system configured to exploit these forces would be capable of adhesion on almost any surface.

In contrast, capillary forces result from the surface tension of a layer of liquid between two contacting surfaces.[35] The liquid layer between the surfaces can act as an adhesive if their wettabilities are high and their contact angles are low. The surface tension bond formed is resistant to perpendicular forces, but will slide along the surface, and thus functions best on rough surfaces, with asperities that can act as anchor points for the fluid. Unlike van der Waals forces, capillary forces do not depend upon contact area, and are not related to the size of the contacting structures.

The intermolecular forces involved in the adhesion of gekkotans have recently been re-examined,[1,36,37] leading to advancements and elaboration upon this theory of attachment. By measuring the attachment forces generated by individual setae, these were found to be consistent with the forces expected to be produced by intermolecular attachment, and were inconsistent with other possible mechanisms of attachment, including suction, friction and microinterlocking,[1] although the importance of frictional forces in attachment was later to be reassessed.[38,39] Specifically, it was shown that a single seta is capable of producing up to 200 μN of force, which is consistent with estimated values of setal force of 40–400 μN obtained from a mathematical model based on the Johnson–Kendall–Roberts theory of van der Waals forces,[40,41] which yields different force values depending on the shape of the contacting surface incorporated into the model.[42] Although these results seemed to be most consistent with van der Waals forces, the possibility that capillary forces are the primary mechanism of adhesion could not be immediately dismissed.

A subsequent set of experiments expanding upon Hiller's findings,[25,32] were conducted in order to ascertain the type of force primarily responsible for adhesion in gekkotans.[36] The inability of geckos to stick to hydrophobic, weakly polarizable surfaces (those with a high water contact angle)[25] could have been due to either reduced capillary adhesion or reduced van der Waals forces.[36] In order to resolve this problem, polarizability was separated from hydrophilicity/hydrophobicity. This is important because van der Waals adhesion depends on the creation of instantaneous dipoles, and thus on the polarizability, rather than hydrophobicity, of a surface.[36] It was found that both whole toes, and single setae were able to attach equally well to both a hydrophilic and a hydrophobic polarizable surface, indicating that van der Waals forces rather than capillary forces are the primary mechanism of attachment in gekkotans.[36] This study provided further support for van der Waals adhesion by revealing that setae are hydrophobic in nature, reducing the effectiveness of capillary adhesion.[36]

It was also suggested that the shape of setae, rather than their material composition, is the primary factor influencing their adhesive capabilities.[36] The chemical composition of setae is, however, important for other setal characteristics, such as their self-cleaning abilities and their nonsticky default state.[8,43]

Whereas such experiments demonstrated that van der Waals forces play an important role in adhesion, it still remained possible that capillary forces, microinterlocking, and frictional forces contribute. Mechanisms of adhesion based on capillary forces are common in a variety of animals including tree frogs,[44,45] feathertail gliders,[46] and a variety of insects.[47–49] Recent measurements of pull-off forces of single gecko spatulae demonstrate that capillary forces may play a significant role in attachment,[50,51] although it has been shown that adhesion is possible without their contribution.[37] Spatular pull-off forces increase with increasing relative humidity and in water, indicating that a water layer between spatulae and the substrate may play a role in enhancing adhesion.[50,51] The results of a theoretical exploration of this possibility[51] were questioned,[52] however, because the model employed did not take into account the hydrophobic nature of setae[8,37,43] suggesting that the results were not representative of the gekkotan system. Furthermore, the measurement of pull-off forces does not provide a realistic representation of the actual forces produced by an intact adhesive system. Pull-off forces are vertically directed and investigations of such forces deploy setae in orientations not likely to be encountered when they are part of an intact setal field.

Huber *et al.*[50] suggested that water layers may contribute to adhesion in a different way, however, theorizing that adsorption of water molecules by the setal spatulae could change the effective short-range interactions, enhancing van der Waals adhesion. The potential role of water and its interaction with adhesive forces in gekkotans is a very complex issue requiring further study.[52]

The role of frictional forces in adhesion has recently been re-examined, further exemplifying the complex mechanism of attachment in geckos.[38,39] Conventionally, friction is described as a contact interaction between two surfaces that opposes their relative motion, thus preventing slippage.[53] Friction at a macroscopic scale is proportional to, and depends upon, the normal force, which acts perpendicular to the surfaces, pushing them together.[54] It also depends upon the coefficient of friction of the two surfaces, a parameter that relates to their surface properties. The force of friction is described by Eq. (1) below.

$$F_F = \mu F_N, \tag{1}$$

where μ is the coefficient of friction and F_N is the normal force or the load. A frictional force may be static (acting on a stationary object) or kinetic (acting on a moving object), and the coefficient of friction varies in relation to this. Friction acts parallel to the surface and is independent of both contact area and speed.[54] On inclined surfaces frictional forces may help an animal maintain traction, but the animal will slip if the component of its weight parallel to the surface exceeds the force of static friction. Furthermore, on perfectly vertical surfaces the force of static friction is zero because the gravitational force has no component perpendicular to the surface.[55] Similarly, as frictional forces act only in shear[56] they cannot aid in adhesion beyond the vertical (on overhangs). It is these characteristics of frictional forces that initially led their rejection as the primary mechanism of adhesion in geckos.

However, the operation of frictional forces at a microscopic scale is very complex, and not fully understood. Amonton's laws state that friction is independent of the surface area of an object. Although this has been found to be true for macroscopic surface area, it does not hold at the molecular level, where it has been shown that the strength of a frictional force is proportional to the true area of contact between two surfaces.[57] At the molecular level surfaces are not smooth, but bear many surface asperities. As a result, true contact between the surfaces is limited to discrete contact points, and the true area of contact is only a fraction of the area that would be in contact if the surfaces were perfectly smooth.[57,58] This also explains why friction is proportional to the normal force; generally, the true area of contact is proportional to the load, or force pressing the two surfaces together.[57,59] Increasing loads will cause the surfaces to deform (either plastically or elastically), thereby increasing the contact area between the surfaces.[57] The discovery that friction is proportional to true contact area led to the suggestion that friction is closely linked to intermolecular bonding.[59] It was suggested that at a molecular level, microscopic points of contact between the surfaces become "cold welded" together as a result of surface adhesion due to intermolecular bonds, and that frictional resistance is associated with these small bonded points of contact.[53,57] For sliding to occur between the two surfaces these bonds would have to be sheared apart.[57]

However, further evidence revealed that this view of friction is not entirely correct. Although adhesion and friction are very closely linked, frictional force is not, in fact, directly correlated to the strength of the adhesive bond.[59,60] Experiments measuring adhesion and friction between atomically smooth mica surfaces have demonstrated that friction is related to the irreversible component of adhesion energy during a loading and unloading cycle.[60] For instance, kinetic friction involves one surface sliding over another. However, instead of sliding, one surface could be detached, moved, and reattached at a new point of contact, which is an adhesion or loading–unloading cycle. The kinetic friction force is thus related to the net energy change involved in the attachment–detachment cycle, and not to the actual strength of the adhesive bond.[60]

Static friction, unlike kinetic friction, can be related to the equilibrium adhesion force or energy.[60] Generally, for two molecularly smooth surfaces, friction can be described by Eq. (2):

$$F = C_1 A + C_2 L, \tag{2}$$

where C_1 and C_2 are constants, A is the adhesion contribution to the frictional force, and L is the load-dependent structural component of friction. On rough surfaces the true contact area between surfaces at a molecular level is very small, and the adhesion contribution to the frictional force is very small and almost nonexistent.[60] The remaining load-dependent component is equivalent to the coefficient of friction in Eq. (1). This term can be equated to the dimensions of surface asperities, and is related to the structural properties of the surfaces.[60] Surface roughness is thus an important aspect of friction, although its exact role is complex. In some cases friction between two surfaces can be reduced if one surface is rougher than the other, and friction can be high between two very smooth surfaces.[62] This seems to depend largely on how the surface roughness influences the actual contact area between the two surfaces. For example, a smooth surface sliding over a rough surface would only come into contact at a few discrete locations. However, two very smooth surfaces would have a larger area of molecular contact, thereby increasing friction. These concepts become very important when considering the role of friction in the adhesive system of gekkotans.

Van der Waals adhesion depends largely upon the closeness of contact between two surfaces, as these forces dominate at very small separation distances (<10 nm).[34] The numerous small, spatular tips of complex setae are thought to allow closer contact between the setae and the surface, thereby increasing the strength of van der Waals adhesion. However, at a molecular level, friction also depends on the true area of contact between two surfaces. Thus setal configuration that enhances van der Waals adhesion will also enhance friction between the setae and a surface. The highly branched nature of setae (Fig. 2(C)) and their very numerous, very small spatular tips provide increased contact area between the animal and a rough surface. This could increase the "adhesion component" of friction (Eq. (2)), thereby enhancing the frictional interaction between the animal and the surface. This suggests that complex, branched setae that contact the surface more efficiently may act better as frictional surfaces than less complex structures.

This clear link between frictional and adhesive forces has recently been reassessed[38,39] and a new model of "frictional adhesion" has been proposed to explain the gecko effect.[38] This model is based on measurements from live geckos and isolated setal arrays, that demonstrate that the force produced by gekkotan setal fields has both frictional (shear) and adhesive (intermolecular bonding) components, and that the strength of attachment is dependent upon the generation of a shear force. This model, therefore, considers adhesive force as a function of shear force, instead of the more conventional view of frictional force as a function of the normal force. As such, this model predicts that shear forces can be generated internally and through the use of opposing foot placement on inverted surfaces, allowing adhesion even in this situation.[38]

In general, a shear force of at least double the adhesive force is required for the adhesive force to be maintained,[38] and measurements from isolated setae of 200 μN of shear force and 40 μN of adhesive force[1,36] suggest that that this factor may often be exceeded. This frictional adhesion model also helps to explain how detachment of the setae occurs. The relationship between the two forces is consistent with a critical detachment angle of 30° (see below). At shear forces less than twice the adhesive force, and an angle of greater than 30°, adhesive forces will no longer be able to be maintained and setae will detach from the surface with no measurable detachment forces. The model, therefore, is defined as:

$$F_A \geq -F_S/\tan\alpha, \tag{3}$$

where F_A is the adhesive or perpendicular force, F_S is the shear or parallel force and α is the critical detachment angle.

Recent discoveries have, therefore, revealed much more about how individual setae behave and how they attach to surfaces.[1,8,43] They are inherently self-cleaning structures, and represent the first self-cleaning adhesive discovered.[52] They are also highly controlled and unidirectional structures and are unique as adhesives in that they possess a nonsticky default state, which prevents inappropriate adhesion to surfaces.[43] To achieve attachment individual setae require both a perpendicular and a parallel preload in order to maximize attachment forces, and this loading must occur in the direction of setal curvature. The perpendicular preload, which involves a small force pushing the seta into the surface, generates an adhesive force between the seta and the substrate via van der Waals bonding. The parallel preload, which involves pulling the seta along the surface, generates the shear force required to establish maximal attachment and to maintain adhesion. Therefore, the perpendicular and parallel preloads are required to generate the forces associated with the frictional adhesion model of setal attachment.[38] Setae will detach from the surface if they are raised beyond a critical angle of approximately 30°,[1] which, as discussed above, is dependent upon the relationship between adhesive and shear forces. Therefore, maximally attached setae are loaded in tension[63] and are held at a low angle (less than 30°) relative to the substrate. An understanding of how setae are brought into contact with surfaces and how attachment forces are generated has allowed the means of deployment of individual setae to be integrated with other elements of the adhesive system.[6]

The highly branched structure of setae is also important for their function as attachment structures. The β-keratin of which setae are composed is a very stiff material with a fairly high elastic modulus. In its bulk form β-keratin would likely be too stiff to function as a tacky material and conform to surface irregularities.[64] However, the fine subdivision of setal tips and the high aspect ratio of the setal shafts results in a considerable decrease in the effective elastic modulus of the β-keratin, allowing setae to act as flexible attachment structures and to conform more easily to irregular surfaces.[64]

3. Structure of the Setal Fields and the Anatomical Hierarchy on Which They Depend

This history of investigation of the mechanism of gekkotan adhesion has been paralleled by a series of studies that have attempted to understand the structure and function of the entire adhesive apparatus.[5,6,22,23,26,27,30,65–71] Other studies have endeavored to investigate the performance capabilities of the adhesive system of geckos in both static clinging[72,73] and dynamic running[74] tests.

Gekkotan setae have been described on numerous occasions,[15,52,75,76] but variation in form is poorly understood, both within and between species.[52] This lack of understanding is partly due to an absence of a well-corroborated pattern of relationships among gekkotan taxa within which comparative observations can be contextualized,[77] and partly due to a lack of comprehensive study of pattern even within a single species.

Our current understanding of setal form is thus rather generalized, with descriptions pertinent to individual species mostly being single examples purported to be "typical" (e.g. Ref. 75) and to have a particular set of dimensions relating to length, stalk diameter, branching pattern, and spatular dimensions.[39,52] Recently we have investigated setal form and variation in the tokay gecko (*Gekko gecko*), and the southern African *Rhoptropus* radiation[3,9] to investigate the pattern expressed across entire setal fields. Entire setal fields were prepared for scanning electron microscopical and

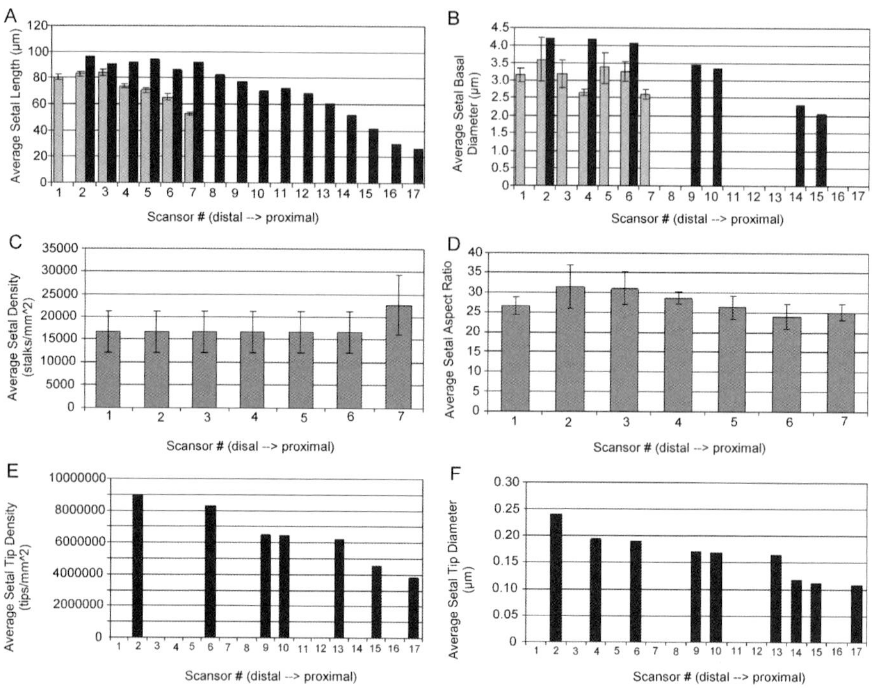

Fig. 3. Setal patterns along digit IV of the pes of *Rhoptropus* cf. *biporosus* (grey) and *Gekko gecko* (black): (A) setal length, (B) setal basal diameter, (C) setal density, (D) setal aspect ratio, (E) setal tip density, and (F) setal tip diameter. Note that *Rhoptropus* cf. *biporosus* has fewer scansors than does *Gekko gecko*, and that comparisons are most effectively made by examining the relative position along the length of the adhesive pad.

histological examination, and were surveyed in a consistent fashion across many individuals for all digits of the manus and pes. Figure 3 provides data for the fourth digit of the right pes of *Rhoptropus* cf. *biporosus* and *G. gecko*, indicating the pattern of variation of setal form in but a single digit. This pattern is essentially identical in its overall conformation between the two species.[9] All scales in the distal segment of the digit are true scansors (plate-like structures bearing sectors of the setal field — Fig. 2(A)) that are underlain by the entire complex hierarchy of control mechanisms associated with setal attachment and detachment[6] (see Fig. 4). It is evident from these data that setal length (Fig. 3(A)) varies considerably from the distalmost to the proximalmost scansor in this region, and that basal diameter (Fig. 3(B)), tip dimensions (Fig. 3(F)), tip area, tip density (Fig. 3(E)), and total spatulate tip area are distributed differently across the subdigital pad. More proximally, in the intermediate and proximal regions of the digit[9] these dimensions change further, with true setae giving way to seta-like elaborations proximally. This transition of form relates to the level of direct control able to be exerted on the setae (see below).

It is evident from these data that there is no typical seta, and that each scansor has a particular pattern that differs in magnitude, if not type, from that of its adjacent neighbors. It is also evident that although individual characteristics, such as the average area of each setal tip (Fig. 3(F)), and average tip density per unit area (Fig. 3(E)), change along the length of the pad, total setal tip area (which relates to the total adhesive force potentially able to be generated) varies much less, indicating that compensatory changes occur that are likely important to functional interactions and the creation and release of the adhesive bond in different regions of the digit. Stalk density remains essentially unchanged in the scansorial region of the digit (Fig. 3(C)), even though stalk diameter decreases (Fig. 3(B)), again indicating compensation on a fine scale. It appears that the mean aspect ratio of the setae (Fig. 3(D)) remains relatively constant[19–23] for all distal scansors, even though dimensions of the setae do not. However, the aspect ratio changes along the length of each scansor in *Rhoptropus* and as a result, all setal shafts do not have the same effective elastic modulus (in terms of bending),[64] but the pattern is consistent for all scansors, and thus the contact fraction potential should be approximately equal per unit area along the length of the scansorial pad.

Such observations, within the confines of single species, and even a single digit, indicate that variation across the pad is real and likely relates to the particular mechanics of the digits of geckos (Fig. 4). Autumn *et al.*[64] have posited that the distance between setal stalks and the stiffness of the shafts relate, in opposite ways, to the avoidance of matting of the tips and the ability to compensate for rough substrate conditions. Variance in aspect ratio is important in the context of its relationship to the material properties of β-keratin, the material of which setae are composed. Such ideas can be investigated across gecko radiations now that basic aspects of setal field design are known.

Consideration of the appropriate configuration of gecko-inspired adhesives has provided a new context for the understanding of the form of setal fields (Fig. 2(A)) seen in geckos themselves. Schargott[78] suggested, based upon general design principles related to fibrillar arrays, that stiffness of the array can be reduced by tilting the beams, resulting in a structured surface with tilted elements. Gekkonid setae are held in a tilted configuration (Fig. 2(A)), affecting their directionality as adhesive structures,[52] and show variation in length along the subdigital pad that may be related to an avoidance of matting when the setae are deployed.[78] Such ideas relating to the configuration of synthetic microfibrillar arrays help to explain the observed pattern of setal fields in geckos, and point to profitable lines of investigation for the future.

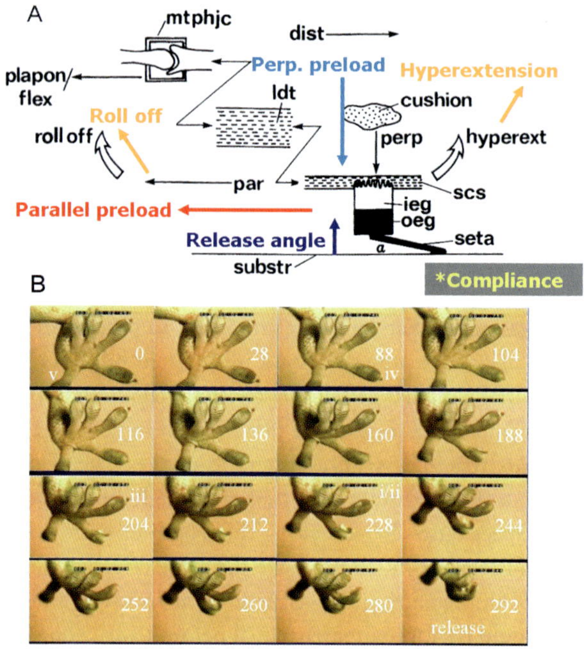

Fig. 4. (A) Diagrammatic representation of the relationships between the seta/substrate contact pattern, the mechanical linkages from seta to skeleton and the application of forces bringing about and releasing the adhesive grip. The setal stalk changes its angle with respect to the substratum and the ventral surface of the foot as substrate contact is made and released. At contact, the stalk angle (α) is depressed below the critical angle for maximal contact to be made. Perpendicular preloading (Perp. Preload) is applied by unfurling of the digit from its hyperextended position distally, or by rolling onto the more proximal scansors proximally, and is assisted by a distal cushion (such as a vascular sinus network). The parallel preload is applied and maintained by a combination of active loading through the tensile skeleton of the digit, mediated by flexor muscles acting through the plantar aponeurosis, and by gravitational shear (on vertical surfaces). Detachment is facilitated by hyperextension distally and roll off proximally, both of which raise the setal shaft above the critical angle threshold. The parallel and perpendicular preloads are diminished as a result of pressure changes and unloading of the tensile skeleton. Abbreviations: α, angle of setal stalk with substratum; dist., distal; ieg, inner epidermal generation; ldt, lateral digital tendon; mtphjc, metatarsophalangeal joint capsule; oeg, outer epidermal generation; plapon/flex, connection to the plantar aponeurosis and crural flexor muscles; scs, stratum compactum of the dermis of the scansor; substr, substratum. (B) Ventral view of the hyperextension and release of the left pes of *Gekko gecko* as viewed through a vertically oriented glass surface, videographed at 500 frames per second. White numbers refer to milliseconds; i–iv refer to digit number. The lizard is detaching its pes from a standing start.

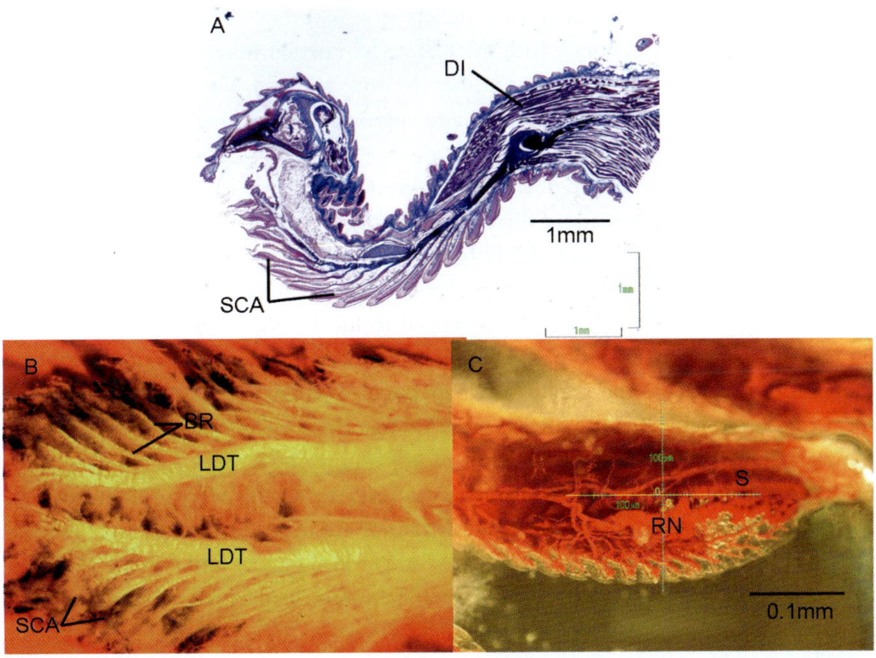

Fig. 5. The anatomy of the adhesive system. (A) Longitudinal section of fourth digit of *Thecadactylus rapicauda* illustrating the dorsal interossei (DI) muscles responsible for hyperextension and the overlapping patterns of the scansors (SCA). (B) The lateral digital tendons (LDT) of *Gekko gecko* that branch (BR) and attach to the dermis of the individual scansors (SCA). (C) Vascular sinuses in the digit of *Gekko gecko*. A large central blood sinus (S) connects with reticular networks (RN) that pervade each scansor.

The entirety of the adhesive system is complex, and many elements, including the muscular (Fig. 5(A)), skeletal, tendon (Fig. 5(B)) and vascular systems (Fig. 5(C)) of the foot, work together to control the actions of the setae and thus the process of adhesion and detachment during locomotion.[5,6,67] The digits of gekkotans undergo active hyperextension (Fig. 4(B)) during locomotion,[5] whereby the digits are rolled off the surface from the distal end proximally. The digits are then rolled back onto the surface from proximal to distal to effect attachment.[5] In gekkotans, hyperextension is controlled by the muscles of the digits (Fig. 5(A)) which either insert directly onto the distalmost phalanges, or by digital tendons.[5] The process of hyperextension is monitored by cutaneous sensilla located on the dorsal surface of the digit[79,80] and is likely to play a major role in the deployment and detachment of setae in gekkotans. During detachment, hyperextension may act to deflect setae past the critical angle of 30°, causing them to break their contact with the surface.[1,6] Furthermore, during attachment, unrolling the digits onto the surface pushes the setae forward (Fig. 4(A)), producing a larger area of contact with the substrate and aligning the setae appropriately with the surface, thereby enhancing adhesion.[5]

In addition to its role in the hyperextension of the digits, the digital tendon system may play another, more direct, role in the attachment of setae. The lateral digital tendons (Fig. 5(B)) are highly branched, and at least one branch inserts on each side of every scansor of the subdigital pad.[5] These tendons, therefore, permit fine control of the scansors, and may be deployed to locally adjust the tension imposed on individual scansors to promote maximal contact with the surface.[5] Furthermore, these tendons provide the parallel preload force required to generate the maximal shear force attainable by individual setae.[1,6] While the parallel preload is contributed to by gravitational forces which pull downwards on the digits when the animal climbs on a vertical surface,[6] gravitational forces can only act in this way on digits aligned with some component of gravity, and thus cannot aid in parallel loading of the setae on surfaces inclined beyond the vertical. As such, both gravity and the lateral digital tendon system may act together in the establishment of maximal attachment force generation.[6]

The vascular system of the digit (Fig. 5(C)) also plays a role in adhesion and in the deployment of setae (Fig. 4).[6,67] Most pad-bearing gekkotans possess a large blood sinus beneath the penultimate phalanx of the digit.[67] It has been suggested that these sinuses allow the subdigital pad to conform to surface irregularities, thus enhancing adhesion.[67] It is also possible that increased pressure in the subdigital sinuses contributes to the perpendicular preload required to engage setae with the substrate.[1,6] The pressure in the digital sinuses may be linked to, and moderated by the hyperextension of the digits.[67] When the digits are hyperextended, the sinuses are drained, which may serve to aid in the detachment of the setae.[67]

Gekkotans possess the unique ability to adhere to surfaces in any orientation, even completely inverted. Thus, their anatomical components must be able to operate to enable contact to be effected in a variety of body configurations and patterns of whole body loading. In an inverted position, geckos are not able to depend upon gravitational forces to aid in the generation of either the parallel or perpendicular preloads required for setal attachment, and must be capable of generating them internally, possibly through the combination of the lateral digital tendon system and subdigital blood sinuses (Fig. 5(B) and 5(C)),[6] and through highly controlled foot placement.[5,38] In inverted situations these animals would also be required to control their muscular system (Fig. 5(A)) so as to enable locomotion in this orientation, and in this altered gravitational state.[81] Gekkotans react differently to microgravity or freefall situations than do other squamates.[82] While most lizards thrash wildly in these circumstances, and grasp at anything available, even their own tails, both arboreal and nonarboreal gekkotans react by spreading the limbs in a "sky diving" posture until a surface is contacted and adhesion in accomplished.[82] A recent study by Jusufi *et al.*[83] revealed that geckos rotate their tail following a fall to achieve this "sky diving" posture, and are also capable of using their tails for manoeuvering in mid-air. These findings indicate that geckos are neurologically adapted to adjust to a variety of body orientations and situations, and that these modifications enhance the ability to employ the adhesive system in environments that present complex three-dimensional and gravitationally variable challenges to the effective employment of their highly modified adhesive system.

The anatomy of the digits and the adhesive system of pad-bearing lizards, and their patterns of neurological control, are complex, and act as an integrated assemblage to bring about efficient adhesion and detachment of the subdigital pads in all orientational situations and on a variety of surfaces, thus enhancing the overall capabilities of the gecko effect.

4. Performance Aspects — Real-World Functional Demands in Relation to the Gecko Effect

Recent research into the adhesive performance of gecko setae has revealed that the perpendicular preload and parallel preload needed for force generation[36] are related to two distinct, but interrelated aspects, of setal attachment to surfaces. Adhesive force output has been shown to be directly related to the shear force applied to the setae. The magnitude of adhesive force measured has been found to be condition dependent, with vertical van der Waals forces[84] providing a significantly smaller proportion of the total than the friction generated through loading in shear.[38,39,86,87] Frictional adhesion has been proposed as a key benchmark for the performance of gecko-like adhesives.[85]

Geckos are able to hold station on vertical surfaces, implying that the frictional loading is able to be maintained for long periods with the animal remaining static. To be effective on surfaces beyond the vertical, this loading must be actively applied in parallel to the locomotor substratum, opposing, rather than with the assistance of, gravitational loading. The intimate contact between integument, dermis, subdermal lateral digital tendons, and joint capsules and limb muscles are implicated in this chain of control (Fig. 4(A)). Evidence gained from calculations based upon isolated setae indicate what the potential of the functional system is, but observation of input forces, and the timing of these, in the intact animal are needed to indicate how geckos control adhesion in the rapid-fire force modulation required for fast locomotion, and in the long-term loading important in station holding.

The recognition that van der Waals forces may be responsible for adhesion at high setal shaft angles (as the setae are brought into contact with the substratum) and that frictional forces may become dominant as the setal shaft angle is lowered to 30° or less[39] provides background information important in the consideration of the evolution of the adhesive system and the inception of successive forces capable of reversibly binding with the substratum. The evolutionary transition from nonadhesive to adhesive functionality in the gecko foot is now beginning to be unraveled as a result of a more complete understanding of the loading vectors of microfibrillar arrays in relation to contact with the substratum.

Setae are arrayed on scansors in ranks and rows (Fig. 2(A) and 2(B)), and their physical orientation changes as the digits alter their configuration through the attachment–detachment cycle. In the hyperextended state, the setae are spread apart in a fan-like arrangement, and are held at their resting angle with respect to their local station on the epidermis of the scansor. As the adhesive pad approaches the surface, it does so with the base of the digit already in contact with the substratum prior to unfurling of the distal part of the digit from its hyperextended state. The setae of each scansor are brought into contact with the substratum sequentially, from the proximal to the distal end of the adhesive pad. It is hypothesized that the unfurling of the digit provides the perpendicular preload, and contact with the surface is made by the distal ends of the spatulae. The contact is likely to close enough in certain regions to permit van der Waals bonding at the tips of the setae, creating relatively weak attractive forces[84] that initiate adhesive contact. Any setae achieving this level of contact are thus candidates for the next phase in the attachment process. Subsequent to the unfurling of the digit, a parallel load is induced through the lateral digital tendons (Fig. 5(B)), imposing tension on the epidermal-dermal interface of the entire scansor. This likely relates to the transition in loading that induces the more powerful frictional interaction[39,86] raising the magnitude of the recorded pull-off forces.[1]

The two stage pattern of contact, via both van der Waals forces and frictional forces, that establishes the totality of adhesive contact, provides information pertinent to the consideration of how such an adhesive system may have originated. This can be approached by investigating the adhesive properties of setal and lamellar arrays of subdigital elaborations that are exhibited by a single species (Fig. 3) and by investigating candidate taxa that exhibit, among their ranks, species that express differential degrees of "setal" expression and control mechanics in conjunction with different patterns of habitat occupancy and substratum exploitation.[66] Among gekkotans, a seta-based adhesive system has evolved independently on many occasions (as well as independently in the polychrotid anoline lineage),[66,88] and it is likely that similar integument–substrate interactions have been the focus for a multitude of integrative evolutionary steps that have led to similar mechanical solutions to the control of adhesion, in both contact and subsequent release, and thus the establishment of the gecko effect.

The setal array acts like a tacky, deformable material, with the effective elastic modulus of the fibre array in bending being four orders of magnitude less than that of the setae in tension.[52] Such properties exploit both the high tensile strength of β-keratin,[89,90] with its proteins polymerized into long cables,[76] in the tensile, friction-dominated mode of loading, and the smaller elastic modulus resulting from its elaboration into cantilevered beams that enhance adhesion via van der Waals interactions. Aspect ratio and setal tip form are important determinants of the potential for making effective, shear-enhanced contact with the surface. Exploration of setal form and aspect ratio in taxa exhibiting incipient development of adhesive pads[66] should reveal aspects important in the transition to the shear-dominated attachment mode of geckos.

The ability of geckos to switch rapidly from attachment to detachment[39] has been related to the phenomenon of digital hyperextension and the lever function of setal shafts. Relaxation of the setae back to their resting state generates a very small release force.[39] The reversibility of attachment has been regarded as being of equal importance to the attachment process in the quest to fabricate microfibrillar adhesive arrays.[91] It has been recognized that the stronger the material anisotropy, the higher the adhesion releasability, with robust attachment occurring in the stiff direction of the directional adhesive and release being promoted by pulling in the soft direction, known as the so-called "stiff adhere–soft release" couplet.[91] Gecko setal arrays operate in this fashion. Release of tension on the setal shafts imposed by the lateral digital tendon system (Fig. 5(B)) likely results in the switching from the "stiff adhere" phase to the "soft release" phase, with digital hyperextension raising the setal arrays to above the critical angle of 30°.[85]

The specific dimensions of setae and their arrangement into setal fields may provide certain functional benefits to enhancing and maximizing adhesive effectiveness. The adhesive system of gekkotans has often been modelled as a mat of fibrillar microstructures that is brought into contact with a surface.[92–94] The reliance of the system on van der Waals forces for initiating adhesion with the surface[36,37] before larger shear forces can be induced[38,39] is predicated upon extremely close contact between the setae and the locomotor surface. The performance of such a system depends primarily on the mechanical control of the contacting structures as well as their geometry and material properties.[94,95] The primary advantage of a fibrillar structure is that it allows for increased compliance between the fibres and the surface, leading to increased contact and adhesion on rough surfaces with a variety of wavelengths and asperity heights.[92–95] This interpretation can be extrapolated for relatively uniform substrata, but is locally applicable for highly irregular and/or highly undulant surfaces as contacted by complete setal fields.[3]

Multiply branched setal tips likely compensate for the ability of only small portions of the adhesive system being able to make effective contact with an undulant locomotor substratum on any given footfall. For geckos, setal fields and their component scansorial subunits are large and deep in comparison to the physical attributes of the terrain traversed,[3] and will bridge across high points rather than conforming to the contours of the undulant surface (Fig. 6). Whereas smooth surfaces provide expansive areas for potential contact, rougher and more undulant surfaces provide only local patches with which intimate setal contact can be made (Fig. 6). In real-world situations the gecko effect is operationalized in circumstances that provide only

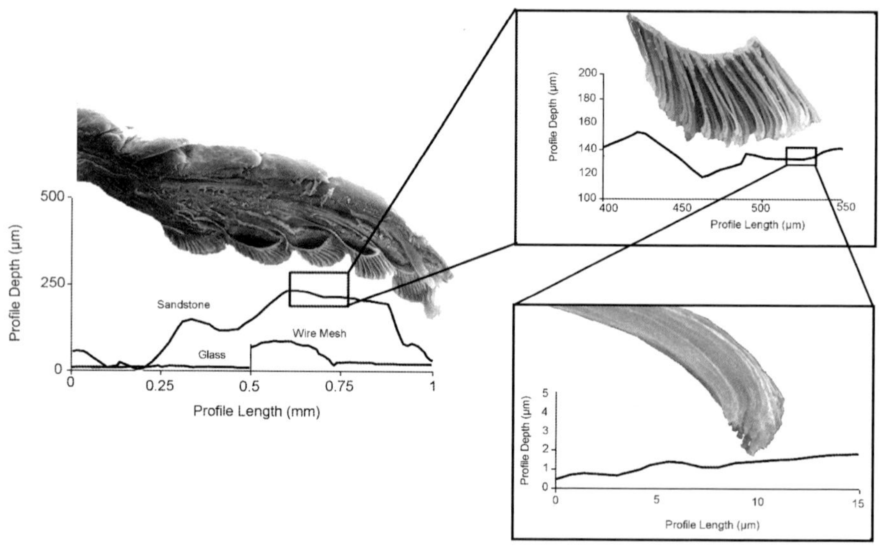

Fig. 6. Longitudinal section of digit IV of *Rhoptropus* cf. *biporosus*, superimposed upon cross-sectional profiles (obtained using the software program MeX 3D–see Sec. 3) of a rock surface (sandstone), a smooth artificial surface (glass), and an undulant artificial surface (wire mesh). This shows the potential restriction of contact to local patches between the setal field of the digit and available portions of these surfaces. The digit and its associated profiles are presented at an identical scale. The upper inset shows individual setae superimposed upon a cross-sectional profile of the sandstone surface illustrating the variability encountered by the setae at this level, and the lower inset shows the interaction of setal tips with a cross-sectional profile of a local area of the sandstone surface. Both the setae and setal tips and their associated profiles are presented at the same scale.

limited opportunities for making such contact, likely accounting for the enormous adhesive potential of the system as a whole.[3] The gecko effect, for geckos in the field, relies upon making sufficient contact to enable support of the body mass on any given footfall, but the enormous capacity of the system is essentially an adaptive response to the unpredictability of suitable contact patches. Exemplification of this is provided by etrapolatory data for the Tokay gecko (*G. gecko*) — the estimated maximum adhesive capacity of this species is 1300 N, nearly 4000 times the force required to support its 50 g mass; this indicates that only 0.04% of its total number of setae are required to effect attachment.[96]

The chemical composition of setae is important in their ability to act as fibrillar mats and to effectively contact undulant surfaces. The β-keratin of which setae are composed is a relatively stiff material, with an elastic modulus too high to allow effective conformation and attachment to surfaces.[64] However, the configuration of the adhesive system into a mat of high aspect ratio shafts of this relatively stiff material allows the setae to behave differently than would bulk keratin, yielding a much lower effective elastic modulus.[64] Setae have dimensions[9] that are conducive to their behaviour as pressure-sensitive adhesive structures, with effective elastic moduli well below the 100 kPa required for them to function effectively in this manner. The ability of setae to act as a field of highly flexible structures, with a low effective elastic modulus, enhances their ability to conform and adhere to microscopically rough surfaces and to comply with surface irregularities.[64,92,95,97–100] However, there is a limit to how long and flexible setae may become, as these properties also promote matting of the setae,[64,93,95,98] which would compromise their adhesive capabilities.[49]

Maderson[30] asked (i) what precise locomotory advantage do setae provide that cannot be adequately met by claws or digital flexure? and (ii) is there any evidence that some or all seta-bearing taxa live on surfaces that have any particular physical characteristics that make setae particularly advantageous? In other words, he asked questions about the specific advantages of what is herein characterized as the gecko effect.

In addressing the first question, studies of *Rhoptropus*[3] have been informative because this taxon is rock-dwelling, lives on undulant, unpredictable surfaces, and lacks claws. The irregularity and randomness of asperities of the scale and distribution evident in the rock surfaces on which it lives (Fig. 6)[3] and the unyielding nature of the locomotor substratum, all indicate that fields of setae may provide a locomotor advantage that individual setae alone might not suggest. Placement of digits onto such a locally unpredictable and unyielding surface would likely mean that adequate purchase provided by claws alone could not be attained without, at best, considerable slippage until an appropriate site of anchorage had been located. Fields of setae, given the random distribution of available support areas, allow islands of suitable substratum to be located and utilized. Furthermore, this will be the case for all body orientations, from vertical to overhanging, to inverted, a situation certainly not so for claws.[55]

In addressing Maderson's[30] second question, it is necessary to consider surfaces naturally exploited by geckos. The greater deformability and consequent increase in real contact area engendered by the extreme levels of splitting of the setal tips seen in geckos[42,98,100–102] enhances the attachment capabilities of a system in circumstances in which only a small portion of the available adhesive contacts can be deployed. For geckos, setal fields are large in comparison to the physical attributes of the terrain they traverse (Fig. 6), and conformation to every irregularity of the surface is not possible. Thus, for localized areas where contact can be established, maximization of contact force generation is of major importance, and extensive contact splitting assists this. Again, this is a particular attribute of the gecko effect at the integrative level of setal field contact with the substratum. Setal fields may be particularly useful on undulant and unpredictable surfaces[3] where contact is likely to be patchy and limited.

The complex configuration of setal fields implies that a simple mat of identical setal structures may not provide adequate attachment to the surfaces these animals move on, and that the microvariation of setae along both scansors and digits allows for optimal contact and adhesion on natural substrates. This region to region variation in setal field configuration may also have implications for the ongoing attempts to fabricate synthetic seta-like adhesives that attempt to mimic the gecko effect.

5. Biomimetics — The Application of Design Principles to Exploitation of the Gecko Effect

The disparate investigations that have been undertaken have brought us to a much fuller understanding of the adhesive processes involved in gekkotan attachment, the anatomical hierarchy through which this is carried out, and the performance capabilities of both gecko and gecko-inspired adhesive systems. We are now at the stage where we can begin to look back over the last 54 million years of evolution of the gekkotan adhesive system and start to apply what we have learned about the phenomenon and control of adhesion to how geckos function in their natural environment, what their capabilities are in those situations, and why the gekkkonid adhesive system is "designed" the way it is.

The impressive adhesive capacity of geckos has led to many recent attempts to fabricate nano-scale adhesive devices that mimic the structure of gecko setae.[93,97,103–107] Researchers have envisioned a wide variety of uses for gecko-like adhesives, including surgical aids, exploratory robots, and even football gloves.[52,96] Evidence suggests that setae may act in a manner similar to pressure-sensitive adhesives (PSAs),[64] but that they differ from conventional PSAs in that they do not adhere spontaneously,[43] are releasable, and are not subject to self-adhesion and degradation.[96] To date, the majority of microfabricated structures have taken the form of fibrillar mats of cylindrical structures composed of various polymer materials.[36,102,104,108,109] Initially these synthetic structures required very high preloads[104,105,109] and generated forces below the capacity of gecko setal fields. A few initial attempts have been made at modelling the hierarchical, branched structure of setae[109,110] but these structures have not been capable of achieving "gecko-like" forces, possibly due to the perpendicular application to the substrate resulting in buckling of the structures,[96,110] or lateral adhesion of the tips, both of which are common problems with fabricated microfiber arrays.[93,104] Synthetic structures capable of producing seta-level forces have been successfully produced at the nano-scale (spatular level), with silicone rubber and polyester materials with dimensions similar to those of setal tips, capable of generating adhesive forces of around 100 nN each, similar to expected values for real spatular tips.[108] Recently, synthetic structures with many characteristics of real setae have been successfully produced.[107] The stiff polymer fibre array produced by Lee *et al.*[107] is capable of generating seta-level forces, with 1 cm^2 of the material being able to support up to 400 g. Furthermore, the adhesive was not applied to the surface perpendicularly, but was pulled along the surface, which resulted in the generation of high shear forces. It is also re-usable, and is the first micro-fabricated adhesive to mimic the nonsticky default state of gecko setae.[107]

Several attempts to produce wall-climbing robots using gecko-like structures have also proved successful. One such is "Mecho Gecko" produced by iRobot in collaboration with UC Berkeley.[103] This robot climbs using a PSA and rolling structures that mimic the rolling of gecko digits onto and off of the surface by hyperextension. Another robot, Stickybot, possesses hierarchical levels of compliance, from the limbs themselves down to microcompliant seta-like stalks. Stickybot uses directional micro-adhesive stalks modelled on gecko setae to effectively climb smooth vertical surfaces at speeds up to 24 cm/s. The stalks are finely controlled by a series of cables which provide the perpendicular and parallel preloads required for adhesion. These impressive robots do not possess the fine setal control and branched setae found in geckos, but their ability to rapidly scale smooth vertical surfaces is remarkable.

Whereas Stickybot[106] possesses some degrees of compliance and control of its attachment devices, the control mechanisms of artificial setal arrays remain crude by gekkotan standards. In depth studies of gecko setal field configuration and control may, therefore, be informative about capabilities that may or may not be realizable in artificial constructs. Our ability to exploit the gecko effect will continue to depend upon insights from biology and engineering reciprocally illuminating one another.

Understanding the real-world capabilities of the adhesive adaptability of gekkotan setal fields[3] may also assist in visualizing solutions to the problems of not only fabricating gecko-inspired microfibrillar arrays, but also to the consideration of how these might be deployed in robotic applications. Selection has guided the scaling and configuration of the entire adhesive pad complex of gekkotans. Mimicking the incredible moment-to-moment and patch-to-patch adjustability of such a system is likely an unattainable goal, but an understanding of these operational capabilities are certainly instructive in guiding our efforts to comprehend and adapt the principles, without necessarily trying to duplicate the intricate details.

6. Conclusions

It is evident that the gecko effect, as manifested by geckos, results in the ability to reversibly attach to a wide array of surfaces in an astonishing array of body orientations. The interaction between the fields of microfibrillar arrays of the digits and the substratum, whether smooth, rough or undulant, depends upon the structural properties of an integument of β-keratin that results in the skin operating as a PSA. The gecko effect is able to be deployed and oscillated rapidly between attachment and release because of a complex structural hierarchy of control mechanisms that allow moment to moment adjustment to topological and material changes in circumstance. The gecko effect, as manifested in the living world, is restricted to animals that although of moderate size, when the entire range of body sizes of animals is considered, are still small by human standards. Attempts to microfabricate adhesives that mimic the properties of gekkotan setae present challenges that are rapidly being met. Employing such synthetic arrays at larger scales, however, may require a deeper understanding of how geckos scale their available setal field surface area as they grow[111] and what the limitations are that seemingly restrict geckos to rather modest sizes. The gecko effect is seemingly quite well understood in terms of the means by which attachment is brought about. The larger aspect of the gecko effect — its dynamic employment in very challenging nanoenvironmental circumstances — promises to present the biomimetic community with significant challenges for the foreseeable future.

References

[1] Autumn K, Liang YA, Hsleh ST, Zesch W, Chan WP, Kenny TW, Fearing R, Full RJ. (2000) Adhesive force of a single gecko foot-hair. *Nature* **405**: 681–685.

[2] Bauer AM, Böhme W, Weitschat W. (2005) An early Eocene gecko from Baltic amber and its implications for the evolution of gecko adhesion. *J Zool* (*Lond*) **265**: 327–332.

[3] Russell AP, Johnson MK. (2007) Real world challenges to, and capabilities of, the gekkotan adhesive system: contrasting the rough and the smooth. *Can J Zool* **85**: 1228–1238.

[4] Bauer AM. (1998) Morphology of the adhesive tail tips of carphodactyline geckos (Reptilia: Diplodactylidae). *J Morph* **235**: 41–58.

[5] Russell AP. (1975) A contribution to the functional analysis of the foot of the Tokay, *Gekko gecko* (Reptilia: Gekkonidae). *J Zool (Lond)* **176**: 437–476.

[6] Russell AP. (2002) Integrative functional morphology of the Gekkotan adhesive system (Reptilia: Gekkota). *Integr Compar Biol* **42**: 1154–1163.

[7] Maderson PFA, Chiu KW, Phillips JG. (1970) Endocrine–epidermal relationships in squamate reptiles. *Mem Soc Endocrinol* **18**: 259–284.

[8] Hansen WR, Autumn K. (2005) Evidence for self-cleaning in gecko setae. *Proc Nat Acad Sci* **102**: 385–389.

[9] Russell AP, Johnson MK, Delannoy, SM. (2007) Insights from studies of gecko-inspired adhesion and their impact on our understanding of the evolution of the gekkotan adhesive system. *J Adhesion Sci Technol* **21**: 1119–1143.

[10] Cartier O. (1872) Studien uber den feineren Bau der Epidermis bei den Reptilien. II. Abteilung. Uber die Wachsthumserscheinugen der Oberhaut von Schlangen und Eidechsen bei der Hautung. *Arb Zool Inst Würzburg* **1**: 239–258.

[11] Cartier O. (1872) Studien über den feineren Bau der Haut bei den Reptilien. I. Abt. Epidermis der Geckotiden. *Verhandlungen der Würzburger phys.-med. Gesellschaft* **1**: 83–96.

[12] Braun M. (1878) Zur Bedeutung der Cuticularborsten auf den Haftlappen der Geckotiden. *Arb Zool Zoot Inst Würzburg* **4**: 231–237.

[13] Schmidt H. (1905) Zur Anatomie und Physiologie der Geckopfote. *Jenaische Zeitschrift für Naturwissenschaften* **39**: 551–580 + plate 551.

[14] Altevogt R. (1954) Probleme eines Fusses. *Kosmos* **50**: 428–430.

[15] Ruibal R, Ernst V. (1965) The structure of the digital setae of lizards. *J Morphol* **117**: 271–294.

[16] Peterson JA. (1983) The evolution of the subdigital pad of *Anolis* 2. Comparisons among iguanid genera related to the anolines and a view from outside the radiation. *J Herpetol* **17**: 371–397.

[17] Wagler J. (1830) Natürliches System der Amphibien, Cotta'sche Buchhandllung, Munich.

[18] Simmermacher VG. (1884) Hafftapparate bei Wirbeltieren. *Zool Garten* **25**: 289–301.

[19] Gadow H. (1901) Amphibia and Reptiles: Geckones. *Camb Nat Hist* **8**: 502–513.

[20] Kunitzky J. (1903) Die Zeit der Entstehung der Borsten und Mechanismus der Bewegung bei den Geckotiden (*Ptychozoon homalocephalum* Crev). *Bull Acad Imper Sci (St. Petersbourg)* **18**: 21–24.

[21] Weitlaner F. (1902) Eine Untersuchung über den haftfuss des gecko. *Verhandlung Zool-Bot Gesell (Wein)* **52**: 328–332.

[22] Dellit WF. (1934) Zur Anatomie und Physiologie der Geckozehe. *Jena Z Naturwiss* **68**: 613–656.

[23] Hora SL. (1923) The adhesive apparatus on the toes of certain geckos and tree frogs. *J Proc Asiatic Society of Bengal* **19**: 138–145.

[24] Maderson PFA. (1964) Keratinized epidermal derivatives as an aid to climbing in gekkonid lizards. *Nature* **203**: 780–781.

[25] Hiller U. (1968) Untersuchungen zum feinbau und zur funktion der haftborsten von reptilien. *Morph Tiere* **62**: 307–362.

[26] Dellit WF. (1949) Zum haftproblem der geckoniden. *Dtsch Aquar Terr Z* **2**: 56–58.

[27] Mahendra BC. (1941) Contributions to the bionomics, anatomy, reproduction and development of the Indian house gecko *Hemidactylus flaviviridis* Rüppel, Part II: The problem of locomotion. *Proc Indian Acad Sci Sect B* **13**: 288–306.

[28] Hiller U. (1976) Comparative studies on the functional morphology of two Gekkonid lizards. *J Bombay Natl Hist Soc* **73**: 278–282.

[29] Maderson PFA. (1966) Some macroscopic and microscopic observations on the foot-pads of the tokay (*Gekko gecko*). *Mem Hong Kong Nat Hist Soc* **7**: 6–10.

[30] Maderson PFA. (1970) Lizard hands and lizard glands: Models for evolutionary study. *Forma et Functio* **3**: 179–204.

[31] Haase A. (1900) Untersuchungen über den bau und die entwicklung der haftlappen bei den geckotiden. *Archiv f Naturgesch* **66**: 321–345.

[32] Hiller U. (1969) Correlation between corona-discharge of polyethylene films and the adhering power of *Tarentola m. mauritanica* (Rept.). *Forma et Functio* **1**: 350–352.

[33] Hiller U. (2000) Walking upside down: The mystery of climbing ability in gekkonid lizards. *Gekko* **1**: 36–40.

[34] Rimai DS, Quesnal DJ. (2001) Fundamentals of Particle Adhesion. Global Press.

[35] Petrucci RH, Harwood WS. (1993) *General Chemistry*, 6th Edn. MacMillan Publishing Co., New York.

[36] Autumn K, Sitti M, Liang YA, Peattie AM, Hansen WR, Sponberg S, Kenny TW, Fearing R, Israelachvili JN, Full RJ. (2002) Evidence for van der Waals adhesion in gecko setae. *Proc Natl Acad Sci* **99**: 12252–12256.

[37] Autumn K, Peattie AM. (2002) Mechanisms of adhesion in geckos. *Int Comp Biol* **42**: 1018–1090.

[38] Autumn K, Dittmore A, Santos D, Spenko M, Cutkosky M. (2006) Frictional adhesion: A new angle on gecko attachment. *J Exp Biol* **209**: 3569–3579.

[39] Tian Y, Pesika N, Zeng H, Rosenberg K, Zhao B, McGuiggan P, Autumn K. (2006) Adhesion and friction in gecko toe attachment and detachment. *Proc Natl Acad Sci* **103**: 19320–19325.

[40] Johnson KW, Kendall K, Roberts AD. (1971) Surface energy and the contact of elastic solids. *Proc Roy Soc Lond A* **324**: 301–313.

[41] Israelachvili JN. (1992) Intermolecular and Surface Forces. Academic Press, New York.

[42] Spolenak R, Gorb S, Gao H, Artz E. (2004) Effects of contact shape on the scaling of biological attachments. *Proc Roy Soc Lond A* **2004**: 1–15.

[43] Autumn K, Hansen WR. (2006) Ultrahydrophobicity indicates a non-adhesive default state in gecko setae. *J Compar Physiol A* **192**: 1205–1212.

[44] Green DM. (1981) Adhesion and the toe-pads of tree frogs. *Copeia* **1981**: 790–796.

[45] Emerson SB, Diehl D. (1980) Toe pad morphology and mechanisms of sticking in frogs. *Biol J Linn Soc* **13**: 199–216.

[46] Rosenberg HI, Rose R. (1999) Volar adhesive pads of the feathertail glider, *Acrobates pygmaeus* (Marsupialia: Acrobatidae). *Can J Zool* 77: 233–248.

[47] Gorb S. (1998) The design of the fly adhesive pad: Distal tenent setae are adapted to the delivery of an adhesive secretion. *Proc Roy Soc Lond B* **265**: 747–752.

[48] Stork NE. (1980) Experimental analysis of adhesion of *Chryosolina polita* (Chrysomelidae: Coleoptera) on a variety of surfaces. *J Exp Biol* **88**: 91–107.

[49] Stork NE. (1983) A comparison of the adhesive setae on the feet of lizards and arthropods. *J Natl Hist* **17**: 829–835.

[50] Huber G, Mantz H, Spolenak R, Mecke K, Jacobs K, Gorb SN, Arzt E. (2005) Evidence for capillarity contributions to gecko adhesion from single spatula nanomechanical measurements. *Proc Natl Acad Sci* **102**: 16293–16296.

[51] Sun W, Neuzil P, Kustandi TS, Oh S, Samper VD. (2005) The Nature of the Gecko Lizard Adhesive Force. *Biophys J Biophys Lett* **89**: L14–L17.

[52] Autumn K. (2006) Properties, principles, and parameters of the Gecko adhesive system. In AM Smith & JA Calow (eds), *Biological Adhesives*. Springer-Verlag, Berlin.

[53] Resnick R, Halliday R, Krane KS. (1992) *Physics*, 4th Edn, Vol 1. John Wiley and Sons Inc., USA.

[54] Cutnell JD, Johnson KW. (1995) *Physics*, 3rd Edn. John Wiley and Sons Inc., USA.

[55] Cartmill M. (1985) Climbing. In M Hildebrand, DM Bramble, KF Liem & DB Wake (eds), *Functional Vertebrate Morphology*, pp. 73–88. Belknap Press, Harvard University Press, Cambridge, Massachusetts.

[56] Bhushan B. (1992) *Introduction to Tribology*. John Wiley and Sons, New York.

[57] Bowden FP, Tabor D. (1966) Friction, lubrication and wear: A survey of work during the last decade. *Br J Appl Phys* **17**: 1521–1544.

[58] Bhushan B, Israelachvili JN, Landman U. (1995) Nanotribology: Friction, wear and lubrication at the atomic scale. *Nature* **374**: 607–616.

[59] Krim J. (1996) Friction at the Atomic Scale. *Sci Am* **275**: 74–80.

[60] Israelachvili JN, Chen YL, Yoshizawa H. (1994) Relationship between adhesion and friction forces. *J Adhes Sci Technol* **8**: 1231–1249.

[61] Israelachvili JN, Chen YL, Yoshizawa H. (1995) Relationships between adhesion and friction forces. In DS Rimai, LP DeMejo & KL Mittal (eds), *Fundamentals of Adhesion and Interfaces*, pp. 261–280. VSP, Utrecht, The Netherlands.

[62] Palmer F. (1951) Friction. *Sci Am* **184**: 54.

[63] Peterson JA, Benson JA, Ngai M, Morin J, Ow C. (1982) Scaling in tensile "skeletons": Structures with scale-independent length dimensions. *Science* **217**: 1267–1270.

[64] Autumn K, Majidi C, Groff RE, Dittmore A, Fearing R. (2006) Effective elastic modulus of isolated gecko setal arrays. *J Exp Biol* **209**: 3558–3568.

[65] Ernst V, Ruibal R. (1966) The structure and development of the digital lamellae of lizards. *J Morphol* **120**: 233–266.

[66] Russell AP. (1976) Some comments concerning interrelationships amongst gekkonine geckos. In A Bellairs & CB Cox (eds), *Morphology and Biology of Reptiles*, Linnean Society Symposium Series 3, pp. 217–244. Academic Press, London.

[67] Russell AP. (1981) Descriptive and functional anatomy of the digital vascular system of the Tokay, *Gekko gecko*. *J Morph* **169**: 293–323.

[68] Russell AP. (1986) The morphological basis of weight-bearing in the scansors of the tokay gecko (*Gekko gecko*). *Can J Zool* **64**: 948–955.

[69] Russell AP, Bauer AM. (1990) Digit I in pad-bearing gekkonine geckos: Alternate designs and the potential constraints on phalangeal number. *Mem Qld Mus* **29**: 453–472.

[70] Russell AP, Bauer AM. (1990) *Oedura* and *Afroedura* (Reptilia: Gekkonidae) revisited: Similarities of digital design, and constraints on the development of multiscansorial subdigital pads. *Mem Old Mus* **29**: 473–486.

[71] Russell AP, Bauer AM, Laroiya R. (1997) Morphological correlates of the secondarily symmetrical pes of gekkotan lizards. *J Zool Lond* **241**: 767–790.

[72] Irschick DJ, Austin CC, Petren K, Fisher RN, Losos JB, Ellers O. (1996) A comparative analysis of clinging ability among pad-bearing lizards. *Biol J Linn Soc* **59**: 21–35.

[73] Bergmann PJ, Irschick DJ. (2005) Effects of temperature on maximum clinging ability in a diurnal gecko: evidence for a passive clinging mechanism? *J Exp Biol* **303A**: 785–791.

[74] Vanhooydonck B, Andronescu A, Herrel A, Irschick DJ. (2005) Effects of substrate structure on speed and acceleration capacity in climbing geckos. *Biol J Linn Soc* **85**: 385–393.

[75] Schleich H, Kästle W. (1986) Ultrastrukturen an Gecko-Zehen (Reptilia: Sauria: Gekkonidae). *Amphibia-Reptilia* 7: 141–166.

[76] Dalla Vale L, Nardi A, Toffolo V, Niero C, Toni M, Alibardi L. (2007) Cloning and characterization of scale β-keratins in the differentiating epidermis of geckoes show they are glycine–proline–serine-rich proteins with a central motif similar to avian β-keratins. *Devel Dynam* **236**: 374–388.

[77] Russell AP, Bauer AM. (2002) Underwood's classification of the geckos: A 21st century appreciation. *Bull Nat Hist Mus Lond (Zool)* **68**: 113–121.

[78] Schargott, M. (2007) Towards biologically inspired hierarchical attachment pads. *Proc. 30th Annual Meeting of the Adhesion Society*, Tampa Bay, FL, 102–104.

[79] Lauff RF, Russell AP, Bauer AM. (1993) Topology of the digital cutaneous sensilla of the tokay gecko, *Gekko gecko* (Reptilia, Gekkonidae), and their potential role in locomotion. *Can J Zool* 71: 2462–2472.

[80] Röll B. (1995) Epidermal fine structure of the toe tips of *Sphaerodactylus cinereus* (Reptilia: Gekkonidae). *J Zool Lond* **235**: 289–300.

[81] Clément G, Rézette D. (1984) Motor behavior underlying the control of an upside-down vertical posture. *Exp Brain Res* **59**: 479–484.

[82] Wasserug RJ, Roberts L, Gimian J, Hughes E, Saunders R, Devison D, Woodbury J, O'Reilly JC. (2005) The behavioral responses of amphibians and reptiles to microgravity on parabolic flights. *Zoology* **108**: 107–120.

[83] Jusufi A, Goldman DI, Revzan S, Full RJ. (2008) Active tails enhance arboreal acrobatics in geckos. *Proc Natl Acad Sci* **105**: 4215–4219.

[84] Huber G, Gorb S, Spolenak R, Arzt E. (2005) Resolving nanoscale adhesion of individual gecko spatulae by atomic force microscopy. *Biol Lett* **1**: 2–4.

[85] Autumn K, Gravish N, Wilkinson M, Santos D, Spenko M, Cutkosky M. (2007). Frictional adhesion of natural and synthetic gecko setal arrays. *Proc 30th Annual Meeting of the Adhesion Society*, Tampa Bay, FL, 58–60.

[86] Pesika NS, Tian Y, Zhao B, Rosenberg K, Zeng H, McGuiggan P, Israelachvili JN, Autumn K. (2007). Geometry-based model of tape peeling applied to the gecko adhesive system. *Proc 30th Annual Meeting of the Adhesion Society*, Tampa Bay, FL, 99–101.

[87] Daltorio KA, Gorb S, Peressadko A, Horchler A, Ritzmann RE and Quinn RD. (2007) A robot that climbs walls using micro-structured polymer adhesive. *Proc 30th Annual Meeting of the Adhesion Society*, Tampa Bay, FL, 329–331.

[88] Russell AP. (1979) Parallelism and integrated design in the foot structure of gekkonine and diplodactyline geckos. *Copeia* **1**: 1–21.

[89] Bonser RHC, Purslow PP. (1995), The Young's modulus of feather keratin. *J Exp Biol* **198**: 1029–1033.

[90] Bonser RHC. (2000) The Young's modulus of ostrich claw keratin. *J Mat Sci Lett* **19**: 1039–1040.

[91] Yao H, Gao H. (2007), Mechanics of bottom-up designed hierarchial adhesion structures of gecko. *Proc 30th Annual Meeting of the Adhesion Society*, Tampa Bay, FL, 111–113.

[92] Jagota A, Bennison SJ. (2002) Mechanics of adhesion through a fibrillar microstructure. *Int Comp Biol* **42**: 1140–1145.

[93] Glassmaker NJ, Jagota A, Hui CY, Kim J. (2004) Design of biomimetic fibrillar interfaces: 1, Making contact. *J Roy Soc Interf* (doi:10.1098/rsif.2004.0004).

[94] Hui CY, Glassmaker NJ, Tang T, Jagota A. (2004) Design of biomimetic fibrillar interfaces: 2. Mechanics of enhanced adhesion. *J Roy Soc Interf* (doi:10.1098/rsif.2004.005).

[95] Spolenak R, Gorb S, Arzt E. (2005) Adhesion design maps for bio-inspired attachment systems. *Acta Biomater* **1**: 5–13.

[96] Autumn K. (2006) How gecko toes stick. *Am Sci* **94**: 124–132.

[97] Campolo D, Jones S, Fearing R. (2003) Fabrication of gecko foot-hair like nano structures and adhesion to random rough surfaces. *Nanotechnology* **2**: 856–859.

[98] Persson BNJ. (2003) On the mechanism of adhesion in biological systems. *J Chem Phys* **118**: 7614–7621.

[99] Persson BNJ, Albohr O, Tartaglino U, Volokitin AI, Tosatti E. (2005) On the nature of surface roughness with application to contact mechanics, sealing, rubber friction and adhesion. *J Phys Condens Matter* **17**: R1–R62.

[100] Persson BNJ, Gorb S. (2003) The effect of surface roughness on the adhesion of elastic plates with application to biological systems. *J Chem Phys* **119**: 11437–11444.

[101] Arzt E, Gorb S, Spolenak R. (2003) From micro to nano contacts in biological attachment devices. *Proc Natl Acad Sci* **100**: 10603–10606.

[102] Peressadko A, Gorb S. (2004) When less is more: Experimental evidence for tenacity enhancement by division of contact area. *J Adhes* **80**: 247–261.

[103] Full RJ. (2001) Using biological inspiration to build artificial life that locomotes. In T Gomi (eds). *8th Int Sypmp Evolutionary Robotics: From Intelligent Robotics to Artificial Life*. Springer, Tokyo, Japan.

[104] Geim AK, Dubonos SV, Grigorieva IV, Novoselov KS, Zhukov AA. (2003) Microfabricated adhesive mimicking gecko foot-hair. *Nat Mater* **2**: 461–463.

[105] Yurdumakan B, Raravikar NR, Ajayan PM, Dhinojwala A. (2005) Synthetic gecko foot-hairs from multiwalled carbon nanotubes. *Chem Comm* **2005**: 3799–3801.

[106] Kim S, Spenko M, Trujillo S, Heynemen B, Mattoli V, Cutkosky MR. (2007) Whole body adhesion: Hierarchical, directional and distributed control of adhesive forces for a climbing robot. IEEE Int Conf Robotics and Automation, Rome, Italy.

[107] Lee J, Majidi C, Schubert B, Fearing R. (2008) Sliding-induced adhesion of stiff polymer microfibre arrays. I. Macroscale behaviour. *J Roy Soc Interf* (doi:10.1098/rsif.2007.1308: published online).

[108] Sitti M, Fearing R. (2003) Synthetic gecko foot-hair micro/nano-structures as dry adhesives. *J Adhes Sci Technol* 17: 1055–1073.

[109] Northen MT, Turner KL. (2005) A batch of fabricated dry adhesive. *Nanotechnology* **16**: 1159–1166.

[110] Kustandi TS, Samper VD, Ng WS, Chong AS, Gao H. (2007) Fabrication of a gecko-like hierarchical fibril array using a bonded porous alumina template. *J Micromech Microeng* 17: N75–N81.

[111] Webster NB, Johnson MK, Russell AP. (2008) Ontogenetic scaling of scansorial surface area and setal dimensions of *Chondrodactylus bibronii* (Gekkota: Gekkonidae): Testing predictions derived from cross-species comparisons of gekkotans. *Acta Zool (Stockholm)* (doi:10.1111/j.1463-6395.2008.00324.x).

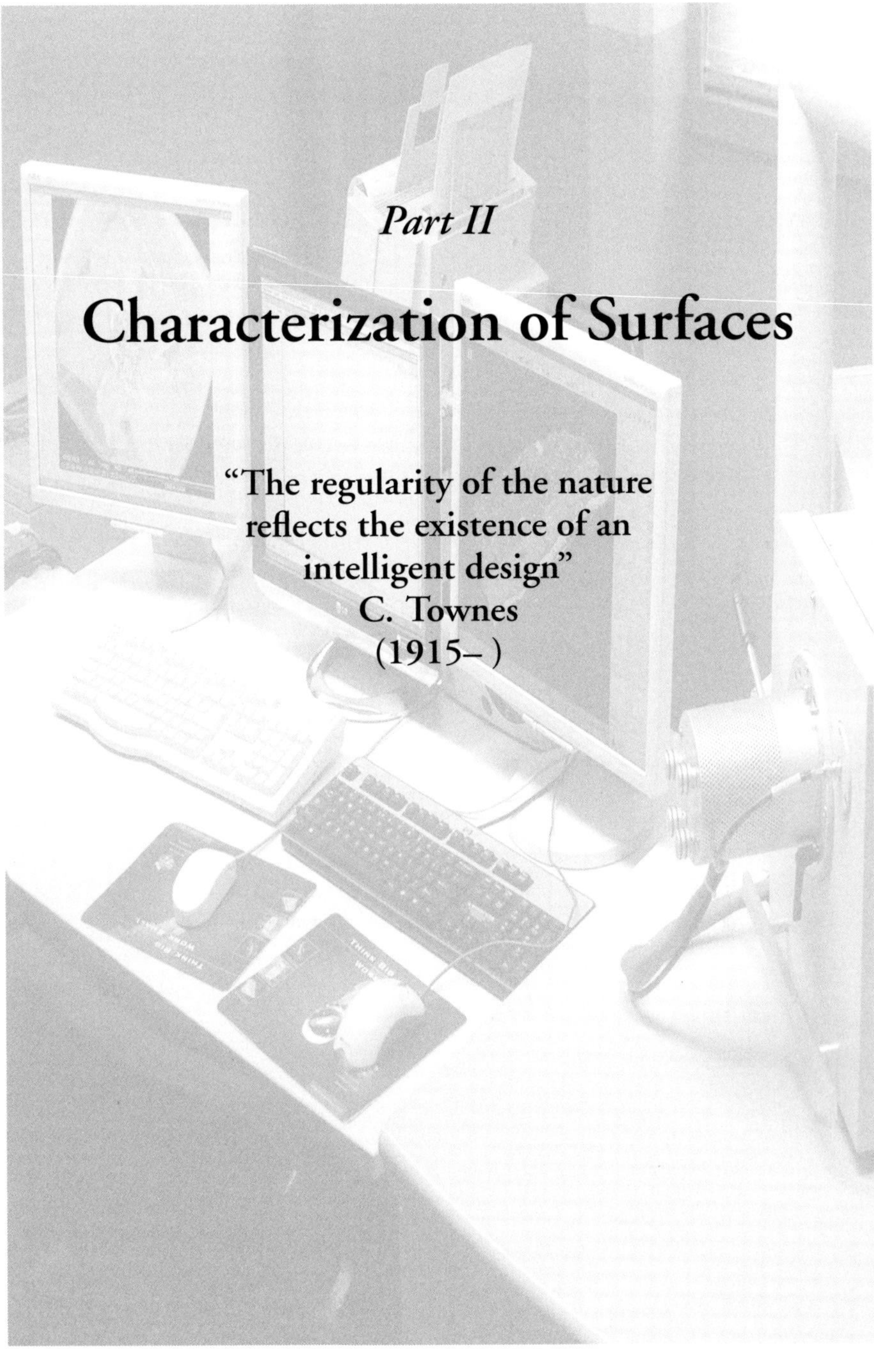

Part II

Characterization of Surfaces

"The regularity of the nature reflects the existence of an intelligent design"
C. Townes
(1915–)

Chapter 6

Micro- and Nano-Scopic Observation of Biological Surfaces

Zhaojie Zhang and Qun Ren

Biological surfaces are the barriers that protect living organisms from the environment, such as cold and dry. Biological surfaces perform many other critical functions, such as providing mechanical support to the organism. Certain specialized surfaces are important for thermoregulation and respiration. Another type of biological surface is that inside the living organism, such as joints of bones (lubrication or friction reduction), inner surface of an intestine (protection and absorption), or surfaces at cellular level, for example, the surface of an egg (Fig. 1). The functions of the biological surfaces rely on the chemical components, as well as their structure, often at micro- or nano-meter level. One critical step towards the understanding of the functions of biological surfaces is the observation and analysis of the surface structure. In this chapter, we discuss the principles and techniques commonly used for observing biological surface structures, including optical microscopy (light microscopy, laser scanning confocal microscopy), scanning probe microscopy, and electron microscopy (scanning electron microscopy, transmission electron microscopy). We describe and contrast the sample preparation of each technique. Quantitative data analyses, as well as the limitations of each technique, are also discussed.

1. Introduction

Broadly speaking, biological surfaces can be categorized into two groups: outer surfaces and inner surfaces. The outer biological surfaces are the barriers that protect living organisms from the environment and provide mechanical support to the body. Besides the main function of protection, many biological surfaces are specialized to provide a variety of functions, such as thermoregulation, respiration, and air or water retention.

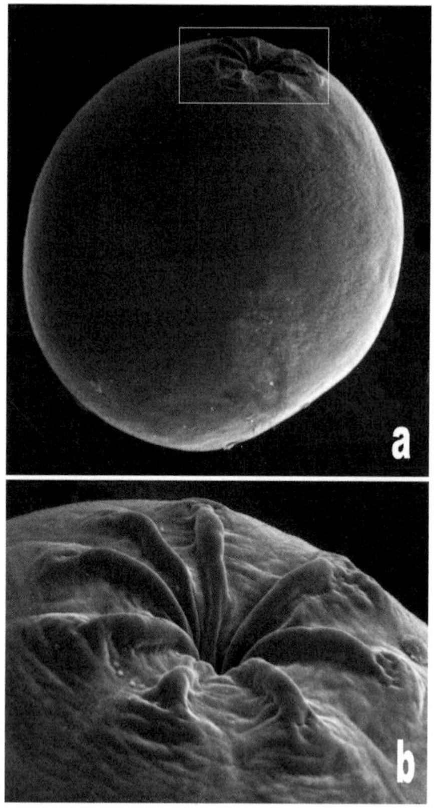

Fig. 1. (a) SEM image of a fish egg. (b) High magnification of the box area in 1(a), showing the fertilization site, where the sperm are attracted (images provided by Maureen McAuliffe).

Some living organisms have developed specialized surface structure to reduce friction (such as sharkskin), while others (such as gecko) have distinct surfaces to attach themselves to a variety of substrates (see Chaps. 2 and 5 in this book for more details related to anti-friction and adhesive surfaces). The inner biological surfaces are those inside the living organism, such as joints of bones (lubrication or friction reduction), inner surface of an intestine (protection and absorption), or surfaces at cellular level, for example, surface of a fish egg, which has a unique structure and chemical components to "attract" sperm for fertilization (Fig. 1).

The functions of the biological surfaces rely on their chemical components as well as their structure, often at micrometer and nanometer level. One prominent example is the so-called lotus-effect (see Chap. 3 in this book for detail).

Biological surfaces not only play important roles in protecting and regulating the living organisms, but also inspire numerous applications in biomimetics. For example, the adhesion mechanism of gecko setae may help design better adhesives commercially.[1] In certain animals, the adhesion system, such as toe pads of tree frog, secretes fluids onto the contact zone, making it "wet".[2] This fluid-based adhesion mechanism may help develop tires that enhance wear performance on dry roads and provide better braking on wet roads.[3]

Traditionally, the structure of biological surfaces has been studied by scanning electron microscopy (SEM). With the advances of technology, especially digital imaging and image analysis, other techniques, such as laser scanning confocal microscopy (LSCM), scanning probe microscopy (SPM) are used as well for understanding the structure and structure–function relation. Image analysis, including three-dimensional reconstruction, topographic analysis, and certain special techniques, such as RIMAPS (see Chap. 7 in this book for detail), also play an important role in interpreting and understanding the structure of biological surfaces.

In the present chapter, we discuss the principles and techniques commonly used for observing biological surface structures, including optical microscopy (light microscopy, laser scanning confocal microscopy), electron microscopy (scanning electron microscopy, transmission electron microscopy), and scanning probe microscopy. We describe and contrast the sample preparation of each technique. Quantitative data analysis as well as the limitations of each technique is also addressed.

2. Surface Observation Using Optical Microscopy

2.1. *Light Microscopy*

One of the most critical aspects of observation using any optical microscope is the specimen illumination. Two illumination systems are commonly used in optical microscopy: transmitted light and reflected light (Fig. 2). Transmitted, also called diascopic illumination, requires the specimen to be "transparent." It is used primarily to examine thin sections of biological or material samples. Reflected light, or episcopic illumination (epi-illumination), is most commonly used for fluorescence microscopy, where fluorphores inside the specimen are excited to produce fluorescent light. The fluorescence is then emitted, or reflected back to the objective and collected by the detector (eyes, or camera). The reflected light is also used to study the surfaces of "opaque" specimens, which is the focus of this session.

One advantage of using an optical microscope with epi-illumination is its simplicity. It requires little sample preparation. The biological samples can be viewed live or fresh with little damage. Taking the plant leaf, for example, the stomata cells

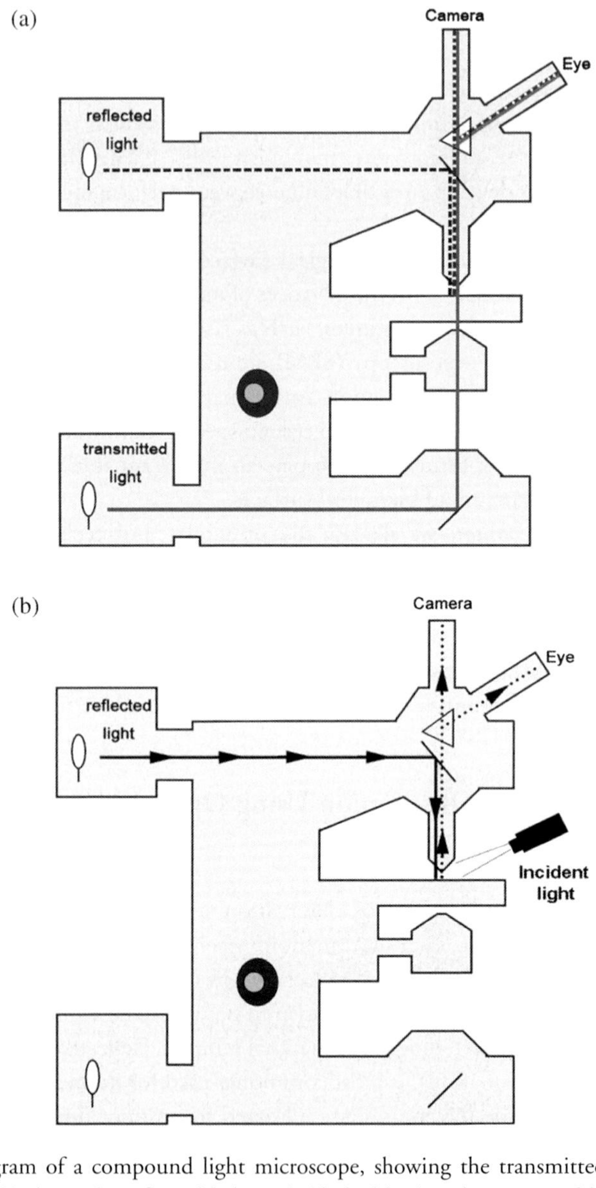

Fig. 2. Diagram of a compound light microscope, showing the transmitted and reflected illumination. (a) shows the reflected light path (dashed line) and transmitted light path (solid line). (b) shows the reflected light path and the addition of an incident light, to improve the image quality.

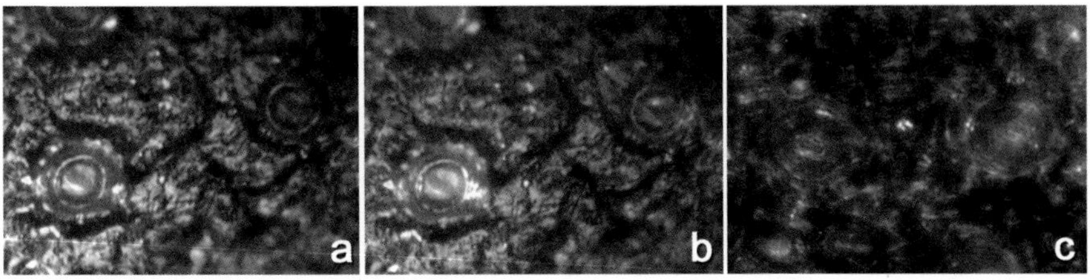

Fig. 3. A plant leaf imaged with epi-illumination using a 20× objective, showing two stomata. (a) Epi-illumination using a halogen light. (b) Same as in (a) except with the addition of an incident light. (c) Epi-illumination with fluorescence light (mercury light bulb without any filter).

open and close according to the physiological and environmental conditions. The light microscopy with epi-illumination can be used to check the instant status of the stomata cells at a specific condition (Fig. 3).

The light source can be a halogen light bulb, or mercury light bulb used on fluorescence microscope, which is available in many biological research laboratories. If a mercury light bulb (for a fluorescence microscope) is used, an excitation (but not emission) filter may be put in place to avoid damages to the eyes. An incident light can also be added (Fig. 2(b)) when necessary to reveal more structural information and to improve the quality of the images. Figure 4 shows a spider head imaged with different illumination. It appears that a combination of epi-illumination plus an incident light (Fig. 4(c)) gives the most balanced illumination for revealing the detailed structure, especially for the feather-like hairs.

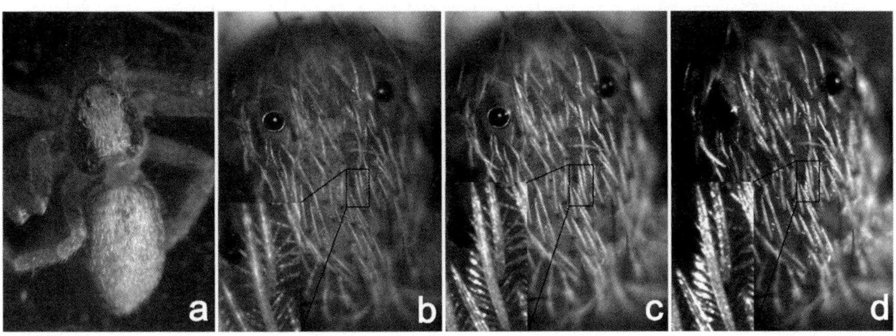

Fig. 4. A spider head imaged with different illumination. (a) Image of the whole spider using a dissecting microscope. (b) Image of the head with epi-illumination (top illumination) using a 10× objective. (c) Same as in (b) except with the addition of an incident light. (d) Image with the incident light only.

The success of surface imaging using epi-illumination relies not only on the illumination, but also the reflectiveness of the sample. Certain samples, such as a plant leaf (Fig. 3), may not be very reflective and thus epi-illumination may not reveal the structure well. The plant leaf appears to be more reflective to the halogen light than the UV light (Fig. 3(c)).

One of the disadvantages of light microscopy, besides its low resolution compared to SEM, is its small depth of field. This problem, however, can be overcome with advanced digital imaging and image editing. A series of images at different focal points can be captured. The images can then be assembled into a single image containing the best focus from each individual image. A variety of softwares are available for the image assembly, such as Adobe Photoshop,[4] or the Extended Depth of Focus plugin[5] from ImageJ (http://rsb.info.nih.gov/ij/). Figure 5 shows a series of images captured at different focal points and the merged image with the extended depth of focus. The topographic view of the image stack can be obtained using the software as well. In order to acquire a good topographic view from the image stack, it is better to use a microscope with a motorized Z-axial stage.

2.2. *Laser Scanning Confocal Microscopy (LSCM)*

The principle of LSCM, in many ways, is similar to that of fluorescence microscopy, except, as the name suggested that (1) it uses a laser beam as the light source; (2) the sample is scanned by the laser beam; and (3) an aperture (or pinhole) is placed in the image plane in front of the detector at a position *confocal* with the in-focus plane. The out-of-focus light is therefore blocked and only the in-focus light goes through the pinhole, resulting in sharper images than those from a conventional fluorescence microscope (Fig. 6). Photomultiplier tubes (PMTs), instead of cameras are often used as the detectors in confocal microscopy. PMTs offer a large dynamic range and better linearity than cameras. They are fast devices and can easily operate in sub-millisecond range. Laser is actually not essential for a confocal microscope. Other light sources are also used in many confocal microscopes, especially in the spinning disk confocal microscopes. The advantages of LSCM include the ability of obtaining sharper images with enhanced resolution, penetrating deep light-scattering tissues, and optical sectioning. Similar to fluorescence microscopy, LSCM is mainly used to observe fluorescently labeled biological samples. Because it usually has a motorized stage with accuracy sometimes at nanometer level, LSCM can be used to study the biological surface structures, especially to study the topology and measure the sample volume. Figure 7 shows the same spider head in as Fig. 4, imaged with a confocal microscope using reflection mode, which collects the emission light in same wavelength as the excitation light. The topological and surface views are easily drawn from the image

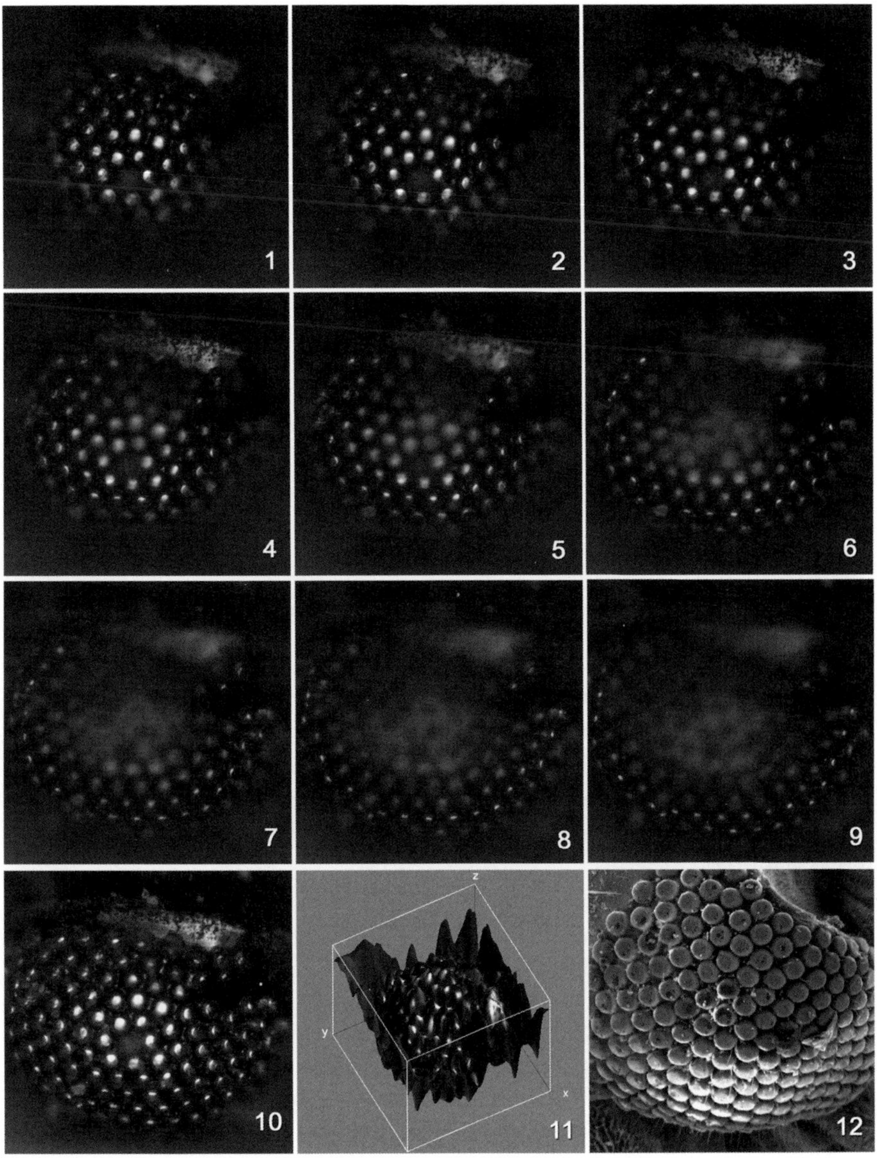

Fig. 5. (1–9) A series of images of a midge eyes taken at different focal points, using an Olympus BX51 microscope with epi-illumination and a 20× objective. (10) The image with extended depth of focus by combining the nine images above using ImageJ. (11) Topographic view generated from the image series. (12) The same midge eyes viewed by SEM.

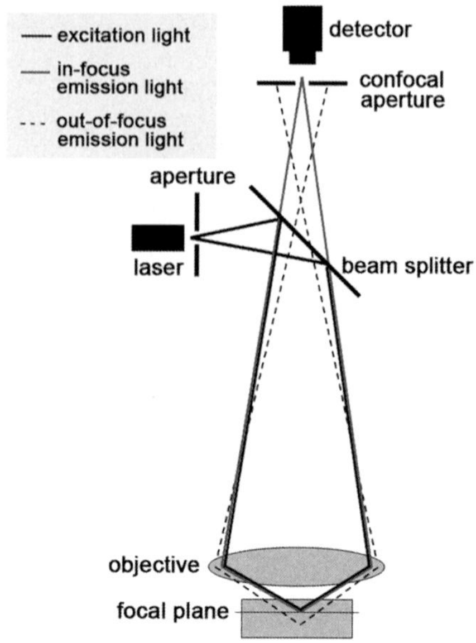

Fig. 6. Diagram showing the light path of a confocal microscope.

series and accurate measurement at 3D level can be done with various softwares. Note however, that the LSCM appears to reveal less morphological details, comparing to traditional microscope with epi-illumination (Fig. 3), due to the poor reflection of the laser by the sample. Different samples may show different reflectiveness to the same laser, while the same sample may also show different reflectiveness to different lasers (wavelength).

Recently, new instruments have been developed, in combination of confocal microscope with atomic force microscope (AFM), enabling topographic and fluorescence spectral imaging simultaneously.[6] The unique combination of these two microscopes not only offers the best resolution at *x*-axis, *y*-axis (confocal), and *z*-axis (AFM), but also provides the ability to study the topographic distribution of certain chemical components, or proteins on the biological surface (see next section for more details).

3. Surface Observation Using Scanning Probe Microscopy (SPM)

SPM is a powerful tool to study surface properties at nanometer scale. Different from the "traditional" microscopy (light or electron microscopy), in which an objective/lens, or a series of objectives/lenses are used to form an image, SPM uses a probe

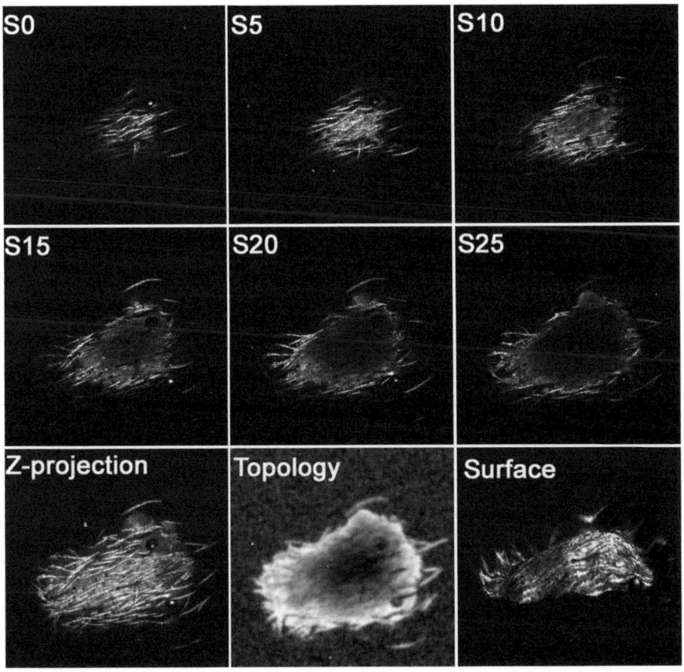

Fig. 7. A spider head (same as in Fig. 4), imaged with a Leica TCS SP LSCM, using a 488 nm laser. Images S0–S25 are representation (S0 = first section from the top, S25 = 26th section) of the 30 images at different *z*-axis. The *Z*-projection, topology, and surface view are all created using imageJ from the 30 serial sections.

to mechanically scan over the surface of the specimen. The image of the surface is generated by recording the probe–surface interaction as a function of position. The resolution of SPM, therefore, is limited by the size of the probe–sample interaction volume, but not by diffraction, as occurred in light or electron microscopy. The probe–sample interaction volume can be as small as a few picometres, giving the SPM much higher resolution.

SPM was invented in 1981 by G. Binnig and H. Rohrer, who won the Nobel Prize in Physics five years later for their invention. There are many kinds of SPMs, such as scanning tunneling microscopy (STM), atomic force microscopy (AFM), scanning near-field optical microscopy (SNOM), and magnetic force microscopy (MFM). All the SPMs are based on similar principles (Fig. 8), although each type of SPM employs a different probe and uses different mode of interaction with the sample surface (for more details about different SPMs, see Ref. 7).

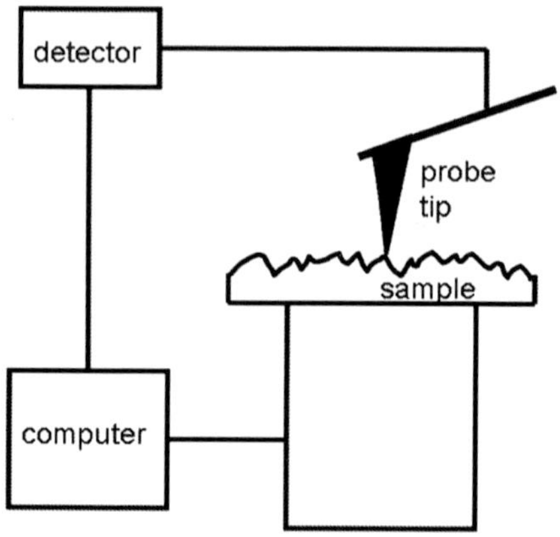

Fig. 8. A simplified depiction of SPM. The probe tip moves on the surface to scan the sample (or sometimes the sample moves). The detector detects the vertical position of the tip; the computer records the vertical position and converts the data into an image.

In this section, we will use AFM as an example, to describe the application of SPM in studying the biological surfaces. Other types of SPMs and their application in studying biological samples, including protein, DNA, and membrane structures can be found in Ref. 8.

3.1. *AFM in Biological Surface Study and Topographic Analysis*

AFM was developed in 1986, to overcome a basic drawback of the scanning tunneling microscope, which can only be used to image conducting or semi-conducting material. In AFM, the probe tip is attached to a tiny leaf spring (called cantilever), which is used to measure the force between the tip and surface, instead of the tunneling current which requires the sample to be conductive. The AFM, therefore, can image almost any type of sample. Because of its versatility, AFM has become the most widely used SPM for studying the surface topology of biological samples. The sample can be examined not only in its native state (no fixation, dehydration, or coating), but also in its physiological condition, such as in liquid culture medium.[9,10] Treatment of biological samples prior to examination is therefore avoided.

AFM generally operates in three different modes: contact, noncontact, and tapping mode. In the contact (also called static) mode, the tip is in contact with the sample surface. The force between the tip and the surface is kept constant during scanning by maintaining a constant deflection, which is used as the feedback signal.

In the noncontact (or dynamic) mode, the probe tip is very close to the sample, but not touching it. The changes to the resonant frequency or amplitude of the cantilever is measured and recorded as the property of the surface. The cantilever used in noncontact mode is usually much stiffer than that in the contact mode. The noncontact mode is very useful for studying biological samples that are either very fragile or in liquid.

The tapping mode, also called intermittent contact mode, is a combination of the contact and noncontact modes. The probe tip is closer to the sample than in noncontact mode and intermittently touches or taps the sample surface. While it eliminates the lateral forces, such as dragging (common in contact mode), the tapping mode improves lateral resolution, especially of soft samples by its intermittent tapping.

Figure 9 shows an example of AFM images of yeast (*Saccharomyces cerevisiae*) cells treated with hydrogen peroxide, which causes cell death.[11] Hydrogen peroxide also causes changes of surface morphology of yeast cells, such as increased roughness of cell surface and collapse of cell wall. These changes can be observed by SEM (Figs. 9(b) and 9(d)), but are difficult to quantify. Certain small changes of the surface may also be masked by the gold conductive coating,[12] which is required for SEM. With AFM, however, the height map (Figs. 9(a) and 9(c)) can be easily obtained and the height and depth can be quantitatively measured on the whole image, or any particular part of the image (Figs. 9(e) and 9(f)). The small protruding area shown in Fig. 9(c), line A, for example, is about 200 nm in length and about 20 nm in height. The depth of the pit in line B of Fig. 9(c), is measured about 100 nm.

3.2. *Combination of AFM With Confocal Microscopy*

AFM provides very high spatial resolution on biological surface, because of its unique mechanism of the tip–sample interaction. This mechanism, however, also limits the AFM's capacity to identify the chemical components of the biological sample. Confocal microscopy or fluorescence microscopy, on the other hand, has been well established to study the location and dynamics of cellular components by using fluorescence labeling of DNA, protein, or lipids. Over the last decade, techniques have been developed by combining the AFM and confocal microscopy to enhance the strength of each microscope.[13–15] The combined AFM and confocal microscope provides both the topographic and optical imaging on the same biological sample. The most recent version of this combination allows not only simultaneous imaging of AFM and fluorescence, but also real time manipulation at nano-meter level.[6,16] A similar technique, called scanning surface confocal microscopy (SSCM), was also developed recently.[17] It combines confocal microscopy with scanning ion conductance microscopy, which uses a nanopipette, instead of a probe tip to scan the cell surface. An ion current that flows to the pipette is used to control the distance between the sample and the nanopipette.[18] The SSCM can simultaneously record high resolution topography and surface fluorescence of living cells at nanometer level.

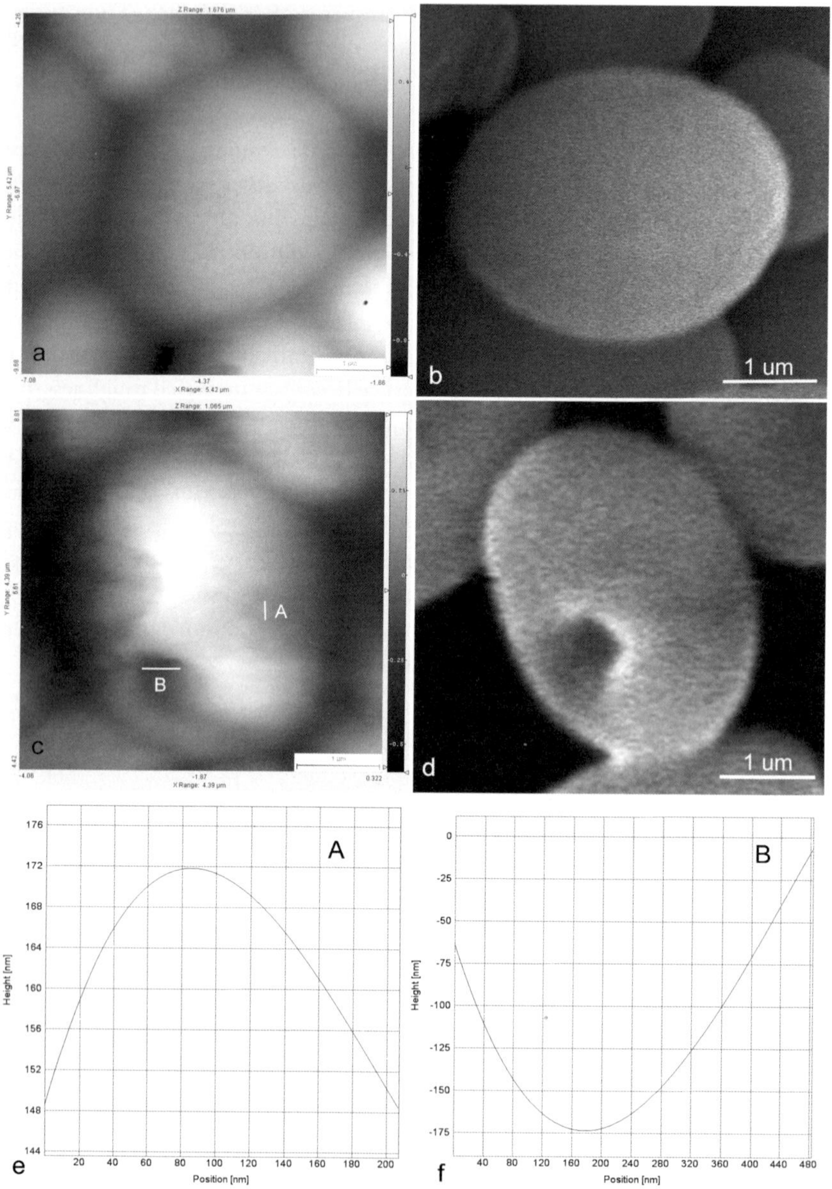

Fig. 9. AFM and SEM images of yeast cells, showing the surface morphology changes after treatment with hydrogen peroxide (a,b). Untreated cell imaged with AFM and SEM, respectively (c,d). Hydrogen peroxide treated cells imaged with AFM and SEM, respectively (e,f). Quantitative measurement of the topographic profiles of lines A and B in (c).

4. Surface Observation Using Electron Microscopy

4.1. *SEM of Biological Surface*

SEM has been the most widely used instrument for biological surface observation. Its advantages include: (1) its small numerical aperture (about 0.002, vs 0.2 on a 10× objective of a light microscope) and large depth of field, which allows a large area of the sample to be in focus at one time (see Fig. 5 for comparison); (2) high resolution; and (3) relatively easy sample preparation, which mainly requires drying the sample and making it conductive by coating it with a thin layer of gold or gold–palladium.

SEM uses an electron beam, instead of light, to scan across the sample to form an image. When the electron beam strikes the surface of a sample, electrons are emitted out from the sample in different forms, such as, secondary electrons, backscattered electrons, transmitted electrons, Auger electrons, and X-rays.[19] In most SEMs, the secondary electrons and backscattered electrons are used to view the sample and generate images. X-rays are often used to determine information about the specimen composition. Secondary electrons possess less energy than backscattered electrons and the secondary electron detector usually sits on one side of the sample (Fig. 10). The closer the sample is to the detector, the more electrons are likely to be attracted to the positively charged detector. This is why most of the SEM images appear to be "side-illuminated" (see Fig. 1 for example).

The backscattered electrons not only possess more energy but also correlates strongly with the atomic numbers, that is, backscattering increases as the atomic number of the specimen increases. This feature is often used by biologists for protein localization on biological surface by labeling the surface with antibodies and immuno-gold, which has a higher atomic number.[20] Figure 11 shows an immuno-gold labeling

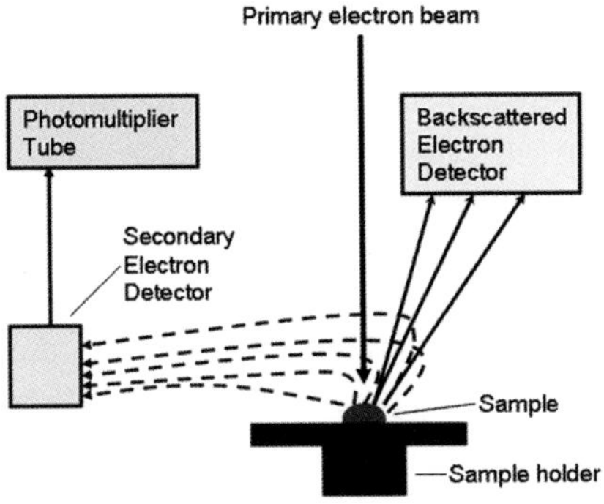

Fig. 10. Diagram showing the relative position of the secondary and backscattered electron detectors in SEM.

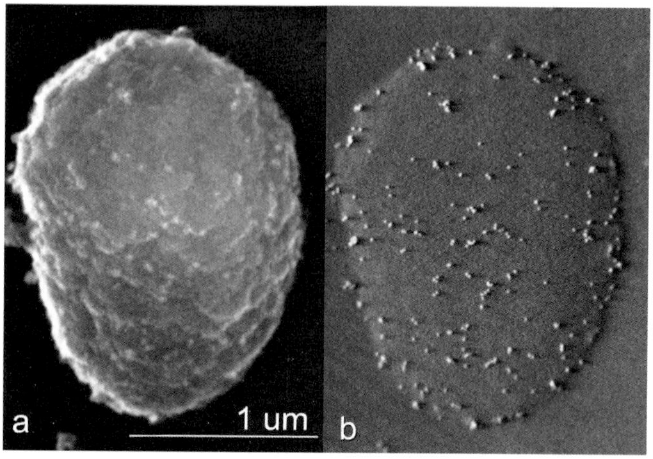

Fig. 11. Immuno-gold labeling of myosin on the surface of tobacco sperm cell. (a) Secondary electron image of an isolated sperm cell; (b) Backscattered electron image of the same field as in (a), showing the 15-nm gold labeled myosin (Courtesy of Zhang *et al.*, Ref. 20, use with permission).

of myosin on the surface of tobacco sperm cells. Sample preparation is critical to the success of labeling to ensure the preservation of both the structure and antigen on the surface (see Ref. 20 for detailed protocol for sample preparation).

The SEM provides a large depth of field. This depth, however, cannot be quantitatively measured due to the fact that the SEM images are not truely three-dimensional. Stereoscopic images had been used to obtain the height, but with certain limitations.[21,22] The recognition and matching of small features in the stereo images, for example, rely on the person who operates and analyzes the data.[23] With the advances of digital imaging and computing technology, various softwares and techniques have been developed recently to overcome the pitfalls.[24] In certain cases, the topology and height map can be obtained at high resolution, comparable to the resolution of AFM.[25] Three-dimensional reconstruction with rendered surface view is also possible through the stereo-pairs of SEM images.[26] These techniques, however, usually work better on smooth, continuous surfaces, but less satisfactory on discontinuous specimens with major occlusions. While few of the software packages are available commercially, freeware as plugin of the ImageJ (http://rsb.info.nih.gov/ij/) can be found on the web (http://www.gcsca.net/IJ/StereoJ.html).

The basic procedure of biological sample preparation for SEM involves in specimen fixation, dehydration, drying, and coating (see Ref. 27 for a comprehensive review of procedures). The purpose of fixation is to immobilize the cellular components, such as liquids and proteins to their original locations so as to preserve the cell morphology as much as possible. Many of the low-molecular weight molecules, including ions, sugars, and amino acids are irretrievably lost during the fixation. Fixation therefore may not be suitable for X-ray analysis and should be avoided when necessary. Many of the fixatives used in SEM, including glutaraldehyde and osmium tetroxide, are derived from TEM studies. Glutaraldehyde immobilizes proteins by forming cross-links among different amino groups. Glutaraldehyde (2%–3%) can be used by itself or supplemented with 1%–4% formaldehyde, which penetrates tissues faster because of its smaller size.

Following the glutaraldehyde fixation, the tissue should be washed thoroughly before post-fixation with osmium tetroxide, which mainly forms cross-links among lipids. Osmium tetroxide can also help to enhance the electron signal and reduce electron charging.[28] Osmium tetroxide (1%–2%) is commonly used and usually the fixation is done at the same temperature used for aldehyde fixation.

Sample dehydration is generally done using a graded series of organic solvent (acetone or alcohol). The goal is to remove the water molecules from the sample and minimize tissue extraction and shrinkage. Once the water inside the sample is replaced by organic solvents, the sample can be dried either by critical point drying or sublimation. The so-called critical point refers to that at certain temperature and pressure (the critical point), the two-phase stage (vapor and liquid) of a volatile liquid disappears; therefore, no surface tension is caused during the process of sample drying.[29] In sublimation, an organic compound (Peldri II for example) that is miscible with dehydrant is added in the final stage of dehydration. The compound becomes solid and then sublimes without producing surface-tension effects.[30]

An alternative of critical point drying or sublimation is liquid substitution, in which a low-vapor-pressure liquid, such as glycerol or triethylene glycol, is used to infiltrate the sample and the sample can be examined by SEM without drying and coating.[31] This technique has been successfully applied to different biological samples.[32]

4.2. *Environmental SEM (ESEM) of Wet Biological Samples*

The biological sample has to be dried and coated with conductive material prior to its observation in a conventional SEM. This makes it impossible for direct observation of the wet, biological sample or any post-manipulation. This obstacle is overcome by the introduction of ESEM. In ESEM, the specimen chamber is isolated in large from the column in terms of vacuum to allow the specimen chamber have a "near environmental" low vacuum. The specimen chamber is then introduced with water vapor, which has two major functions. One function is to produce more secondary electrons when struck by secondary electrons from the sample. The second function of water vapor is to neutralize the negative charge on the sample surface by the positive ions produced when the water molecules lose their secondary electrons. A specialized, gaseous secondary electron detector, which is very efficient for collecting the water-molecule-amplified secondary electrons, is used in ESEM.

Recently, a small tabletop SEM was introduced by Hitachi and FEI. Despite its relatively low resolution (about 30 nm), the tabletop SEM is capable to view and image directly certain wet biological samples, such as fresh plant leaves (Fig. 12). Surface coating with gold or gold–palladium is not necessary, due to the usage of an advanced backscattered electron detector. Coating with a thin layer of gold, however, does provide more contrast and enhanced details of the surface structure (Fig. 13).

4.3. *TEM Observation of the Inside of the Biological Surfaces*

Transmission electron microscopy (TEM) is primarily used to examine the internal structure of biological samples. It can be used, however, to reveal certain surface characteristics that may otherwise be difficult to view by other types of microscopic methods. The thickness of cell wall, which could be related to the physiological or growth condition of the cell, can be best revealed by ultra-thin sections and TEM, which provides higher resolution than light microscopy (Fig. 14). The roughness of the surface can also be examined in detail, although at two-dimensional level.

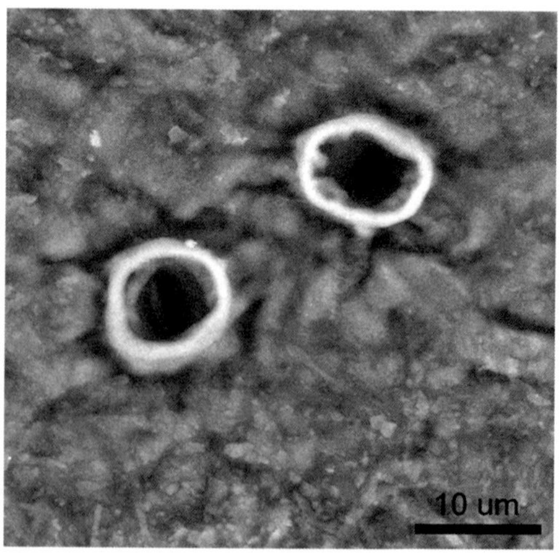

Fig. 12. A fresh plant leaf (same as in Fig. 3) imaged with a Hitachi TM-1000 tabletop SEM without any preparation.

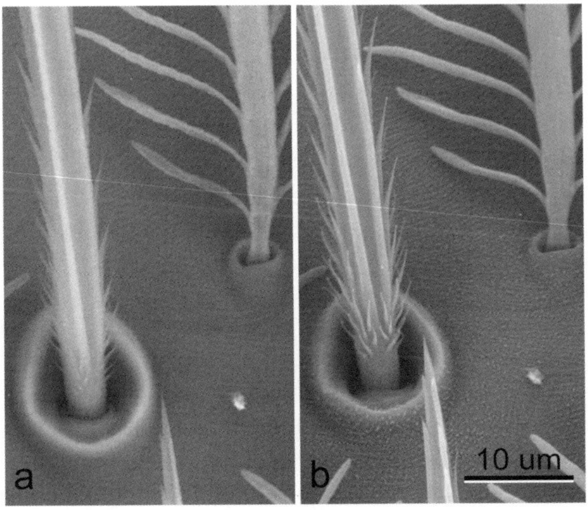

Fig. 13. Hair structure on a spider head (same as in Fig. 4) imaged using the Hitachi TM-1000 tabletop SEM without coating (a) and with gold coating (b). The spider was fixed with 75% alcohol and dried in the air.

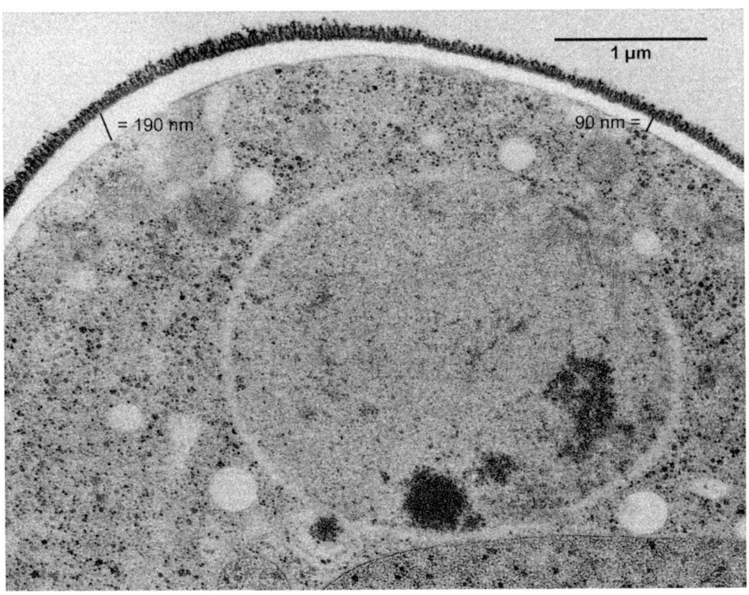

Fig. 14. TEM image of a yeast cell during early meiosis, showing the thickness of the cell wall and the roughness of the surface (see Ref. 34 for detailed protocol).

The principle of TEM resembles a light microscopy, in which light is transmitted through the sample to form an image. TEM requires ultra-thin sections (around 50–100 nm) in order for electrons to be transmitted. The thin sections are usually stained with heavy metals to enhance the contrast. In order to obtain the ultra-thin sections, the biological sample needs to be fixed and embedded in resin. The TEM sample preparation is probably the most difficult one among those for microscopic observation. It is time consuming and requires extensive training for the operator (see Ref. 33 for details).

Acknowledgments

This research is supported by the National Institutes of Health, Institute for Research Resources, and grant P20 RR15640 to the Neuroscience Center of Biomedical Research Excellence and NIH-NCRR grant RR-16474 (University of Wyoming's IDeA Networks for Biomedical Research Excellence). The authors thank Dr. Dan Buttry and Poonam Singh for their assistance using the atomic force microscope. The authors also thank Dr. Francis Flynn for his critical reading and comments of the manuscript.

References

[1] Autumn K, Sitti M, Liang YA, Peattie AM, Hansen WR, Sponberg M, Kenny TW, Fearing R, Israelachvili JN, Full RJ. (2002) Evidence for van der Waals adhesion in gecko setae. *Proc Natl Acad Sci USA* **99**: 12252–12256.

[2] Scherge M, Gorb SN. (2001) *Biological Micro- and Nanotribology: Nature's Solutions*. Springer, Berlin.

[3] Gorb, SN. (2006) In Y Bar-Cohen (ed), *Biomimetics — Biologically Inspired Technologies*. CRC Press, Florida, pp. 381–416.

[4] Wilson TC, Charlton WA, Postuszny U. (2006) Increased image depth for epi-illumination microscopy. *Can J Bot* **84**: 508–513.

[5] Forster B, Van De Ville D, Berent J, Sage D, Unser M. (2004) Complex Wavelets for Extended Depth-of Field: A New Method for the Fusion of Multichannel Microscopy Images. *Microsc Res Tech* **65**: 33–42.

[6] Kassies R, van der Werf KO, Lenferink A, Hunter CN, Olsen JD, Subramaniam V, Otto C. (2005) Combined AFM and confocal fluorescence microscope for applications in bio-nanotechnology. *J Microsc* **217**: 109–116.

[7] Meyer E, Hug HJ, Bennewitz R. (2004) *Scanning Probe Microscopy: The Lab on a Tip*. Springer-Verlag, Berlin Heidelberg.

[8] Marti O, Amrein M. (1993) *STM and SFM in Biology*. Academic Press, San Diego.

[9] Bustamante C, Rivetti C, Keller DJ. (1997) Scanning force microscopy under aqueous solutions. *Curr Opin Struct Biol* 7: 709–716.

[10] Hansma PK, Cleveland JP, Radmacher M, Walters DA, Hillner P, Bezanilla M, Fritz M, Vie D, Hansma HG, Prater CB, Massie J, Fukunaga L, Gurley J, Elings V. (1994) Tapping mode atomic force microscopy in liquids. *Appl Phys Lett* **64**: 1738–1740.

[11] Madeo F, Fröhlich E, Ligr M, Grey M, Sigrist SJ, Wolf DH, Fröhlich KU. (1999) Oxygen stress: A regulator of apoptosis in yeast. *J Cell Biol* **145**: 757–767.

[12] Pereira RS, Geibel J. (1999) Direct observation of oxidative stress on the cell wall of Saccharomyces cerevisiae strains with atomic force microscopy. *Mol Cell Biochem* **201**: 17–24.

[13] Horton M, Charras G, Ballestrem C, Lehenkari P. (2000) Integration of atomic force and confocal microscopy. *Single Mol* **1**: 135–137.

[14] Lal R, Proksch R. (1997) Multimodal atomic force microscopy: Biological imaging using atomic force microscopy. *Int J Imag Syst Tech* **8**: 293–300.

[15] Noy A, Huser TR. (2003) Combined force and photonic probe microscope with single molecule sensitivity. *Rev Sci Instrum* **74**: 1217–1221.

[16] Gradinaru CC, Martinsson P, Aartsma TJ, Schmidt T. (2004) Simultaneous atomic-force and two-photon fluorescence imaging of biological specimens in vivo. *Ultramicroscopy* **99**: 235–245.

[17] Gorelik J, Shevchuk A, Ramalho M, Elliott M, Leik C, Higgins CF, Lab MJ, Klenerman D, Krauzewicz N, Korchev Y. (2002) Scanning surface confocal microscopy for simultaneous topographical and fluorescence imaging: application to single virus-like particle entry into a cell. *Proc Natl Acad Sci USA* **99**: 16018–16023.

[18] Hansma PK, Drake B, Marti O, Gould SA, Prater CB. (1989) The scanning ion-conductance microscope. *Science* **243**: 641–643.

[19] Becker HC. (1975) Scanning electron microscopy, Lubrication, Texaco, Inc. 61, 37–56.

[20] Zhang Z, Tian HQ, Russell SD. (1999) Localization of myosin on sperm-cell-associated membranes of tobacco (*Nicotiana tabacum* L.). *Protoplasma*, **208**: 123–128.

[21] Boyde A. (1973) Quantitative photogrammetric analysis and qualitative stereoscopic analysis of SEM images. *J Microsc* **98**: 452–471.

[22] Howell PGT, Boyde A, Ross HF. (1986) A three axis stereocomparator for scanning electron microscopic photogrammetry — RS3. *Scanning* **8**: 182–186.

[23] Boyde A. (1986) Stereological stereometry: Simplified approaches to volumetry by SEM. *Scanning* **8**: 240–249.

[24] Minnich B, Leeb H, Bernroider EWN, Lametschwandtner A. (1999) Three dimensional morphometry in scanning electron microscopy: A technique for accurate dimensional and angular measurements of microstructures using stereopaired digitized images and digital image analysis. *J Microsc* **195**: 23–34.

[25] Raspanti M, Binaghi E, Gallo I, Manelli A. (2005) A vision-based, 3D reconstruction technique for scanning electron microscopy: Direct comparison with atomic force microscopy. *Microsc Res Tech* **67**: 1–7.

[26] Manelli A, Sangiorgi S, Binaghi E, Raspanti M. (2007) 3D analysis of SEM images of corrosion casting using adaptive stereo matching. *Microsc Res Tech* **70**: 350–354.
[27] Goldstein J, Newbury D, Joy D, Lyman C, Echlin P, Lifshin E, Sawyer L, Michael J. (2003) *Scanning Electron Microscopy and X-Ray Microanalysis*, Third edition, Kluwer Academic/Plenum Publishers, pp. 591–646.
[28] Wood PS, Ledbetter MC. (1976) Cell organelles at uncoated cryofractured surfaces as viewed with the scanning electron microscope. *J Cell Sci* **21**: 47–58.
[29] Cohen AL. (1979) Critical point drying-principles and procedures. *Scan Electron Microsc.* **2**: 303–322.
[30] Kenney JR, Williams RW, Gray JP. (1989) Use of Peldri II (a fluorocarbon solid at room temperature) as an alternative to critical point drying for biological tissues. *J Electron Microsc Tech* **11**: 117–125.
[31] Ensikat HJ, Barthlott W. (1993) Liquid substitution: A versatile procedure for SEM specimen preparation of biological materials without drying and coating. *J Microsc* **172**: 195–203.
[32] Favret EA, Fuentes NO, Molina AM. (2006) Quantitative study of Xanthosoma violaceum leaf surfaces using RIMAPS and variogram techniques. *Microsc Res Tech* **69**: 684–688.
[33] Bozzola JJ, Russell LD. (1992) *Electron Microscopy: Principles and Techniques for Biologists.* Jones and Bartlett Publishers, Massachusetts.
[34] Ren Q, Yang H, Rosinski M, Conrad MN, Dresser ME, Guacci V, Zhang Z. (2005) Mutation of the cohesin related gene PDS5 causes cell death with predominant apoptotic features in *Saccharomyces cerevisiae* during early meiosis. *Mutat Res* **570**: 163–173.

Chapter 7

RIMAPS and Variogram Characterization of Micro-Nano Topography

Néstor O. Fuentes and Eduardo A. Favret

The utility and acceptability of images obtained from different microscopes, with no prior knowledge of the type of microscopy (optical or electronic) used to form the image of three-dimensional objects, depend on the direct interpretation that can be made from them by any observer. But this simple "eye-inspection" is not enough to obtain all the characteristic topographic patterns from biological or technological surfaces, and advanced imaging techniques are imposed.

Answers to the unavoidable questions of "what can be seen" in an image are closely related to the amount of information about the whole specimen that can be detected in a small region, and to what extent they can give an adequate representation of the specimen characteristics. It is within this framework that two imaging techniques, capable of overcoming these difficulties in topographic pattern determination, are reported.

1. Introduction to RIMAPS and Variogram Analysis

Visual observation has played a major role in all areas of natural science. Human beings perceive most of the information about their environment through their visual sense. However, quantitative evaluation of images never found widespread application until the born of personal computers. Applications in image processing have now been applied to all the sciences. Many books have been written on the subject of image analysis, including linear and nonlinear filter operations, fast algorithms, morphological operations to detect the shape of objects, segmentation and classification, reconstruction of three-dimensional objects, and analysis of stereo images.[1–5] Therefore we present in this chapter two image techniques, RIMAPS and Variograms, which usually do not appear in reference books and have shown to be helpful tools for characterizing biological and technological surfaces.

1.1. *Basic Concepts of RIMAPS Technique*

1.1.1. *Theory*

RIMAPS is a new imaging characterization technique independent of the class of microscopy and of conditions used for observation as long as they remain constant.[6–11]

This technique involves rotating the image using available algorithms of commercial software and computing one space-variable integral of the two-dimensional Fourier transform for each value of the other variable.[12–22]

The mathematical procedure is essentially the rotation of the digitized image $I(x, y)$ a certain angle α, and the calculation of the x-step of the two-dimensional Fourier transform for each y-line of the new image $I_\alpha(x, y)$ obtained after rotation:

$$\mathsf{I}_\alpha(\nu_x, y) = \int_{-\infty}^{\infty} I_\alpha(x, y) \exp(-i\nu_x x) dx. \tag{1}$$

Results from Eq. (1) are averaged over all y-coordinate of the image and an average power spectrum (APS) is obtained for each angular position:

$$\mathsf{A}_\alpha(\nu_x) = |\langle \mathsf{I}_\alpha(\nu_x, y)\rangle^* \langle \mathsf{I}_\alpha(\nu_x, y)\rangle|, \tag{2}$$

where $\langle\ \rangle$ means average value and with $\langle \mathsf{I}_\alpha(\nu_x, y)\rangle^*$ being the complex conjugate of $\langle I_\alpha(\nu_x, y)\rangle$. If the set of maximum values, in arbitrary units (a.u.), of all Average Power spectra (MAPS):

$$\mathsf{M}(\alpha) = \max \mathsf{A}_\alpha(\nu_x), \tag{3}$$

is plotted as a function of the angle of rotation (typically steps of 1°) of original Rotated Image (RI), valuable information can be obtained from the surface pattern under study.[6,7] The peaks appearing in the resulting plots indicate surface pattern orientation.

In the present chapter, we focus our attention on both the tendencies of the curve and on the intensity and shape of the peaks. To clarify the use of this technique, a few images from ideal geometrical shapes are compared to experimental surface patterns obtained on different materials.

1.1.2. *Examples of using RIMAPS for ideal geometrical forms*

The results of the analysis obtained with the RIMAPS technique for a straight line and a square are shown in Figs. 1 and 2, respectively.[6,7] Figure 1 shows: (a) a line forming an angle of 138° with horizontal, (b) the relationship of MAPS versus the RI in Cartesian coordinates, and (c) the relationship of MAPS versus the RI in polar

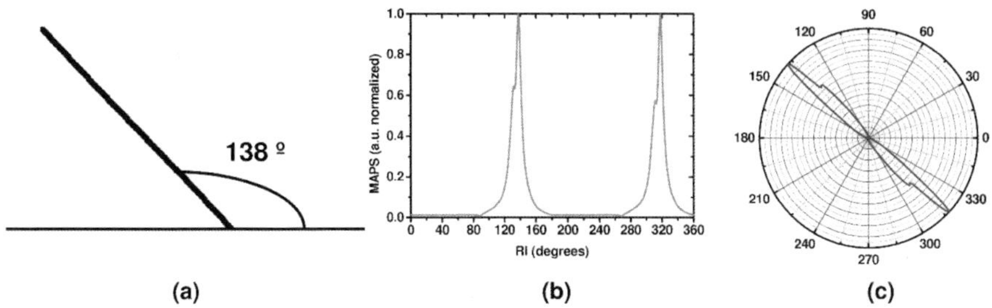

Fig. 1. (a) A line at an angle of 138° with a horizontal axis; (b) MAPS vs RI in Cartesian coordinates; and (c) the graph MAPS vs RI in polar coordinates, obtained using RIMAPS, typical for the linear-type pattern.

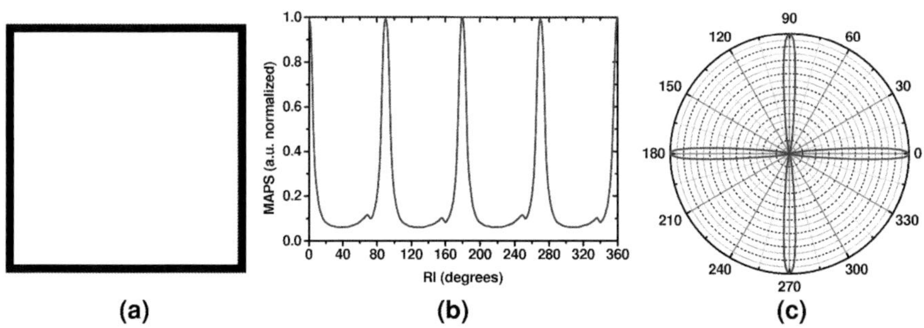

Fig. 2. (a) A square; (b) MAPS vs RI in Cartesian coordinates; and (c) MAPS vs RI in polar coordinates, obtained using RIMAPS, typical for the square-like pattern.

coordinates. Figure 2 shows: (a) a square, (b) the relationship of MAPS versus the RI in Cartesian coordinates, and (c) the relationship of MAPS versus the RI in polar coordinates. Note that the peaks in Figs. 1(b), 1(c), 2(b), and 2(c) indicate the angular location for the directions corresponding to the line position and to the four sides of an ideal square. These figures clearly reveal symmetric shapes of graphs MAPS versus the RI and a possibility of characterizing simple geometrical forms, using the RIMAPS technique.

The simple combination of straight lines is shown in Fig. 3.[6,7] This figure shows: (a) a square and a line forming an angle of 138° with horizontal, (b) the relationship of MAPS versus the RI in Cartesian coordinates, and (c) the relationship of MAPS versus the RI in polar coordinates. It is clear from the example given in Fig. 3 that the RIMAPS spectra corresponding to cases representing the superposition of simple geometrical forms can be obtained as the combination of the individual spectrum of each single geometrical shape represented in Figs. 1 and 2.

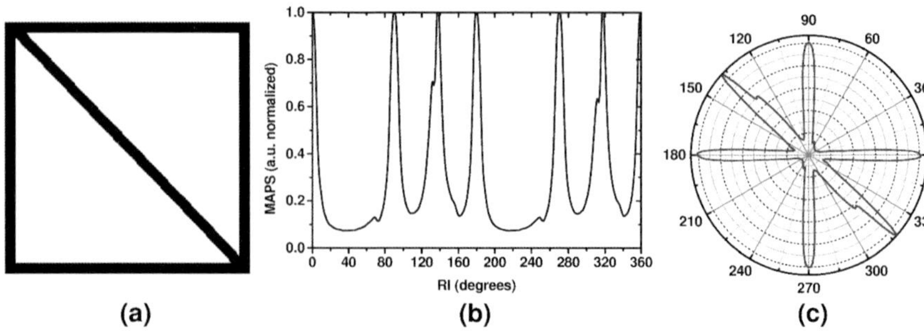

Fig. 3. (a) A square with a line inside it forming an angle of 138° with a horizontal axis; (b) MAPS vs RI in Cartesian coordinates; and (c) MAPS vs RI in polar coordinates, obtained using RIMAPS, showing the results of the combination of the square-like and linear patterns.

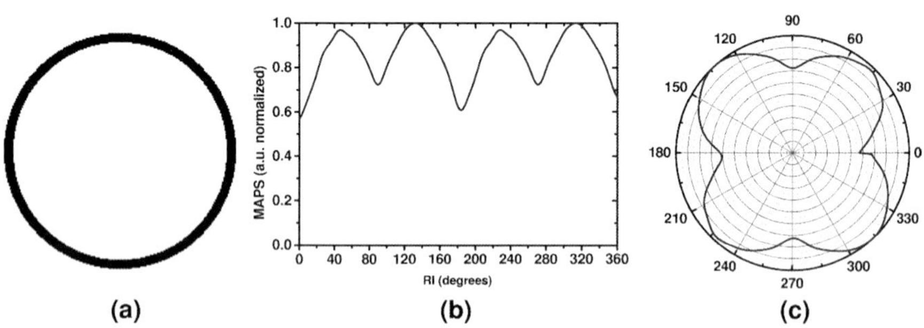

Fig. 4. (a) A circle; (b) MAPS vs RI in Cartesian coordinates; and (c) MAPS vs RI in polar coordinates, obtained using RIMAPS.

Examples of the analysis obtained with the RIMAPS technique for a circle and the combination of circle and straight lines are shown in Figs. 4 and 5, respectively.[6,7] Figure 4 shows: (a) a circle, (b) the relationship of MAPS versus the RI in Cartesian coordinates, and (c) the relationship of MAPS versus the RI in polar coordinates. Figure 5 shows: (a) a circle and a straight line inside forming an angle of 138° with horizontal, (b) the relationship of MAPS versus the RI in Cartesian coordinates, and (c) the relationship of MAPS versus the RI in polar coordinates.

In all the examples presented here, the figures clearly reveal symmetric shapes of graphs MAPS versus RI and a possibility of characterizing simple geometrical forms, using the RIMAPS technique.[23–25] It must be remembered that any RIMAPS spectrum can be considered as the combination of the spectra of simple geometrical forms. If it is assumed that a physical process determines a distinctive pattern on a

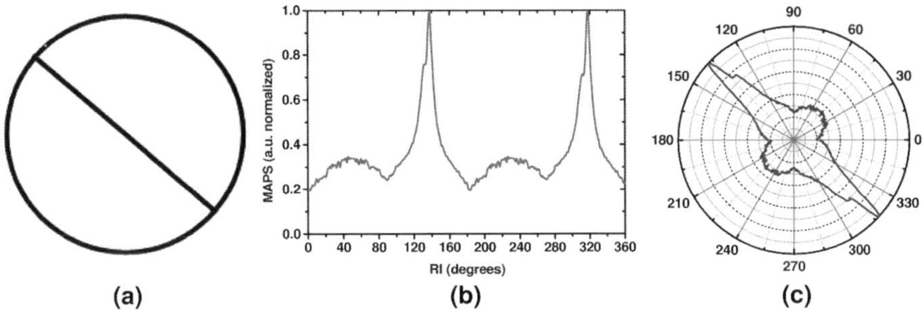

Fig. 5. (a) A circle with a line forming an angle of 138° with horizontal direction; (b) MAPS vs RI in Cartesian coordinates; and (c) MAPS vs RI in polar coordinates, obtained using RIMAPS, showing the results of the combination of the circular and linear patterns.

surface, the combined action of several processes may be described by identifying the corresponding RIMAPS spectra.[23–25]

1.1.3. *Characterization of simple experimental surfaces using RIMAPS*

Different topographic patterns generated on metallic plates were used to show examples of RIMAPS applications on real experimental surfaces, creating a surface roughness and complex topographic characteristics that it becomes nearly impossible to visually analyze the corresponding surface images to characterize geometrical patterns.

After being chemically etched with different time of exposure to a modified Palmerton reagent,[25] different zinc-plate surfaces, shown in Figs. 6(a), 7(a), and 8(a), were analyzed using the RIMAPS technique. Figures 6(b), 6(c), 7(b), 7(c), 8(b), and 8(c), show the graphs MAPS versus RI, which contain a series of well-defined peaks located at the angular positions of the lines. Comparison of the results shown in Figs. 6–8, with those corresponding to ideal patterns given in Figs. 1, 3, and 5 allows us to conclude that all the surface patterns produced on the zinc-plates could be represented as the combination of a series of simple geometrical figures.

In some cases (Fig. 6), surface pattern can be identified by simple observation of the micrograph and can be associated with a simple geometrical figure. In this case, it can be thought that the characteristic pattern is formed by a series of dashed lines with a principal orientation at approximately 138° with horizontal. On other sample surfaces, the topographic pattern seems to be a superposition of lines oriented in different directions. In the micrograph of a Zn grain (Fig. 7) obtained with Bright

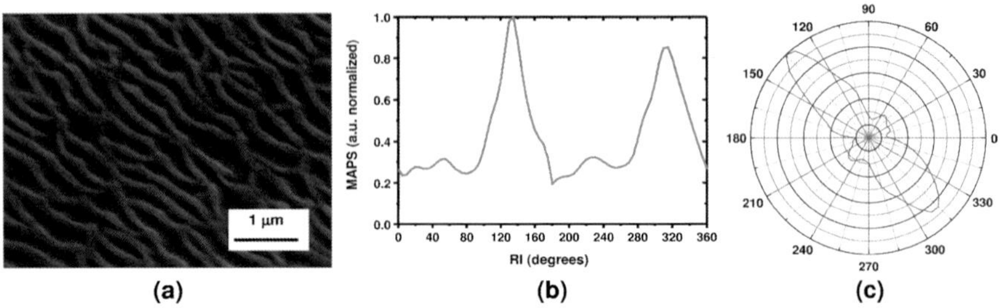

Fig. 6. (a) Digitized SEM micrograph of a Zn-surface after chemical treatment with a modified Palmerton reagent (reprinted with permission from Fuentes and Favret,[7] Wiley); (b) and (c) MAPS vs RI in Cartesian and polar coordinates, respectively, which correspond to a linear-type pattern of surface topography presented in Fig. 1.

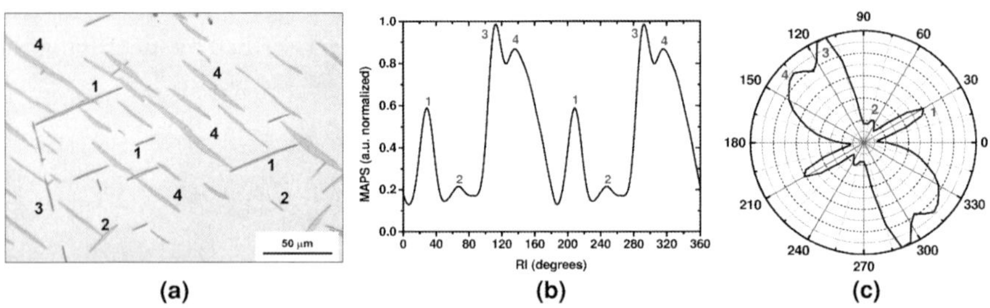

Fig. 7. (a) A light micrograph of a Zn grain surface after chemical etching using a modified Palmerton reagent, obtained with Bright Field Illumination (reprinted with permission from Fuentes and Favret,[7] Wiley); (b) and (c) MAPS vs RI in Cartesian and polar coordinates, which correspond to a combination of segmented lines that form a square-like pattern of surface topography similar to that presented in Fig. 3.

Field Illumination (light microscopy) RIMAPS results give well-defined peaks located at the angular positions at which the lines are oriented.

Directions indicated as 1, 3, and 4 formed a four-sided figure while orientation 2 gives a contribution in a different direction. This combination determines a typical pattern very similar to the ideal situation presented in Fig. 3. With different time of exposure to the modified Palmerton reagent, one of the possible combinations of circles and lines that are presented in Fig. 8 can be obtained. In the micrograph 8(a) obtained with Dark Field Illumination, the intersections of curved lines form facets oriented in a certain direction on the surface. It is noted that in Figs. 8(b) and 8(c) that all the peaks of RIMAPS represent the superposition of the angular positions of circles alignment and the directions of facet orientation. This situation can be compared to the simplified ideal pattern of only one circle and a line shown on Fig. 5. It can be

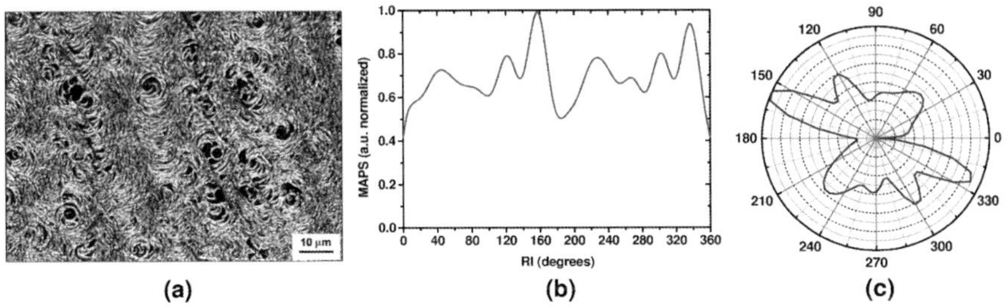

Fig. 8. (a) A light micrograph of a Zn grain surface after chemical etching using a modified Palmerton reagent, obtained with Dark Field Illumination (reprinted with permission from Fuentes and Favret,[7] Wiley); (b) and (c) MAPS vs RI in Cartesian and polar coordinates, which correspond to the superposition of the angular positions of circles alignment and the directions of facet orientation. This situation can be compared to the ideal pattern of Fig. 5 formed by the combination of one circle and a line.

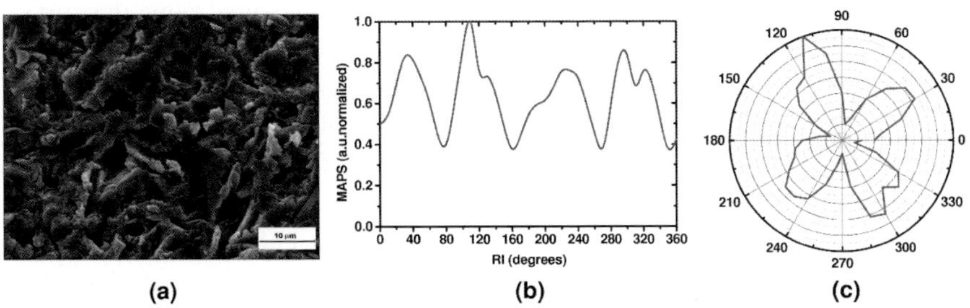

Fig. 9. (a) Digitized SEM micrograph of the Ti-surface after the Alumina particle bombardment (reprinted with permission from Fuentes and Favret,[7] Wiley); (b) and (c) MAPS vs RI in Cartesian and polar coordinates, which correspond to a square-like pattern of surface topography presented in Fig. 2.

noted that the presence of circular shapes gives smoothed curves that represent the base values for the peaks corresponding to well-defined directions associated to straight lines or a particular alignment of geometrical forms.

A titanium-plate surface was subjected to the Alumina particle bombardment[7] to create a surface roughness with no clear surface orientation (Fig. 9(a)). Using RIMAPS, it was determined a function MAPS versus RI in Cartesian and polar coordinates (Figs. 9(b) and 9(c)), which is characterized by peaks typical for a four-sided geometrical pattern, including two main orthogonal directions along the surface. These results can be compared with those given in Figs. 2(b) and 2(c). Note that the peaks in Figs. 2(b) and 2(c) indicate the angular position for the main directions of the surface topography, which can generally be seen in Fig. 2(a). The width of each peak is related to scattering around the main orientations.

1.2. *Basic Concepts of Variogram Method*

1.2.1. *Variogram analysis*

The Variogram method is based on a log–log representation of a characteristic roughness parameter versus the length scale. This algorithm is applicable to fractal and nonfractal surfaces characterization. It is well known that the geometric structure of rough surfaces is random, and that roughness features are found at a large number of length scales between the length of the sample and atomic scales. To understand the topography-dependent phenomena, the surface geometry must be described with parameters which take into account different length scales.

The root mean square average (RMS), σ, is one of the parameters most often used to characterize the surface roughness from a profile measured along this surface. While such parameters are useful for many applications, they do not cover information on the range of length scales over which different topographic features exist. Indeed, conventional parameters depend only on a few particular length scales, such as the instrument resolution or the sample length, while rough surfaces contain roughness at a large number of length scales.

During the last decade, various methods based on fractal analysis have been proposed to characterize surface roughness at different length scales. Most of the methods are aimed at calculating fractal parameters, in order to characterize the roughness at several length scales. For this purpose, the log–log representation of the variance σ^2 versus the sample size is used to determine the fractal dimension from the slope of the resulting curve.[26–28] However, not all surfaces show fractal behavior. In these cases, the surface topography may not be appropriately described using a fractal dimension.

Therefore, we propose a new method for the quantification of length scale-dependent topography.[10,29,30–36] From a given set of windows of different sizes of one digitized image, the RMS parameter is calculated for different window areas on a surface. The calculated roughness parameter is then represented on a log–log plot as function of the length scale (window area). Intersections of different slopes in the plot give crossover lengths that characterize the surface.

1.2.2. *Examples of using Variogram for ideal geometrical forms*

Using the same ideal geometries presented in a previous section, the results obtained using the Variogram methods are presented. Simple geometrical shapes were drawn on

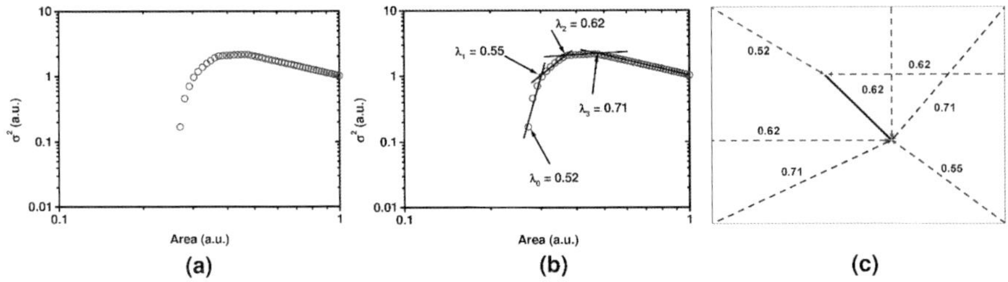

Fig. 10. (a) Log–log representation of the Variance σ^2 vs the windows areas; (b) the intersections of slopes give the typical length scales for one line pattern; and (c) a straight line with the indication of the different length scales.

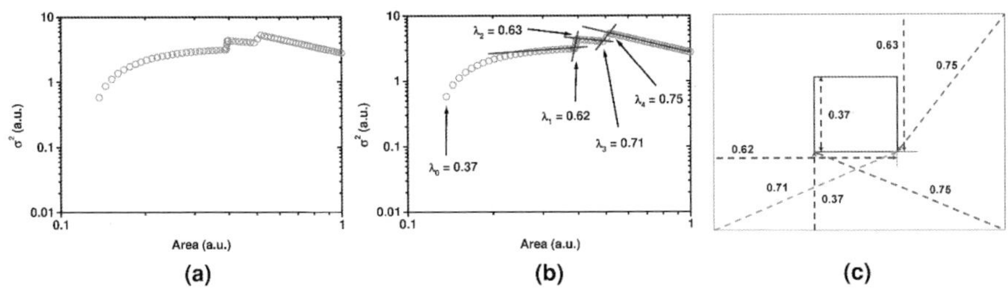

Fig. 11. (a) Log–log representation of the Variance σ^2 vs the windows areas; (b) the intersections of slopes give the typical length scales for one square pattern; and (c) a square with the indication of the different length scales.

a dimensionless unit area. The analysis corresponding to a straight line and a square are shown in Figs. 10 and 11. Both figures present: (a) the log–log representation of the variance (in arbitrary units) versus the different observation-window areas, (b) the slopes superimposed to the Variogram curve indicating the typical length scales, and (c) a line (Fig. 10) and a square (Fig. 11) with the different length scales indicated on the images.

In these figures, all length scales represent different reference points for positioning and sizing of the simple patterns under study.

The combination of these two situations is shown in Fig. 12. This figure gives: (a) the log–log representation of σ^2 versus the observation-window areas, (b) the typical length scales indicated at the intersections of slopes superimposed to the Variogram curve, and (c) a line inside a square indicating all the length scales.

The example given in Fig. 12 clearly demonstrates that the Variogram results corresponding to cases representing the superposition of simple geometrical forms include all the length scales obtained from each one of the single geometries.

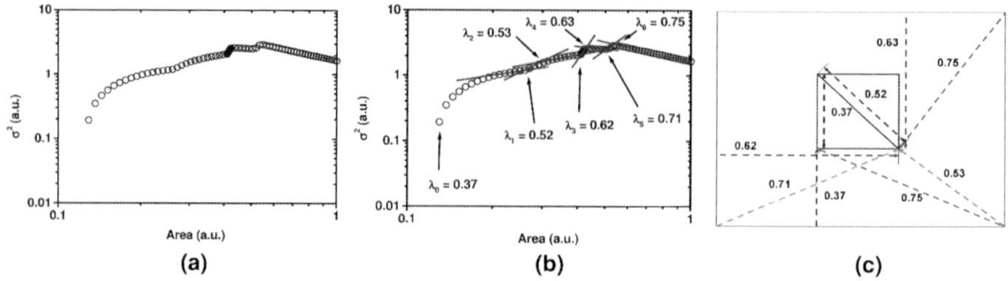

Fig. 12. (a) Log–log representation of the Variance σ^2 vs the windows areas; (b) the intersections of slopes give the typical length scales for the combination of straight lines pattern; and (c) a line inside a square with the indication of the different length scales.

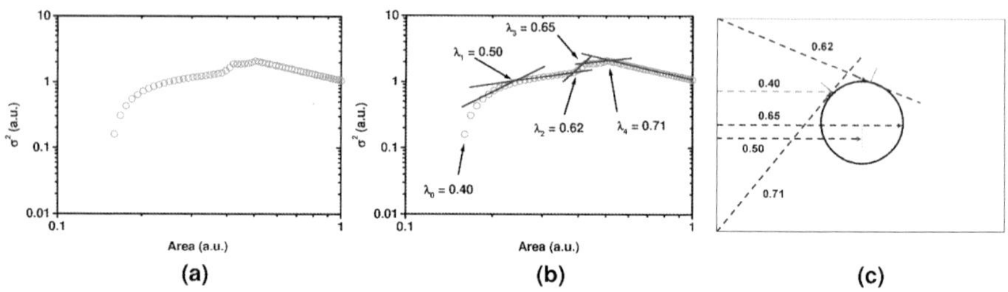

Fig. 13. (a) Log–log representation of the Variance σ^2 vs the windows areas; (b) the intersections of slopes give the typical length scales for a circle; and (c) a circle with the indication of the different length scales.

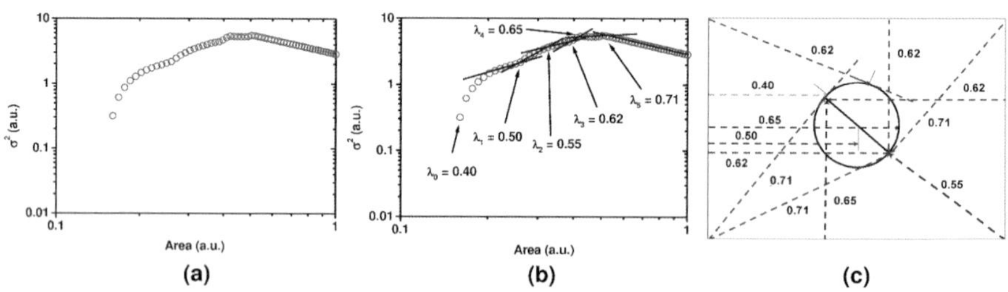

Fig. 14. (a) Log–log representation of the Variance σ^2 vs the windows areas; (b) the intersections of slopes give the typical length scales for the combination of straight line and a circle; and (c) a line inside a circle with the indication of the different length scales.

Figures 13 and 14 show two examples of the Variogram analysis obtained for a circle and the combination of circle and a straight line inside of it. As in the previous examples, Figs. 13 and 14 present: (a) the log–log representation of the variance versus the different observation-window areas, (b) the slopes superimposed to the Variogram curve indicating the typical length scales, and (c) a circle (Fig. 13) and the combination of a straight line inside a circle (Fig. 14) showing the different length scales indicated on the images.

All the previous examples conclude the fact that Variogram analysis will clearly reveal all the length scales that characterize a surface under study. As it was stated before with RIMPAS technique, if several physical processes act on a surface producing different distinctive patterns, the Variogram method is able to identify the corresponding length scales.[29,31–37]

1.2.3. *Characterization of simple experimental surfaces using Variogram*

The same topographic patterns generated on metallic plates and were studied with RIMAPS technique (Figs. 6–9), are used here to give examples of the Variogram analysis. In all cases, the figures present: (a) the log–log representation of the Variance σ^2 versus the windows areas; (b) the intersections of slopes giving the typical length scales for the surface pattern, and (c) the digitized image of a metallic surface and the first length scales indicated on the image.

Zinc-plate surfaces with different time of exposure to chemical etching, are shown in Figs. 15–17. Three distinctive patterns cover the whole surface under study in each case. In the example of a linear-type pattern of Fig. 15, Variogram analysis gives the width of the facets, the length of its different segments and the separation between them.

The case of the pattern that results after chemically etching and resembles a combination of straight line-segments, is shown in Fig. 16. From a Bright Field

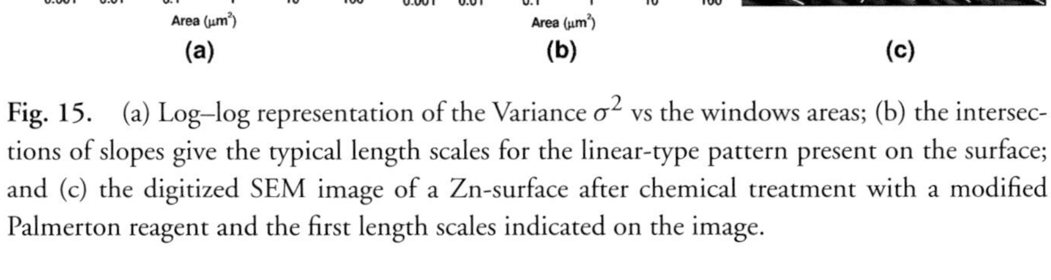

Fig. 15. (a) Log–log representation of the Variance σ^2 vs the windows areas; (b) the intersections of slopes give the typical length scales for the linear-type pattern present on the surface; and (c) the digitized SEM image of a Zn-surface after chemical treatment with a modified Palmerton reagent and the first length scales indicated on the image.

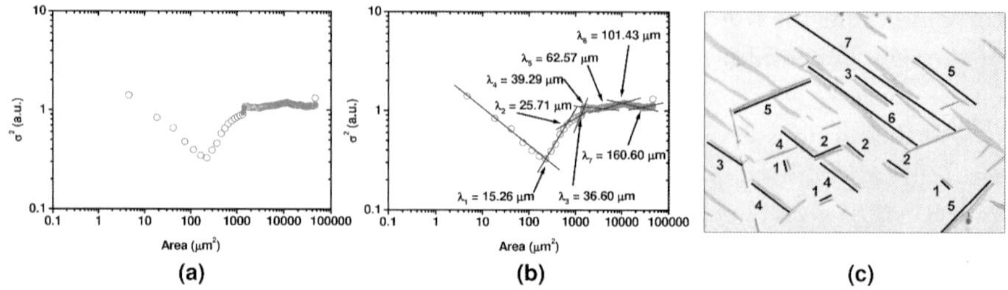

Fig. 16. (a) Log–log representation of the Variance σ^2 vs the windows areas; (b) the intersections of slopes give the typical length scales corresponding to a combination of segmented lines that form a square-like pattern present on the surface; and (c) the light micrograph of a Zn grain surface after chemical etching using a modified Palmerton reagent, obtained with Bright Field Illumination, and the first length scales indicated on the image.

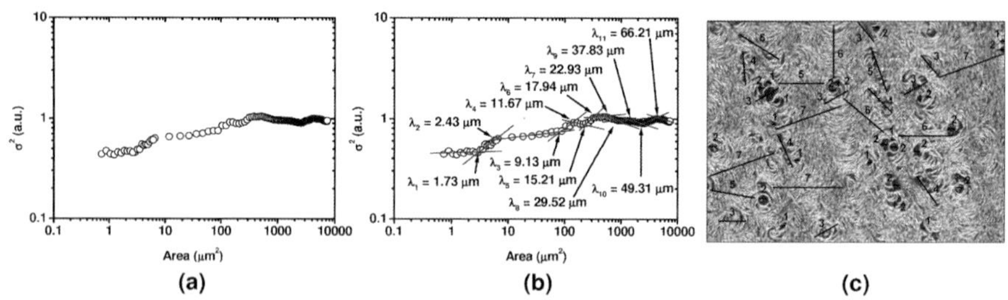

Fig. 17. (a) Log–log representation of the Variance σ^2 vs the windows areas; (b) the intersections of slopes give the typical length scales for the circular shape size and the distances between circles alignment and the directions of facet orientation in the pattern present on the surface; and (c) the light micrograph of a Zn grain surface after chemical etching using a modified Palmerton reagent, obtained with Dark Field Illumination and the first length scales indicated on the image.

Illumination image, Variogram detects the size of each segment that can be identify on the surface.

The Zinc surface image obtain with Dark Field Illumination, Fig. 17, gives another example of the application of the Variogram method. Here, the length scales obtained represent the size (approximate diameter) of the circular shapes of the surface pattern, and the distances between these circles, their alignment directions and the orientation of facets that compose the topographic pattern on the surface. As in the examples presented before, only the first length scales are indicated on the surface image.

In the case of the surface pattern obtained on a Titanium-plate after Alumina particle bombardment, Fig. 18, the Variogram analysis brings the length scales representing

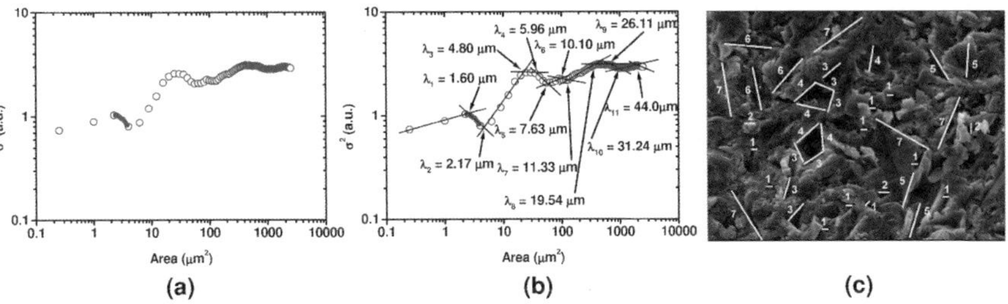

Fig. 18. (a) Log–log representation of the Variance σ^2 vs the windows areas; (b) the intersections of slopes give the typical length scales corresponding to four-sided shape pattern present on the surface; and (c) the digitized image of the Ti-surface after the Alumina particle bombardment and the first length scales indicated on the image.

the line-segments determining the four-sided shape of the simplified geometry that characterizes the surface topography.

These examples given in Figs. 15–18 can be compared with the Variogram results obtained from the ideal geometrical shapes of Figs. 10–14. It can be noted that in all cases, the resulting length scales characterize the size and relative location of the topographic pattern on the surfaces under study. In the case of real experimental surfaces, the density of the pattern covering the whole surface increases the number of length scales to be detected on the surface.

2. Micro-Nano Topography Characterization

2.1. *Biological Surfaces*

Plant surfaces, optimized in their function as protective layers and according to several demands which has been put upon them throughout evolution has led to the development of multifunctional surfaces. In search of the typical patterns that determine the different surfaces behavior, it is a common practice for botanists and taxonomists to describe and characterize leaf surfaces using electronic microscopy.[38,39] Scanning electron microscopy (SEM) allows the characterization of biological surfaces in a qualitative manner, based only on visual observation. These are primarily qualitative descriptions of micrographs. If the magnification of the different images is known, observers can measure cell sizes or distances between any of the structures present on the surface. This is an observer-dependent technique for describing botanical material.

On the other hand, the combined use of RIMAPS and Variogram techniques for analyzing biological surfaces from digitized images is a strong user-independent procedure that can be used to obtain valuable information on the topography and microstructure of the leaf surface from both fresh and dried herbarium material.

In this section, RIMAPS and Variogram analysis of two botanical species with superhydrophobic behavior are presented (see Chap. 3). A surface is considered superhydrophobic when its static contact angle is higher than 150° and its contact angle hysteresis is below 10°. Two *Colocasia esculenta* leaves (adaxial faces), one fresh (low magnification, 400×) in Fig. 19 and the other dried (high magnification, 6400×) in Fig. 20, and fresh rice leaf showing the adaxial epidermis (magnification of 1000×) in Fig. 21, are given as examples.

Even though the surface condition of leaves presented in the micrographs of Fig. 19(a) (fresh) and Fig. 20(a) (dried), and the magnifications used in each case are very different, the RIMAPS analysis detect the same topographic patterns on the

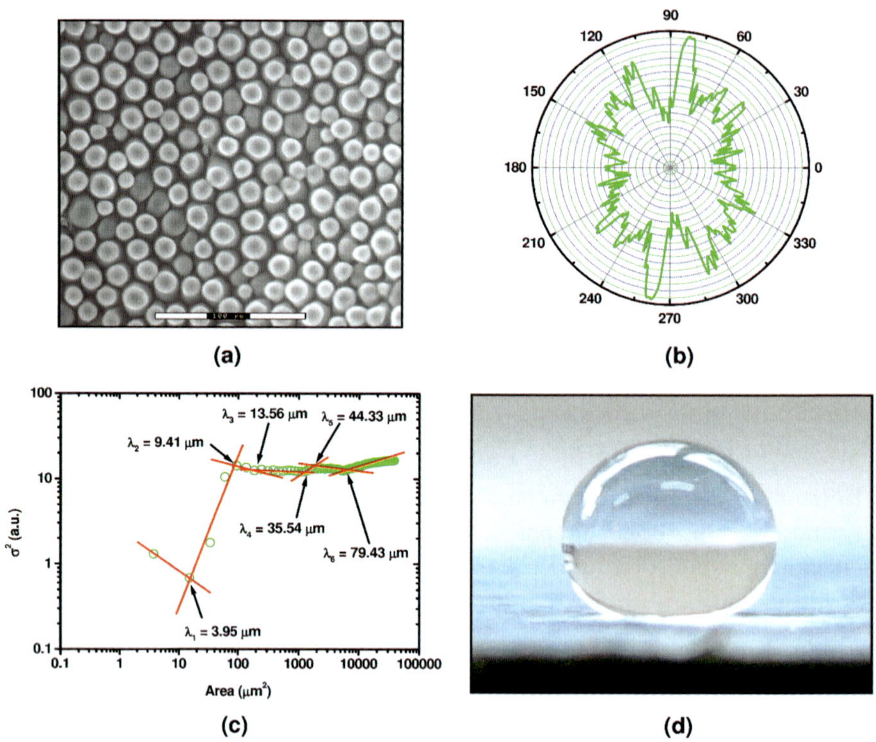

Fig. 19. (a) SEM micrograph of the adaxial face of a *Colocasia esculenta* fresh leaf, observed with a magnification of 400× (reprinted with permission from Favret and Löthman,[11] Wiley); (b) MAPS vs RI in polar coordinates, obtained using RIMAPS; (c) Log–log representation of the Variance σ^2 vs the windows areas with the intersections of slopes giving the typical length scales for the surface pattern; (d) the shape of a water drop of 10 μl on the leaf surface with a static contact angle of approximately 150°.

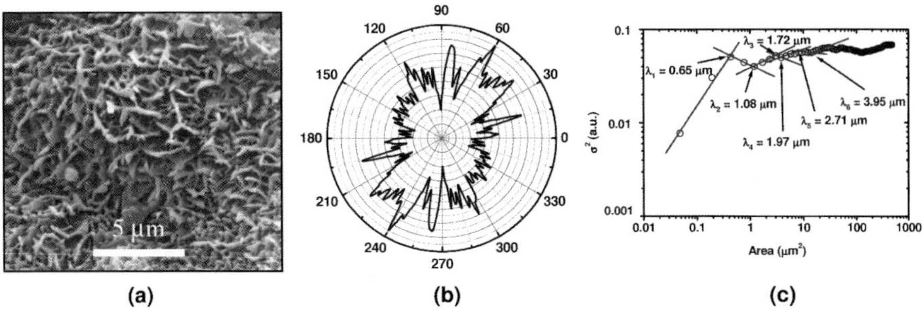

Fig. 20. (a) SEM micrograph of the adaxial face of a *Colocasia esculenta* dried herbarium leaf, observed with a magnification of 6400×, showing the pattern formed by wax crystals (reprinted with permission from Favret and Löthman,[11] Wiley); (b) MAPS vs RI in polar coordinates, obtained using RIMAPS; (c) Log–log representation of the Variance σ^2 vs the windows areas, where the intersections of slopes indicate the typical length scales of the pattern corresponding to the wax crystals distribution.

leaves surfaces. The pattern of the micro- and nanostructures has coincident principal directions and distribution.

The simple eye-inspection of their respective images cannot give enough information to confirm that both micrographs correspond to the same botanical species.

The general shape of the polar representation of RIMAPS spectra (Figs. 19(b) and 20(b)) corresponds to the case of the combination of circular shapes and straight lines, and their respective alignment, previously presented in Figs. 5 and 8. The circular shapes correspond to the papillae and stoma, and the straight lines correspond to the distribution pattern of the wax seen in between the papillae and the alignment of papillae. In both cases, a main alignment direction is represented by a peak at 85°. As the magnification is increased, the peaks representing the linear-type wax pattern become higher in intensity. These peaks are located between 0° and 75°.

The Variogram results obtained from both micrographs, give all the length scales characterizing the leaves patterns. It can be said that the topography determined by papillae and stoma has typical lengths that are greater than 3.95 μm, while the pattern due to the crystals of wax has length scales lower than 3.95 μm. These results are illustrated in Figs. 19(c) and 20(c).

Wetting behavior of the *C. esculenta* leaf is illustrated in Fig. 19(d). Water drops with a volume of 10 μl adopted a spherical shape denoting the hydrophobic characteristics of the leaves of this plant.

The fresh rice leaf was observed with a magnification of 1000×, and the corresponding SEM micrograph of Fig. 21(a) shows the adaxial epidermis of the leaf.

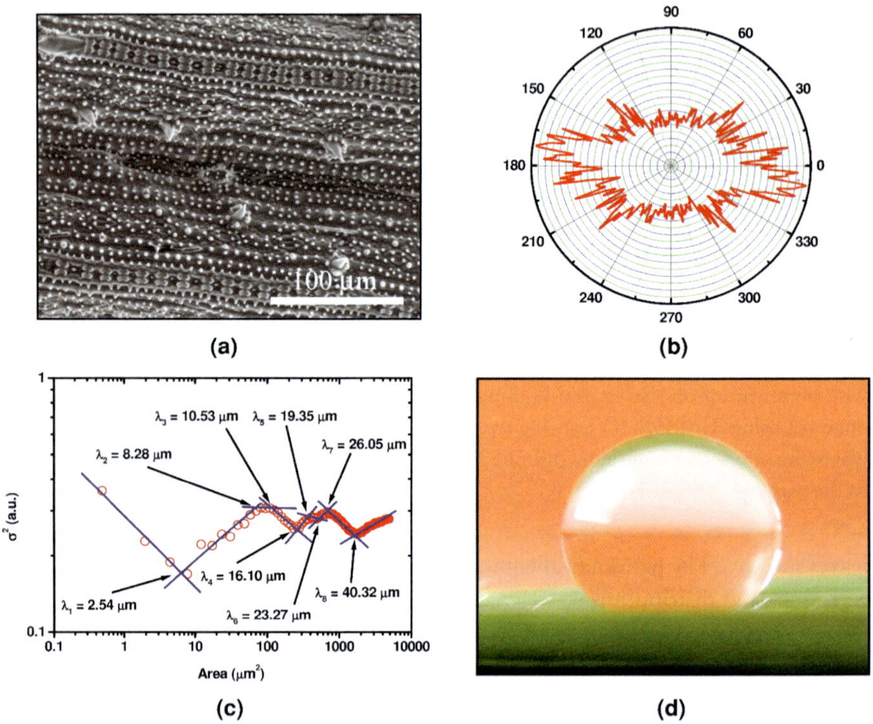

Fig. 21. (a) SEM micrograph of the adaxial surface from a fresh rice leaf, observed with a magnification of 1000×; (b) MAPS vs RI in polar coordinates, obtained using RIMAPS; (c) Log–log representation of the Variance σ^2 vs the windows areas with the intersections of slopes giving the typical length scales for the leaf surface pattern; (d) the shape of a water drop of 10 μl on the leaf surface with a static contact angle of approximately 150°.

The topographic pattern detected by using RIMAPS (Fig. 21(b)), reveals a prevailing direction of alignment very close to the horizontal direction (170°). The shape of RIMAPS spectrum resembles the characteristics of the linear-type patterns introduced in Figs. 1, 6, and 7. The existence of low-intensity secondary peaks superimposed to the main direction is the consequence of the alignment of undulating-line patterns and elliptic shapes present on the surface. Even though this pattern is different from the one observed on the *C. esculenta* leaves, the existence of length scales (Fig. 21(c)) of the same order of magnitude than in the mentioned case, seems to indicate that the topographic pattern will verify the nonwetting condition on the leaf surface as it is illustrated in Fig. 21(d) with the image of a 10 μl water drop on the surface.

2.2. *Technological Surfaces*

Early description of wetting phenomena was formulated in the 19th century from the basis of Young's equation with the balance of surface tensions acting on a water drop of millimeter or micrometer sizes. A drop fulfilling this equation is set to be partially wetting the substrate and has a contact angle close to 90°. For angles smaller than 90°, the drop will flatten out to form a film. On the other limit, if the contact angle increases almost until 180° the surface will remain dry (see Chaps. 3 and 8). It was in the first half of the 20th century, with the works of Wenzel[40] and Cassie and Baxter[41] that the effect of surface structure has come into play for the first time. Since then, many scientific investigations have been devoted to explaining how surface roughness can induce wetting or nonwetting behavior.[42–44] The importance of understanding these complex mechanisms relies on the possibility to design functional surfaces with direct applications in fields like biotechnology, corrosion-free medical devices, lubrication, and use of adhesives on space qualified composite structures.

The nonwetting, self-cleaning, and anti-adhesive properties are due to the combination of surface architecture and surface chemistry. Hierarchical surface structures, i.e. a surface composed of a microstructure superposed by a nanostructure, together with a hydrophobic surface chemistry, are believed to be responsible for the properties of the surface. Those superior nonwetting properties are often called super or ultrahydrophobic.[45]

Technological development of different functional surfaces, has been made by studying and manufacturing topographic patterns using physical and chemical procedures, for example by two step plasma treatment[46] or by microwave plasma-enhanced chemical vapor deposition.[47] Biomolecules are currently being investigated for their use as hydrophobic coatings as well as a component of multifunctional technical surfaces.[48] Applications of self-cleaning or anti-adhesive surfaces can be seen in different fields such as facade paint, hemocompatible coronary stents, and implant materials which show an altered initial osteoblast attachment.[49]

In this section, three examples of different man-made topographic patterns are presented (Figs. 22–24). Different materials such as copper and plates made of glass fibers or carbon fibers reinforced composites were used to characterize the fluid behavior in response to the different pattern characteristics. In all cases, the surfaces were coated with gold by sputtering.

The first surface corresponds to a rough copper sample (Fig. 22(a)), prepared by electrochemical methods generating a dendrite-like spheroid rough structure, with different topographic features. To a minor degree, some copper-oxide might be present in the structure. In principle, the roughness and the chemical composition (due to

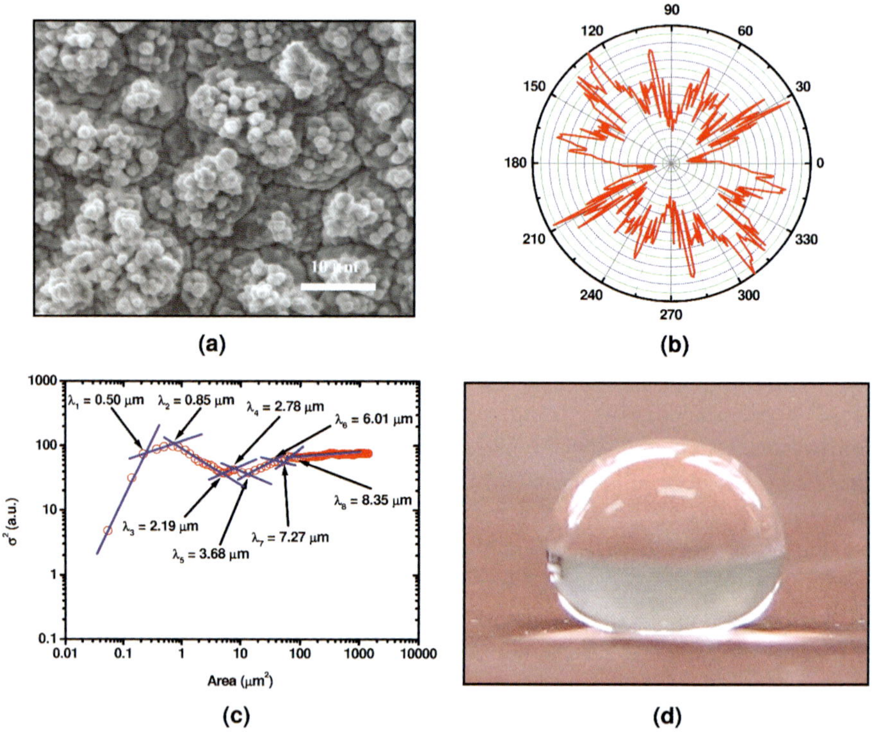

Fig. 22. (a) SEM micrograph of rough copper sample prepared by electrochemical methods, observed with a magnification of 2560× (reprinted with permission from Favret and Löthman,[11] Wiley); (b) MAPS vs RI in polar coordinates, obtained using RIMAPS; (c) Log–log representation of the Variance σ^2 vs the windows areas, where the intersections of slopes indicate the typical length scales corresponding to the pattern present on the surface; (d) spherical shape of a water drop of 10 μl on the copper surface forming a static contact angle of 120°–130°.

hydrophobic residues) were responsible for the nonwetting properties as seen in the high static contact angle around 120°–130° and a sliding angle of less than 10° for a micro-drop of water of 10 μl. However, the same conditions were found in the case presented here, when the sample was coated with gold (Au), which means that roughness and not chemical composition was the key to the nonwetting condition (Fig. 22(d)) as was observed in the case presented in Fig. 19(d). Comparison of RIMAPS results given in Fig. 22(b) with the spectrum corresponding to the analysis of *C. esculenta* (Figs. 19(b) and 20(b)) shows similar but rotated shapes. It can be observed that the peaks appearing between 120° and 180° (Fig. 22(b)) are the consequence of a

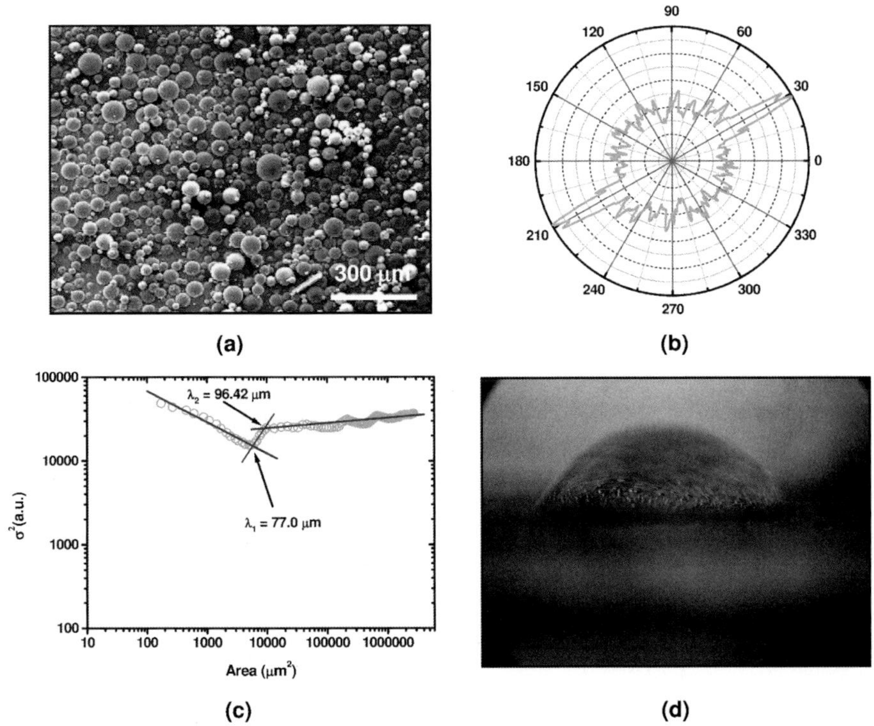

Fig. 23. (a) SEM micrograph of rough surface obtained by the deposition of glass spheres, with a magnification of $400\times$; (b) MAPS vs RI in polar coordinates, obtained using RIMAPS; (c) Log–log representation of the Variance σ^2 vs the windows areas, where the intersections of slopes indicate the typical length scales corresponding to the pattern present on the surface; (d) image of a water drop of 5 μl on the surface forming a static contact angle of 80°.

topography distribution similar to the structures formed by the wax crystals on the *C. esculenta* leaves and correspond to the peaks detected by RIMAPS located between 0° and 60° (Figs. 19(b) and 20(b)). Variogram analysis detects length scales with values lower than 10 μm (Fig. 22(c)). Both circumstances, similar topographic pattern on surface and the same order of magnitude of length scales, are indicating that the technological surface obtained will respond in a hydrophobic way similar to the botanical species introduced in the preceding section.

In the other two examples, the samples were prepared starting from surfaces with smooth topography, and then several random rough patterns were obtained with a combination of spray adhesives and the deposition of glass spheres and carbon particles and then they were coated with gold by sputtering (Figs. 23–24). The base surface on

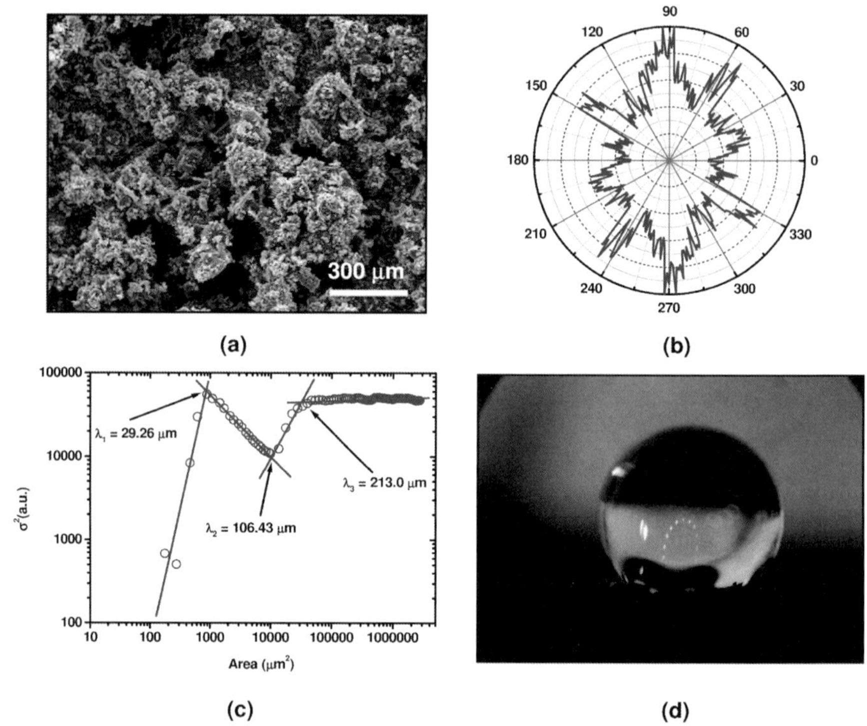

Fig. 24. (a) SEM micrograph of rough surface obtained after the deposition of carbon particles on a glass fiber composite plate, with a magnification of 400×; (b) MAPS vs RI in polar coordinates, obtained using RIMAPS; (c) Log–log representation of the Variance σ^2 vs the windows areas, where the intersections of slopes indicate the typical length scales corresponding to the pattern present on the surface; (d) image of a water drop of 5 μl on the surface forming a static contact angle of 130°.

sample plates was prepared using glass fibers with epoxy/polyphenylene oxide resin.[50] The external laminates of glass fibers composites were in contact with peel plies[51] that determine the topographic patterns of the plate surfaces. In all cases, the samples were covered with bleeders[51] before vacuum bagging for polymerization in autoclave.

Two examples of surface topography are given in Figs. 23(a) and 24(a). The SEM micrograph of Fig. 23(a) shows a topographic pattern obtained with the adhesion of glass spheres using spray adhesive on the base surface. This micrograph looks similar to the image obtained from the *C. esculenta* at low magnification (Fig. 19(a)). Even though the RIMAPS spectrum of Fig. 23(b) corresponds to a topographic arrangement of circular shapes and their alignment, as in the *C. esculenta* case (Fig. 19(b)), the length scales are similar or greater than the greatest length scale of the plant leaf (Fig. 19(c)). The absence of characteristic lengths in the order of a few tenths of microns or lower (Fig. 23(c)), seems to determine a behavior very different from the nonwetting condition. It can be observed in Fig. 23(d) that the shape of a drop of 5 μl flattens on the surface until reaching 3 mm, with a static contact angle of 80°.

The last example in Fig. 24 corresponds to the case of carbon particles deposited on a glass fiber plate surface by means of spray adhesive. In this case, the SEM micrograph (Fig. 24(a)), the RIMAPS spectrum (Fig. 24(b)) and the Variogram analysis (Fig. 24(c)) completely coincide with the results obtained from the plants leaves study described previously (Figs. 19–21). Furthermore, fluid behavior on this surface is represented by a static contact angle of 130° and a spherical shape of drop shown in Fig. 24(d), where the water volume is of 5 μl and its diameter of 1.5 mm. However, the sliding angle is far higher than 10°, avoiding a superhydrophobic condition.

It must be remembered that in all cases of the man-made surfaces, the plates had received a coating of gold in order to avoid the incidence of chemical characteristics on the wetting behavior of the surfaces. Then, it must be emphasized that the surface behavior is only the consequence of geometrical features of the topographic patterns. In general, the contact angles increases when smaller scale lengths appear on surface and, as a consequence, transition is observed to dewetting condition.

3. Conclusions

Revealing hidden surfaces structures which are likely to be responsible for the function or ideally multifunction of the surface, has been a challenge in surface science and also for the applications and implementations of biomimetically derived technical surfaces. RIMAPS and Variogram are user-friendly tools that can be of value for this purpose. The results obtained by the different research works show the robustness of RIMAPS and Variogram analysis to distinguish and characterize leaf surfaces. The use of RIMAPS and Variogram opens a wide spectrum of possibilities by providing tools for the quantitative description of leaf surfaces. The combined use of both techniques will allow researchers to distinguish similarities and differences of species that occur in different environments, and the quantitative parameters that represent an organismal lineage. Thus, enabling the identification of simple geometrical patterns[52] with the corresponding length scales to be reproduced on technological surfaces to obtain distinctive functional properties of man-made materials.

RIMAPS technique has formerly been used to characterize the micro-relief of metallic surfaces and is now also being applied to biological surfaces. In all cases, it can be stated that the RIMAPS technique can be used to analyze digitized images of any kind of surfaces providing the principal topographic characteristics that are frequently extremely difficult or impossible, to obtain from a visual examination of surface images.

This chapter includes several examples of the RIMAPS analysis of simple ideal geometrical forms. Comparison between the ideal cases and the real topography made on different metallic surfaces demonstrates that the surface geometry can always be represented as a combination of ideal geometrical patterns. The analysis of different leaves surfaces and their patterns similarities to the simple geometrical shapes previously given are also presented. With the use of this technique, the main directions of a surface pattern can be identified by obtaining their respective angular location on the surface. The existence of secondary directions is represented by low-intensity peaks and the eventual pattern dispersion increases the width of the peaks.

All the results included in the present chapter emphasize the possibility of using RIMAPS and Variogram to obtain an accurate quantitative description of biological and bioinspired surfaces where the Variogram analysis provides the typical scale lengths that characterize a surface.

The studies of the wetting conditions of different leaves and materials surfaces reveal the existence of three ranges of length scales that characterize the topographic patterns and determine the applicability of the wetting or nonwetting conditions when fluid flow on surfaces is considered. In those surfaces with typical lengths of the order of 100 μm or greater, the static contact angles are lower than 80° and drops flatten out to wet the surface. This is a hydrophilic situation where the no-slip condition of fluid is applicable. In the case of topography having typical lengths between 1 and 100 μm, the water drops have spherical shapes and partially wet the surface with static contact angles greater than 90°. The wetting condition becomes more inapplicable as the contact angle tends to 180°. The extreme situation corresponds to hydrophobic surfaces with length scales lower than 1 μm and contact angles close to 180°.

The RIMAPS analyses of biological and man-made surfaces show the existence of different topographic patterns having the same wetting condition. No evidence was found on the existence of a unique surface pattern that was able to guarantee the applicability of the ultrahydrophobic condition. On the other hand, the Variogram analysis shows that on those surfaces showing the same wetting behavior, the same length scales are also present. This circumstance puts in evidence that the wetting conditions are closely related to a combination of microtopography given by the pattern geometry on surface and the nanotopography expressed in the existence of length scales in the nanometer range.

Acknowledgments

The authors are very grateful to Patricia B. Bozzano, Ramón A. Castillo Guerra, Ricardo J. Montero, Lorena Setten, Natalia Andrada, Adriana Dominguez, and Sara

L. Novas for the preparation and observation of the different samples. Special thanks should be given to Ana M. Molina, Christoph Neinhuis, and Per Löthman for providing the biological and copper surfaces.

References

[1] Smart P. (2007) *Image Analysis*. Glasgow University Press. Glasgow.
[2] Umbaugh SE. (2005) *Computer Imaging: Digital Image Analysis and Processing*. Taylor and Francis, Boca Raton, Florida.
[3] Russ JC. (1999) *The Image Processing Handbook*. CRC, Boca Raton, Florida.
[4] Jähne B. (2002) *Digital Image Processing*, 5th ed. Springer Verlag, Germany.
[5] Bow ST. (1992) *Pattern Recognition and Image Preprocessing*. Marcel Dekker, New York.
[6] Fuentes NO, Favret EA. (2002) A new surface characterization technique: RIMAPS (Rotated Image with Maximun Average Power Spectrum). *J Microsc* **206**: 72–83.
[7] Fuentes NO, Favret EA. (2002) Surface characterization of materials using the RIMAPS technique. *Microsc Anal* **56**: 5–7.
[8] Favret EA, Fuentes NO, Molina AM. (2003) Identification of botanical species using RIMAPS analysis on images from leaf surfaces. *Microsc Microanal* **9** (Suppl. 2): 1338–1339.
[9] Favret EA, Fuentes NO. (2004) RIMAPS and Variogram analysis of barley leaf surfaces. *Microsc Today* **12**(3): 24–26.
[10] Favret EA, Fuentes NO, Morrás HJM. (2006) RIMAPS and Variogram characterization of soil microstructure patterns. *Microsc Microanal* **12** (Supp. 2): 1688–1689.
[11] Favret EA, Löthman P. (2007) RIMAPS image analysis of biological and technical non-wettable surfaces. *Microsc Anal* **21**(1): 7–9.
[12] Elliot D, Rao K. (1982) *Fast Transforms Algorithms, Analyses, Applications*. Academic Press, New York.
[13] Gorcester J, Millhauser GL, Freed JH. (1989) Two dimensional and fourier transform EPR. In AJ Hoff (ed), *Advanced EPR. Applications in Biology and Biochemistry*, pp. 105–125. Elsevier Science, Amsterdam.
[14] Mc Donough R, Whalen A. (1995) *Detection of Signals in Noise*, 2nd ed. pp. 24–25; 52–54; 96–101. Academic Press, New York.
[15] González-Velasco E. (1996) *Fourier Analysis and Boundary Value Problems*. Academic Press, New York.
[16] Felinger A. (1998) *Data Analysis and Signal Processing in Chromatography*, pp. 19–40; 167–173; 241–247; 279–286; 385–406. Elsevier Science, New York.
[17] Ozaktas HM, Alper Kutay M, Mendlovic D. (1999) Introduction to the fractional fourier transform and its applications. In B Kazan and T Mulvey (serial eds), P Hawkes (Editor in Chief), *Advances in Imaging and Electron Physics*, Vol. 106, pp. 239–286. Academic Press, New York.
[18] Takeda M. (2000) Fourier transform profilometry. In PK Rastogi and D Inaudi (eds), *Trends in Optical Non-Destructive Testing and Inspection*, pp. 576–600. Elsevier Science, New York.

[19] Jeffrey A. (2001) *Advanced Engineering Mathematics ISE*, International Edition, pp. 545–622. Academic Press, New York.

[20] Smith S. (2002) *Digital Signal Processing: A Practical Guide for Engineers and Scientists*, pp. 141–260; 311–318; 567–580. Newnes, New York.

[21] Palm A. (2003) Capillary isoelectric focusing developments in protein analysis. In G Marko-Varga and P Oroszlan (eds), *Emerging Technologies in Protein and Genomic Material Analysis. Journal of Chromatography Library*, Vol. 68, pp. 118–122; 209–210. Elsevier Science, New York.

[22] Palmblad M, Bergquist J. (2003) Identification and characterization of peptides and proteins using Fourier Transform ion cyclotron resonance mass spectrometry. In G Marko-Varga and P Oroszlan (eds), *Emerging Technologies in Protein and Genomic Material Analysis. Journal of Chromatography Library*, Vol. 68, pp. 209–210. Elsevier Science, New York.

[23] Favret EA, Fuentes NO. (2003) RIMAPS characterization of etch pits on deformed aluminium surfaces. *Microsc Microanal* **9** (Suppl. 2): 766–767.

[24] Favret EA, Fuentes NO. (2003) RIMAPS detection of incipient damage on metallic surfaces. *Mater Characterization* **49**(5): 387–393.

[25] Favret EA, Fuentes NO. (2003) RIMAPS detection of underlying topography. *Practical Metallography* **40**(11): 556–563.

[26] Babadagli T, Develi K. (2003) Fractal characteristics of rocks fractured under tension. *Theoret Appl Fracture Mech* **39**: 73–88.

[27] Vázquez L, Vara JM, Herrasti P, Ocon P, Salvarezza RC, Arvia AJ. (1995) Methods of fractal analysis applied to STM imaging. *Chaos, Solitons Fractals* **6**: 569–573.

[28] Williams JM, Beebe TP Jr. (1993) Analysis of fractal surfaces using scanning probe microscopy and multiple-image variography. 1. Some general considerations. *J Phys Chem* **97**: 6249–6254.

[29] Favret EA, Fuentes NO, Álvarez MR. (2004) RIMAPS and Variogram analyses of microwear traces in experiments with stone tools. *Microsc Microanal* **10**(Supp. 2): 968–969.

[30] Favret EA, Fuentes NO, Molina AM. (2006) Quantitative study of *Xanthosoma violaceum* leaf surfaces using RIMAPS and Variogram techniques. *Microsc Res Tech* **69**: 684–688.

[31] Fuentes NO, Faybishenko BA. (2002) RIMAPS characterization of surface in the variable aperture model and determination of the main paths for water flow. *Proc XXIX Annual Meeting of AATN*, Buenos Aires, Argentina, **036**: 1–8.

[32] Fuentes NO, Faybishenko BA. (2003) Study of the fracture surface characteristics that determine the main paths for liquid flow and contaminant transport in fractured-porous media. *Proc VI Annual Meeting of SETAC – LA*, Buenos Aires, 20–23 October 2003, 179–180.

[33] Fuentes NO, Faybishenko BA. (2003) Evaluation of the conditions imposed by the fracture surface geometry on water seepage through fractured porous media. *Proc XXX Annual Meeting of AATN*, Buenos Aires, Argentina, **029**: 1–6.

[34] Fuentes NO, Faybishenko BA. (2004) RIMAPS and Variogram characterization of water flow paths on a fracture surface. *Proc of the 2nd Int Symposium "Dynamics of Fluids in Fractured Rock"*, Berkeley, CA, USA, 120–123.

[35] Fuentes NO, Faybishenko BA. (2005) Determination of the liquid flow paths imposed by the geometry in a fractured basalt rock Vadose zone. *Proc XXXII Annual Meeting of AATN*, Buenos Aires, Argentina, **015**: 1–7.

[36] Fuentes NO, Favret EA. (2006) Variogram characterization of length scales of topographic patterns. *Microsc Microanal* **12**(Supp. 2): 82–83.

[37] Favret EA, Fuentes NO, Fayou Y. (2004) RIMAPS and Variogram analysis of the surface topography induced by laser interference micropatterning. *Appl Surf Sci* **230**: 60–72.

[38] Barthlott W, Neinhuis C. (1997) Purity of the sacred lotus, or escape from contamination in biological surfaces. *Planta* **202:** 1–8.

[39] Neinhuis C, Barthlott W. (1997) Characterization and distribution of water-repellent, self-cleaning plant surfaces. *Annals Bot* **79**: 667–677.

[40] Wenzel RN. (1936) Surface roughness and contact angle. *Ind Eng Che* **28**: 988–994.

[41] Cassie ABD, Baxter S. (1944) Wettability of porous surfaces. *Trans Faraday Soc* **40**: 546–551.

[42] Shibuchi S, Onda T, Satoh N, Tsuji K. (1996) Super water-repellent surfaces resulting from fractal structure. *J Phys Chem* **100**: 19512–19517.

[43] Herminghaus S. (2000) Roughness-induced non-wetting. *Europhy Lett* **52**(2): 165–170.

[44] Seemann R, Mönch W, Herminghaus S. (2001) Liquid flow in wetting layers on rough substrates. *Europhys Lett* **55**: 698–704.

[45] Blossey R. (2003) Self-cleaning surfaces. Virtual realities. *Nature Mat* **2**: 301–306.

[46] Fresnais J, *et al.* (2004) Polyethylen ultrahydrophobic surface: synthesis and original properties. *Eur Phys J Appl Phys* **26**: 209–214.

[47] Hozumi A, Takai O. (1998) Preparation of silicon oxide films having a water-repellent layer by multi-step microwave plasma-enhanced chemical vapour deposition. *Thin Solid Films* **334**(1–2): 54–59.

[48] Löthman P, *et al.* (2005) Nanostructuring of monomolecular tetraetherlipid surfaces. *Proc of EUROMAT*, pp. 5–8.

[49] Zhang YM, *et al.* (2003) Surface analyses of micro-arc oxidized and hydrothermally treated titanium and effect on osteoblast behavior. *J Biomed Mat Res* **68A**(2): 381–391.

[50] FR-4Product Data (1999) GE electromaterials, *General Electric Company*.

[51] Vacuum bagging materials, tooling prepregs & accessories (2004) ISO 9002 Quality Systems, *Airtech International, Inc.*

[52] Favret EA, Fuentes NO, Molina AM, Setten L. (2008) Description and interpretation of the bracts epidermis of Gramineae (Poaceae) with rotated image with maximum average power spectrum (RIMAPS) technique. *Micron* **39**: 985–991.

Chapter 8

Capillary Phenomena

Gerardo Callegari and Adriana Calvo

Interfacial and capillary phenomena are present in multiple biological processes. Some examples are duck's feathers impermeability, spiders' sticky traps, and Lotus leaf's effect. The last subject is considered in a separate chapter due to its important technological applications. The basis to understand all those processes is the focus of the present chapter, divided into three subsections. The first one addresses the fundamentals of interfacial tension and wetting conditions as thermodynamical concepts. In the second, capillarity effects under dynamical conditions are considered. The third section is devoted to liquid films, their stability, and the spontaneous retraction in simple geometries.

1. Introduction

Several aspects of physico-chemistry of surfaces have big impact in biology. For example, insects and arachnids make use of capillarity and adhesive effects in several beneficial ways. Some of them exploit those features as a secure shield against their predators, this is the case of the beetle (*Hemisphaerota cyanea*) which utilizes only a small portion of the total pads to walk, but when it is under attack, it strongly sticks to the surface by activating all the pads at the same time.[1] In this way, the beetle can seize a force 60 times its weight for 2 min and up to 200 for shorter periods. This allows him to stand while predators, many times his weight, want to flip him to attack his vulnerable tummy, as can be seen in Fig. 1.

Other animals create powerful weapons against their preys: spiders are sticky traps master-builders. As an example, the Golden Orb Weaver spins a tough web which remains as a permanent fixture, after that she spins a spiral of sticky insect-ensnaring silk[2] and then she just waits and collects her reward. Many spiders wrap up their victims with silk, but they kill them by venom injection. There are some rare spiders

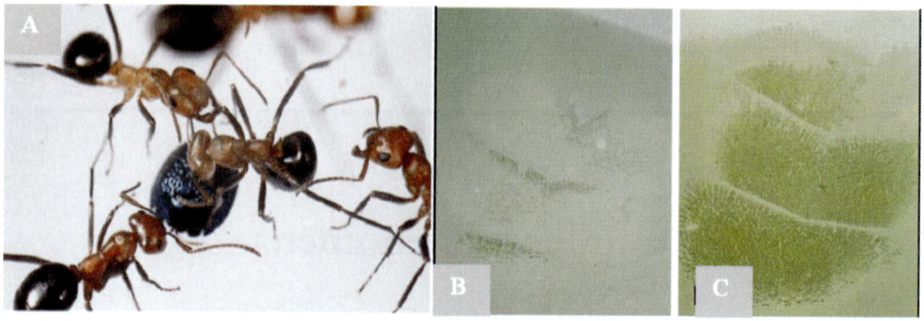

Fig. 1. (A) Beetle under attack by *F. Exsectoides* on a glass surface. (B) Tarsal contact during ordinary walking: only a few bristles touch the glass. (C) Tarsus pressed down flatly, in defensive response while the beetle is subjected to an electromagnetic pull.[1] Reprinted from Ref. 1 with permission of National Academy of Sciences. Copyright 2000, National Academy of Sciences, USA.

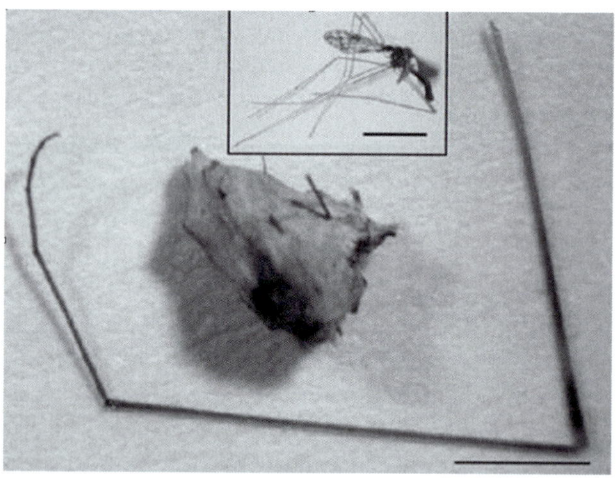

Fig. 2. A tipulid fly from which one leg was removed before it was wrapped and fed on by a *P. vicina* spider.[3] Reprinted from Ref. 1 with kind permission of Springer. Copyright 2006 Springer.

(like the Uloboridae) which does not have venom and wraps the prey with a big amount of silk.[3] The victims die due to strangulation while the silk is drying and contracting (Fig. 2).

For other animals, the use of capillary properties serves as a mean to walk on water surface (see Fig. 3). Gorb[4] wrote a review on how insects make use of adhesion in

Fig. 3. Common water-walking arthropods, ordered roughly by size. (a) Broad-shouldered water strider Microvelia, (b) water treader Mesovelia, (c) springtail *Anurida maritima*, (d) water measurer *Hydrometra stagnorum*, (e) water strider Gerris (f) fisher spider *Dolomedes triton*. Scale bars, 1 mm, (g) the fisher spider on the free surface (scale bar 1 cm), (h) hairs protrude from the leg and about the free surface, (i,j) scanning electron microscope images of a strider's leg. While the leg has thickness comparable to that of a human hair, it resembles a brush, whose thick carpet of hairs is tilted relative to the leg surface. Hairs are typically 30 mm long, 1–3 mm thick at the base and tapered; their density is 12,000–16,000 hairs mm^2, (j) a closer view of the hairs shows that their tips are bent inwards, towards the leg which discourages their piercing the water surface; moreover, each hair is patterned with grooves of characteristic width 400 nm that run its length.[5] Reprinted from Ref. 5 with permission of Elsevier. Copyright 2008, Elsevier.

different ways to get the same result: climb any surface at any angle. Most of the insects use wet adhesion: secreting a sticky liquid to assure their pads to almost any surface, but the most striking of all the ceiling walking creatures is the gecko.

The ability of a gecko to climb vertical walls and walk through ceilings was recently discovered to be due to adhesion originated in van der Waals attractions[6] which will be discussed in this chapter and also in Chap. 5.

Another example of application of interfacial physico-chemistry in biology is the wetting property of some natural surfaces as the super-hydrophobicity of the Lotus leafs, which will be discussed in detail in Chap. 3.

Other hydrophobic example in nature is duck's feather. Both preen oil and feather structure have been regarded as important for the ability of adult plumage to shed water droplets and resist water penetration while swimming.[7–9] Again a hierarchical structure barbs and barbules in the down coat makes the feather impermeable and prevents the skin of the duck to be in contact with water.[9] This not only helps ducks floatability, but also provides thermal insulation.

Last but not least, it is worth to mention the studies of cell adhesion to multiple surfaces. This is a broad topic and we cannot even refer to thousands of papers in the field. Here, we will only mention two reviews from the biochemical[10] and physical[11] points of view as well as a more recent review on the relation between wetting properties of implantable materials and its biocompatibility.[12] Biocompatibility of implantable material[13] is probably the hottest topics in material science with application in medicine.

All those phenomena mentioned above are strongly related to the wetting properties of real surfaces. In this chapter, we will present a short review on this extensive topic.

2. Wetting Properties: Surface Energy and Tension

Even when much effort was invested by the scientific community to understand wetting, surface energy and other apparently simple concepts, lots of questions still remain open in which we will discuss in this chapter. The exposition will be focused on some of those controversies with the hope to contribute to both the reader's knowledge acquisition and inquisition.

Both wetting and adhesion are manifestations of the attractive forces that exist between molecules. A molecule in the bulk of a phase suffers a zero net force exerted by their neighbors, but when a molecule is on the surface, the surrounding molecules exert an unbalanced net force which pulls the surface molecule towards the interior of the phase (Fig. 4). This net force is manifested as a surface tension. As some of the bonds present in the molecule when it is in the bulk of the phase are cut when it is

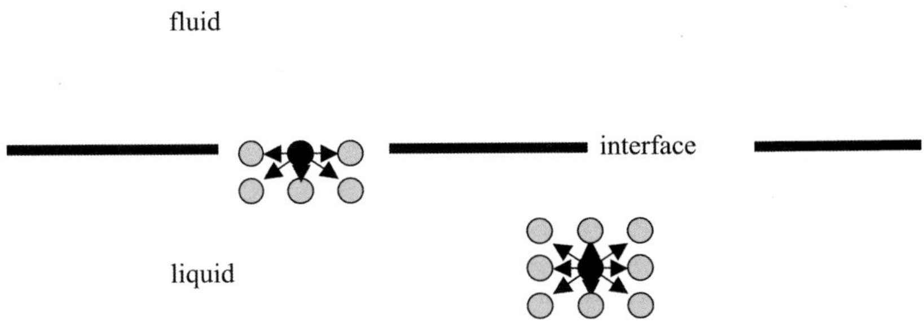

Fig. 4. Liquid–fluid interface. Diagram of the forces on a molecule of liquid.

placed at the interface, the energetic configuration changes. The surface energy, γ is defined as excess of free energy (per unit area) due to the existence of the interface.

2.1. *Molecular Interactions*

Molecular interactions can be classified into: non-polar (np), polar (p), and a third kind of forces called associative forces, which include Lewis acid basic interactions like hydrogen bonding (*ab*).

Non-polar forces arise from instantaneous dipoles produced by the motion of electrons within the molecule. Those interactions are called London dispersion forces.[14] Its variation with the distance between the centers of the molecules is well described by the negative of the gradient of the Lennard-Jones potential[15]

$$U_{12}(r) = A(r^{-6} - r^{-12}). \tag{1}$$

Polar interactions are divided into: Keesom forces,[16] arising from molecules with permanent dipoles and Debye forces[17,18] caused by a molecule with permanent dipole inducing a dipole in a neighboring molecule by polarization. Later, Good and Girifalco[19] showed that their contributions in condensed phases are not important. Nevertheless, they can also be well described by a Lennard-Jones' potential though mathematically associated with non-polar interactions.

We lack of a general description of the electron acceptor and electron donor abilities of the molecules (like in a hydrogen bond interaction) in terms of a potential energy. Considering that the exact quantum mechanics analysis of those interactions is an extremely complicated task, there is a need for simpler approaches. The application of those simple ideas is presented below, with experimental evidence of the problems that each of such approaches faces. It is likely that the complex nature of the interaction cannot be satisfactorily solved within a simple approach. Hydrogen bonds are perhaps the most important bond in biology: they are not only responsible for the particular properties of the most common substance in all biologic systems: water, but also give to proteins the particular structure which built up their specificity, key of their function. It is also worth to note that even when hydrogen bonds are hundreds of time weaker than metallic bonds, ropes made of spider silk are as strong as steel and most metals. This is because hydrogen bonds come in triplets or quadruplets and are not broken sequentially, but all at the same time.[2]

If the forces between two different molecules are "of similar type", the molecules behave as "indistinguishable" and they mix. Those substances in liquid state are miscible. If the forces are not "of a similar type", a phase separation will happen and the liquids are called immiscible.

Glycerol and water are examples of miscible liquids because both molecules express hydrogen bond interactions. Oils and fats are also miscible because they interact through dispersive forces. Of course, water–oil and water–fat are immiscible because the kind of interactions between dislike molecules is different from that of the like ones. The magnitude of the force in the water-phase direction is larger than in the opposite sense. As a result of this difference, there is a phase separation and a net interfacial tension.

2.2. *Adhesion and Cohesion*

Adhesion and cohesion are other manifestations of molecular interactions. The work of adhesion W_A, refers to the free energy difference between two states of the system: in the first one, phase 1 is in contact and in equilibrium with phase 2 (surface free energy per unit area: γ_{12}) (Fig. 5(a)) and in the second one, phase 1 and phase 2 are in contact and in equilibrium with their own vapors (surface free energy per unit area: $\gamma_1 + \gamma_2$) (Fig. 5(b)). Therefore, W_A is

$$W_A = \gamma_1 + \gamma_2 - \gamma_{12}, \tag{2}$$

relation known as Dupre equation.

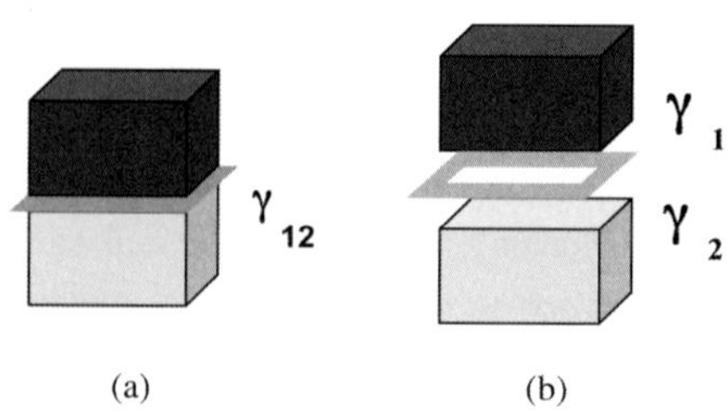

Fig. 5. Work of adhesion: (a) the two phases are in contact and (b) the two phases are separated.

Work of cohesion refers to the same difference when phase 1 is equal to phase 2. In that case, the first system has zero surface energy while for the second one it is

$$W_C = 2\gamma_1. \tag{3}$$

It is still a controversy if it is possible to directly determine the surface energy and work of adhesion of solid materials. Churaev and Sobolev,[20] for example, argue that due to the irreversible character of the separation of surfaces that is accompanied by the emission of energy during the rupture of chemical bonds, it is not possible to achieve those measurements. This point will be discussed later.

2.3. *Wettability and Contact Angle*

When a portion of liquid is placed in contact with an un-dissolvable, smooth, homogeneous, and un-deformable solid surface in the presence of a second fluid, the liquid can either spread on the solid surface or keep as a droplet on it. In this last situation, the equilibrium contact angle, θ_0, formed between the liquid/fluid interface and the solid (measured through the liquid) determines the wetting property of the system. The condition when the liquid completely spreads on the solid is known as *totally wetting* situation and corresponds to $\theta_0 = 0$. A *nonwetting situation* is achieved when the contact angle is larger than 90° and a *partially wetting* one, is intermediate: $0° < \theta_0 < 90°$.

The difference between the partial and the nonwetting situations will become clearer when discussing capillary effects. For now, we will just mention that the names are associated with what happens when a capillary or a porous material (the solid surface) is placed in contact with the liquid. In total and partially wetting cases, the liquid will spontaneously penetrate into the material while in the nonwetting situation the liquid will not penetrate without an external force.

The spreading coefficient (S) quantifies the wetting or spreading property of a solid–liquid–fluid system. It is defined as the difference in the Gibbs free energy per unit area between the two situations schematized in Fig. 6. In the left-hand side of the figure, the solid surface is only in contact with the environmental fluid: $G_1 = \gamma_{sf}$. In the right-hand side, a suitably thick liquid film covers the solid surface: $G_2 = \gamma_{lf} + \gamma_{sl}$, where γ_{sf} is the solid–fluid, γ_{lf} the liquid–fluid and γ_{sl} the solid–liquid interfacial energies of the system. The spreading coefficient is, then

$$S = G_1 - G_2 = \gamma_{sf} - \gamma_{lf} - \gamma_{sl}. \tag{4}$$

If $S > 0$ ($G_2 < G_1$) the second situation is energetically more favorable than the first one, i.e. the surface is "more comfortable" when it is covered by the liquid. Then, a liquid droplet deposited on the solid will spread until the liquid covers it

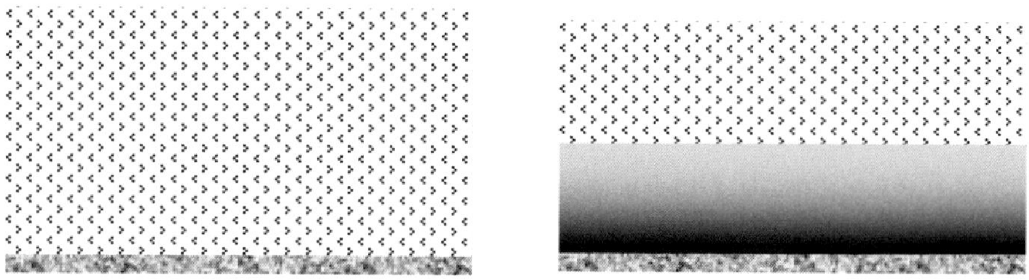

(a) (b)

Fig. 6. (a) The solid surface is in contact with the environmental fluid; $G_1 = \gamma_{sf}$; (b) an adequately thick liquid film covers the solid surface: $G_2 = \gamma_{lf} + \gamma_{sl}$. The spreading coefficient is $S = G_1 - G_2$.

completely. In the opposite case: $S < 0$, the droplet will remain covering a finite area of the solid.

It is worth to note that, when describing the second state, we raised the point that the liquid film should be thick "enough." This condition, safely fulfilled for thickness above 100 nm, is imposed to avoid the interaction of the solid molecules with the fluid through the liquid. If the film thickness is less than 100 nm the long range interaction between solid and fluid molecules cannot be neglected and another term known as disjoining pressure (π_{sf}),[21] should be added

$$\pi_{sf} = -\frac{1}{A}\left.\frac{\partial G}{\partial x}\right|_{A,T}, \tag{5}$$

where x is the film thickness, A the covered area, and T the temperature. This term vanishes for $x > 100$ nm and the sign is such that inhibits the spreading of the film. This means that if a liquid totally wets a solid ($S > 0$) it will not spread up to the monolayer scale because previously the spreading driving force per unit length will be compensated by the effects of the disjoining pressure. Indeed, the equilibrium thickness of the film can be calculated through the described balance.[22]

If $S < 0$, the most favorable situation for the solid is to be in contact with the environmental fluid. So, a liquid forced to cover it will "retract" to form a droplet with the shape of a spherical cup if gravitational effects are negligible. This phenomenon is known as dewetting and will be addressed later in this chapter.

Young[23] understood the interfacial tensions as forces per unit length acting parallel to each interface and reducing the corresponding contact area, as shown in Fig. 7.

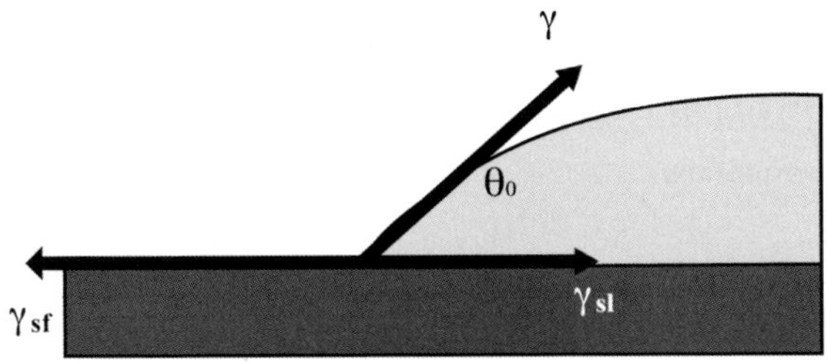

Fig. 7. Forces acting in the triple contact line.

The balance of the components of those forces parallel to the solid surface is called Young's equation

$$\gamma_{\mathrm{sf}} = \gamma \cos\theta_0 + \gamma_{\mathrm{sl}}. \tag{6}$$

Note that the angle θ_0 is not defined if $S > 0$; on the contrary, if $S < 0$, $\cos\theta_0 < 1$. The vertical component of the force γ is balanced by the elastic force in the solid producing a deformation in it which even when small, always exists. In some cases, this deformation of the solid can influence the behavior of the contact line in dynamic situations (and even contribute to the hysteresis of the contact angle, as will be discussed later). Another feature that was not taken into account in Young's equation is that, even with "ideal" solid surfaces, at the triple contact line (TCL) the three interfacial tensions may be influenced by the presence of the third phase due to its effect on the interaction between the other two phases. This possibility was recognized by Gibbs who suggested that this three-phase mutual interaction must be accounted for by a "line tension".[24] The value of the line tension has been a controversial issue for many years; however, it is clear now that it may affect the contact angle of drops only when they are extremely small.[25]

2.4. *Liquid vs Solid Surface Energy: Measurement Techniques*

It was shortly noticed before that measuring the surface energy of a solid is a challenging task. We will broadly discuss here this point and compare the solid with the liquid case.

Liquid–fluid interfacial tension can be directly measured because the interfaces can be deformed through the application of small (and measurable) pressures in bulk and liquids can retain a high degree of mobility at the molecular level. Thus, liquids have the ability to spontaneously reorganize at the surface to minimize the total surface free energy and present a reproducible, smooth, and well-defined surface configuration.

When a liquid–fluid interface is curved, the pressure on both sides is not the same: it is higher in the fluid situated on the concave side. The mean curvature of the interface (C) is related with the pressure difference (ΔP) and the surface tension (γ) through Laplace equation

$$\Delta P = \gamma C. \tag{7}$$

The mean curvature is a mathematical description of the deformation of a surface and is defined as the sum of the inverse of the maximum and minimum radii of curvature of the surface in a given point. For all points of a spherical surface, both radii are equal to the sphere radius R and $C = 2/R$; for a cylindrical surface $C = 1/R$ (in all points, the maximum radius is infinite and the minimum is the radius of the cylinder, R), and for a plane $C = 0$.

There are two most classical methods to measure interfacial tensions of liquids: to generate liquid droplets in the presence of another fluid and to measure the force necessary to introduce or to extract a solid of a liquid bath. There are multiple variations for both of them.

The advantage of the first method is that there is no need of an "external" solid material to probe the liquid–fluid interface. On the other hand, techniques based on the second idea are normally easier to implement and one of them (see below Wilhelmy method) entails no corrections and can be used to measure contact angles on solid surfaces, as well.

The two most used approaches of the droplet technique are described below.

2.4.1. *Drop weight or drop detachment*

The method is based on Tate's law[26]

$$W = 2\pi R\gamma, \tag{8}$$

where W is the weight of the droplet, R the radius of the tip from where the droplet detaches, and γ the interfacial tension. In 1919, Harkins and Brown[27] correlated and measured a correction factor which is a function of $R/V^{1/3}$ where V is the volume of the detaching droplet

$$V(\rho_1 - \rho_2)g = f\left(\frac{R}{V^{1/3}}\right) 2\pi R\gamma, \tag{9}$$

where ρ_1 and ρ_2 are the densities of the heavy and light phases, respectively, and g is the gravitational acceleration. They report correction factor values in the range $0.3 < R/V^{1/3} < 1.6$. Polynomial approximations for different ranges of $R/V^{1/3}$ were proposed for practical applications of the above expression.[28,29]

Even when the determination of the interfacial tension through this method is empirical, the results obtained proved to be very well correlated with the ones obtained by other measuring techniques as sphere and cylinder tensiometry.[30] The main factors restricting the general applicability of the drop detachment technique are the lack of sufficient experimental drop correction factor data for $R/V^{1/3} < 0.25$ and situations in which the volume of satellite droplets, formed during the pinch off process, cannot be ignored when compared with that of the detached drop. This situation is typical for liquid–liquid systems exhibiting low interfacial tension.

2.4.2. *Shape of the droplet*

It is usually called axisymmetric drop shape analysis.[31] The interfacial tension and contact angles are determined from the shape of the axisymmetric menisci of both sessile and pendant drops. The employed strategy is to fit the shape of an experimental drop to the theoretical drop profile according to the Laplace equation, using surface tension as adjustable parameter. Details of the methodology together with a program to implement it can be found elsewhere.[31]

As stated above, the other most used methods to determine interfacial tension are based on measuring the force necessary to introduce or to extract a solid of a liquid bath. Normally, the material used is platinum but the particular technique depends on the shape of the solid. Ring and fiber or plate cases are described below.

2.4.3. *Ring*

This method was developed by du Nuoy[32] and is based on the balance between the force needed to detach a ring from a liquid–fluid interface and the capillary force that the liquid exerts on the ring. Equipments that use this principle are commercially available. Care should be taken of the circularity and planarity of the ring as well as the straight mounting for a perfect coplanar detachment. However, there are some corrections to perform to the classic formula

$$F = f\gamma P, \tag{10}$$

where F is the measured force, P the mean perimeter of the ring, and f a correction factor.

2.4.4. *Fiber or plate*

Developed by Wilhelmy in 1863[33], this method involves measuring the force on a plate or a fiber while they are immersed in or withdrawn from a liquid. Zero contact angle must be ensured through intensive cleaning of the solid. The advantage is that it allows continuous monitoring of the force $F(t)$ and thus the result does not rely on only one point but on a curve. This curve should be a straight line with the slope related to the buoyancy effect of the solid object.

The Wilhelmy technique is simple and reliable and it has another important advantage: it also allows measurement of contact angles when γ is known. As will be seen below the most widespread methodology to determine the surface energy of a solid, which is not an easy task, is by measuring its contact angle with different liquid–fluid systems and using Young and/or Dupre equations.

Solid interfaces cannot be deformed as liquid–fluid ones, therefore, it is not possible to measure directly the interfacial force with techniques similar to those just described. The stillness of solid molecules is also responsible for the apparition of defects (voids, inclusions, dislocations, impurities, etc.) which can induce local tensile and compressive stress fields that alter the surface energy of solids. This topic will be discussed in more detail in the next subsection.

A different approach yielding values of work of adhesion involves sensitive measurements of detachment force and contact radius when two surfaces are carefully moved towards and apart from another. The instrument is known as surface force apparatus (SFA) and was first developed by Tabor and Winterton[34] for direct measurement of van der Waals forces between molecularly smooth surfaces. Further improvements were carried on by Israelachvili and Tabor.[35]

It is noticeable that when only dispersive forces are present both Young and SFA techniques give similar results (for example, in polyethylene).[36] But when polar molecules are present, polyethylene terephthalate (PET) is an example, the results varies.[36]

2.5. *Components*

So, in principle, it is possible to calculate the surface energy of a solid. Indeed, molecular interactions have been modeled and the density function (average position of molecules) can be either measured or assumed and the results averaged and extended. However, some researchers went on further: they expressed the work of cohesion and adhesion as a sum of components associated with different types of forces acting across an interface and calculate the interfacial energies using the components of the surface energy of each phase. In doing that, they are assuming that the internal configuration of molecules (electronic clouds energies, distance and angular position of atoms, etc.) is not affected by the mutual interaction, which is not valid in general.

Fowkes[37,38] was the first to propose this approach in the 60s by writing the total interaction as having three components: the Lifshitz–van der Waals dispersive (d), the polar (p), and the hydrogen bond (h) terms. The expression for the surface tension given by work of cohesion is

$$\gamma = \frac{W_c}{2} = \gamma^d + \gamma^p + \gamma^h. \tag{11}$$

Then, he used Berthelot's rule[39] for each type of interaction

$$\gamma^i = \sqrt{\gamma^i \gamma^i}, \tag{12}$$

and he proposed to combine the polar (*p*) and hydrogen bond (*h*) interactions into the acid–base component, γ^{ab}. So

$$\gamma = \sqrt{\gamma^d \gamma^d} + \sqrt{\gamma^{ab} \gamma^{ab}}. \tag{13}$$

Twenty years later, Van Oss and coworkers[40] proposed that the polar component γ^p could be combined into the dispersive term and that the acid–base components could be calculated as

$$\gamma^{ab} = \sqrt{\gamma^+ \gamma^-} + \sqrt{\gamma^- \gamma^+}, \tag{14}$$

where γ^- is the surface acidic or hydrogen-bond-donating parameter and γ^+ is the surface basic or hydrogen-bond-accepting parameter. This proposition is known as VOCG approach. So, the total surface energy is

$$\gamma = \gamma^d + 2\sqrt{\gamma^+ \gamma^-}. \tag{15}$$

It is worthy to note that in the last expression γ^d is now the non-polar dispersion component originally defined by Fowkes plus the polar terms (Keesom and Debye components).

The same idea was also applied to the work of adhesion of two materials, which gives

$$W_a = \gamma_1 + \gamma_2 - \gamma_{12} = 2\sqrt{\gamma_1^d \gamma_2^d} + 2\sqrt{\gamma_1^+ \gamma_2^-} + 2\sqrt{\gamma_1^- \gamma_2^+}. \tag{16}$$

Using this expression for water–hydrocarbons interfacial tension (γ_{wH}) the dispersive surface energy component of water may be found. As basic components for hydrocarbons should be zero, the dispersive component is the surface tension of the hydrocarbon. Measuring γ_H, γ_w, and γ_{wH} the only unknown is γ_w^d which turns to be

$$\gamma_w^d = 21.8\,\mathrm{mN/m}. \tag{17}$$

With the experimental value for the surface tension of water (72.8 mN/m), using Eq. (15) and assuming that water has the same ability to donate than to accept electrons, so the components are equal, the other components were calculated

$$\gamma_w^+ = \gamma_w^- = 25.5\,\text{mN/m}. \tag{18}$$

The same methodology was proposed to obtain the components and the total surface energy of solids from the following equation:

$$W_a = \gamma_1(1+\cos\theta_1) = 2\sqrt{\gamma_s^d\gamma_l^d} + 2\sqrt{\gamma_l^+\gamma_s^-} + 2\sqrt{\gamma_l^-\gamma_s^+}, \tag{19}$$

and using three different liquids, due to the existence of three unknowns for the solid components.

In the late 90s, Kwok and collaborators[41] showed that the method did not give consistent values of the surface free energy components using different sets of liquids for the same tested solid. Kwok and coworkers[41] recommended the use of the so-called "equation of state".[42] But this last approach has also been criticized.[43]

Della Volpe and Siboni[44] refuted the VOCG components model using Kwok's own experimental values. Della Volpe and Siboni argued that the determination of the solid surface energy components by the VOCG model could be ill-conditioned if any set of liquids were used to measure the contact angle with the solid. For example, if only pure dispersive liquids are employed, the accurate result of the acidic or basic components will never be obtained. The same would happen if liquids with almost no basic components were used.

But this was not the only critique to the VOCG acid–base model. The method seems to consistently give higher basic than acidic components. Moreover, for some polymers like cellulose, PVA, PMMA, and PVC,[45] which are believe to be acidic in the surface, the method gave much larger basic than acidic components. Lee[46] and Della Volpe and Silboni[44] also proposed to vary the acidic–basic scale by changing the relation between γ_w^+ and γ_w^-. This suggestion smoothens the problem a bit, but does not solve it totally.

As a conclusion: there is a good agreement between all the methods to characterize the surface energy of a solid surface when only dispersion forces are present (in the sense of VOCG model: with the weak polar interactions added). When either the liquid or the solid (or both) have a strong polar component (hydrogen bond, for example), disagreement among the results are evident. This result may suggests that hydrogen-bond like interactions are still not very well described by VOCG model. Moreover, some inconsistencies appear between the acidic or basic character of certain solids and the component values obtained with the same approach.

2.6. *Hysteresis of the Contact Angle*

Up to this point, only the thermodynamic contact angle (defined by Young's equation) was discussed. In real systems, a droplet laying in a solid surface may have a different shape from a spherical cup and display different local contact angles along the TCL at the same time. Moreover, adding liquid to a still droplet may not necessary mean a movement of the TCL. The fact that several different contact angles are possible in the stationary situation is called hysteresis of the contact angle.

Two critical contact angles can be experimentally determined, the larger one is called advancing contact angle θ_a: above this value, the contact line advances (the liquid through which the angle is being measured displaces the other fluid). The smallest one is known as the receding contact angle θ_r: below this value the liquid through which the angle is being measured recedes displaced by the external fluid. Hysteresis is defined in two different ways, as: $\Delta = \theta_a - \theta_r$ and as: $\Delta = \cos\theta_r - \theta_a$, while the first is a pure definition, the second one has some physical sense in terms of surface energies differences.

Contact angle hysteresis has been extensively studied for the last two decades due to the technological interest in super-hydrophobic systems. Classical approaches, experimental evidence, and understandings of the topic will be presented in the following sections. Much more will be discussed on this subject in Chap. 3 where the Lotus effect will be described.

2.6.1. *Classical approach: roughness and chemical heterogeneities*

The phenomenon of the hysteresis of the contact angle has been mostly associated with physical or chemical heterogeneities on the solid surface or with impurities in any of the components of the system. It was thought that Young's equation was still applicable locally; this statement was only proved for particular situations, for example: choosing the shape of the liquid–fluid interface as a spherical cup, but not in general due to the mathematical difficulty to describe all the possible states of the system.

Wenzel[47] extended Young's equation to describe rough surfaces when the area covered by liquid is much larger than the roughness scale and if the liquid completely penetrates into the roughness grooves, obtaining

$$\cos\theta_w = R\cos\theta_m, \tag{20}$$

where θ_m and θ_w are the Young and the apparent contact angles of a drop on a smooth and a rough surface, respectively, and $R > 1$ is the Wenzel roughness factor defined as the ratio of the total surface area to the geometrically projected area.

On the other hand, Cassie and Baxter[48] considered a heterogeneous surface made up of two species with area fractions ϕ_1 and $\phi_2 = 1 - \phi_1$ and contact angles θ_1 and θ_2. When the individual areas are much smaller than the droplet size, they found that the apparent contact angle (θ_{C-B}) is given by

$$\cos\theta_{C-B} = \phi_1 \cos\theta_1 + (1 - \phi_1)\cos\theta_2. \tag{21}$$

Equation (21) can be also applied to a homogeneous and rough solid surface when the liquid of the droplet does not penetrate into the grooves and the base of the droplet is in contact with the solid (with contact angle θ_1) and with the air trapped in the grooves (with contact angle $\theta_2 = 180°$). The transition between this and Wenzel's cases is discussed by Lafuma and Quere.[49] These approaches are discussed in Chap. 3.

2.6.2. *Metastable configurations*

Some theoretical studies have investigated the structure of the mechanical energy functional of ideal three-phase systems on surfaces with ordered patterns of wettability.[50–53] Energy barriers in that functional arise from the heterogeneity of the surface and trap the contact line in metastable positions. This gives rise to contact angle hysteresis.

Neumann and Good[51] developed a contact angle model for smooth, but heterogeneous vertical plates; they demonstrated the existence of a large number of metastable states.

A fundamental study by Huh and Mason[54] also revealed the effect of surface roughness on wetting. They considered a liquid drop growing or retracting concentrically on a solid surface with axisymmetric grooves.

Marmur and coworkers[53] proposed a model for two-dimensional (cylindrical) drops on a periodically heterogeneous but smooth solid surface. Their calculations revealed a possibility of the dependence of advancing and receding contact angles on the drop volume.

2.6.3. *Vibrations and the global energy minimum (GEM)*

Johnson and Dettre[55] produced roughness elements in the micrometer range on paraffin wax and found that the hysteresis depends on roughness and on the applying energy when the liquid used for wetting and dewetting experiments was water. They acknowledged that the systems behave as composite surfaces similar to those observed by Cassie and Baxter.[48] They also suggested that the advancing and receding contact angles might converge to a common value if sufficient energy were supplied to overcome the energy barriers between the metastable states. This was demonstrated by Stokes *et al.*,[56] by Andrieu *et al.*[57] for drops on heterogeneous surfaces, and by Decker *et al.*[58] for vertical heterogeneous plates penetrating into a liquid. The latter two groups applied vibrations to supply the necessary energy. Marmur and coworkers[59] assume that this common value may be the contact angle corresponding to the global energy minimum (GEM is defined as the lowest energy out of all possible metastable states[60]). To assess the validity of this assumption, an indication that results from a mathematical analysis of the Wenzel equation may be used. It was proven that when the droplet is sufficiently large compared with the roughness scale (which is a necessary condition for the validity of the Wenzel equation), it becomes axisymmetric as it reaches the GEM.[61] Marmur and coworkers[59] suggested that even when the opposite statement has not been proven, a practical working hypothesis may be made: when a large droplet on a rough surface becomes round following vibrations, it is at the global minimum in energy. He tested this idea on surfaces made of the same material, but with different roughness. All those systems showed a practically identical ideal contact angles, as calculated from the Wenzel equation.

2.6.4. *Other sources of hysteresis: smooth and homogeneous surfaces*

Progress made in the past few years with respect to the preparation and characterization of solid surfaces has facilitated the fabrication of extremely smooth and homogeneous surfaces. Even though, remarkable hysteresis (up to 35°) of water on some low energy polymers was observed,[62] where contact angles are larger than 90°. Polymeric smooth and homogeneous surfaces have been useful in identifying other causes for hysteresis: liquid absorption and/or retention (swelling) on smooth surfaces of polymer films[63]; molecular rearrangements or reorientation occurring at solid–liquid interfaces.[62,64,65] In this line, the work done by Rangwalla and coworkers[66] who explored the configuration of polymeric brushes when in contact with water and in air using IR–visible sum frequency generation spectroscopy (SFG) has to be remarked. Their analysis indicated that in contact with water, the polymer surface is heterogeneous, showing areas of different surface energies. The degree of heterogeneity showed by SFG was in accordance with hysteresis of the contact angle measured with the sessile droplet technique on the different substrates. Churaev and Sobolev[20] suggested also that the vertical component of Young's force can locally deform or at least stress the solid under the contact line, causing microhills[67] (in the first case) or faults (in the second) which can also lead to hysteretic behavior.

3. Capillarity

The term capillarity is used to designate the physical process by which a liquid penetrates into a single capillary tube or in multiple pores in a porous material. A pore does not necessarily has to be a closed tube: the space between three or even two parallel rods or fibers is also considered as a pore[68]; indeed, this is the kind of pore normally found in fibrous materials.[69–71] So, it is not the cross-sectional shape but the size which determines the definition of a capillary tube: its radius must be smaller than the capillary length $\kappa^{-1} = \sqrt{\frac{\gamma}{\rho g}}$, which is the length scale above which gravitational exceeds capillary forces. As a result, if all the pores inside a sponge have a radius larger than κ^{-1} (for water it is 2.7 mm), liquid will never climb through them and only the pores below the liquid level will be filled. On the other hand, if some pores are capillaries (their radius below the capillary length), liquid is wicked "by capillarity" into the part of the sponge which is above the liquid level.

Consider a vertical plate partially immersed in a wetting liquid bath in Fig. 8. The liquid spreads on the solid surface and climbs through the vertical wall. The interface curves and a pressure difference which obeys Laplace law is built up across it. It can be shown that the maximum height of the liquid is: $h = \sqrt{\frac{2\gamma}{\rho g}} = \sqrt{2}\kappa^{-1}$ when the liquid totally wets the solid.

This result is obtained by balancing pressures in points M and N of Fig. 8. In equilibrium the pressure in N is equal to the pressure in the outer fluid, P_0, and the pressure in M is $\rho g h + \Delta P + P_0$ where ΔP is the Laplace pressure difference. The order of magnitude can be obtained by assuming that the meniscus mean radius of curvature is of the order of κ^{-1}, the pressure difference is $\Delta P \cong -\frac{\gamma}{\kappa^{-1}}$ and $h \cong \kappa^{-1}$.

When the liquid is partially wetting ($0 < \theta \leq \pi/2$) the maximum height is smaller and can be written as: $h = \sqrt{2}\kappa^{-1}(1 - \sin\theta)^{1/2}$ (22). When $\theta = \pi/2$, $h = 0$ and the

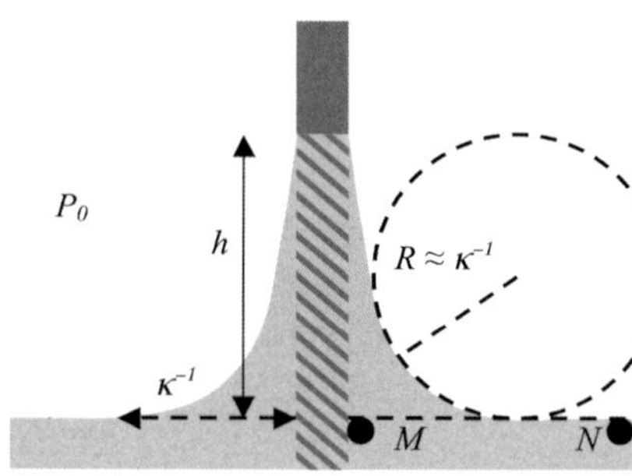

Fig. 8. Capillary rise in a plane surface.

free surface of the liquid remains plane, if $\theta > \pi/2$ the liquid does not climb on the solid but sinks on it. The reason why the liquid is called partially wetting in the first case is now obvious: it behaves as a wetting liquid in the sense that it spontaneously climbs through walls and as we see below, it also penetrates into capillaries without being forced.

If two vertical plates are now immersed in a wetting liquid bath, the distance d between them will define the rise height. If d is larger than κ^{-1} the menisci do not "interact" and the situation is as depicted in Fig. 9(a): the curvature in point N is still zero, its pressure is P_0 and the rise height is the same as before.

Differently, if d is smaller than κ^{-1}, the interface at N will have a curvature with radius $R \approx d/2$ (see Fig. 9(b)). The pressure at that point will initially be smaller than P_0 and in consequence, smaller than the pressure in point M. Equilibrium is only achieved by some extra pressure in point N, so, the meniscus should climb a certain height H, called Jurin's length, until the pressures in both points become equal. At that state, H must satisfy: $\rho g H \cong \frac{2\gamma}{d}$ and increases as the distance between plates decreases. In fact, if $d \gg \kappa^{-1}$, the capillary rise between the plates is much larger than the outside one: $H \gg h$.

If instead of two plates a cylindrical tube of diameter d is immersed in the same liquid, the meniscus will take the shape of a spherical cup of radius $d/2$. So, the Laplace pressure difference across the interface is $\frac{4\gamma}{d}$ and the capillary rise inside the

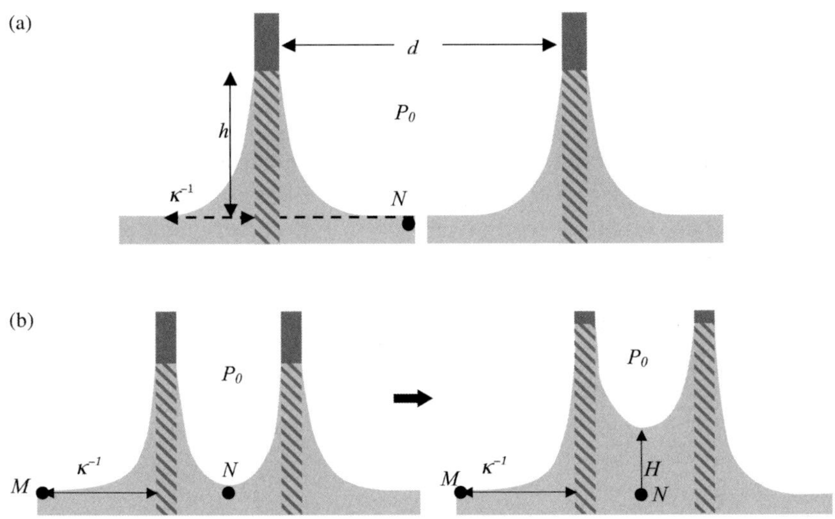

Fig. 9. Capillary rise in two vertical plates immersed in a totally wetting liquid when the distance between them is: (a) larger (b) smaller, than the capillary length.

tube gives: $h = \frac{4\gamma}{d\rho g}$. If the liquid partially wets the tube, the meniscus radius will be $\frac{d}{2\cos\theta}$ and $h = \frac{4\gamma\cos\theta}{d\rho g}$.

Note that up to this point nothing has been said on the dynamics of the processes: only static configurations have been taken into account. In fact, in all the above expressions, the static contact angle θ is the advancing one if the dry solids have been immersed in the liquid, but the receding contact angle must be used if the solids have been extracted out of the liquid bath.

The dynamics of capillary rise, as the description of any flow involving an interface, is very complicated and there is no full agreement on how to solve the problems that arise when a TCL is in motion. This topic will be discussed later when dealing with the problem of dynamic contact angle. Before going deeper in details, it must be pointed out that many approximations work very well in most of cases.

In 1917, Lucas[72] determined the position of the front of liquid being wicked by a porous material as a function of time. He found that it propagates as $t^{1/2}$; a behavior which was previously reported by Bell and Cameron.[73] A few years later, in 1921, Washburn[74] explained Lucas' observations assuming: (a) that the porous material can be seen as a system of parallel capillaries; (b) that the flow in each capillary tube is stationary and axisymmetric; and (c) that the flow is well described by a Poiseuille profile with the pressure difference across the interface given by the Laplace equation. Under these conditions, the position of the meniscus inside a capillary tube varies with time as

$$x = \sqrt{\frac{\gamma R\cos\theta}{2\eta}t}, \tag{22}$$

where R is the radius of the capillary tube and η the dynamic viscosity of the liquid. Although the model needs strong assumptions, it has two robust advantages: its simplicity and that it reproduces most of the available experimental data.

In 1923, Bosanquet[75] pointed out that in Lucas–Washburn equation the velocity goes to infinity for $t = 0$. This physically impossible situation was a result of neglecting inertial effects. He described the motion of the meniscus with the following differential equation (obtained from the one-dimensional Navier–Stokes equations)

$$\frac{d}{dt}\left(R^2\rho x\frac{dx}{dt}\right) + 8\eta x\frac{dx}{dt} = 2\gamma R\cos\theta, \tag{23}$$

where the first term on the left side is due to inertia, the second one takes into account the viscous effect (and gives the Pouseuille flow) and the term on the right side is the driving capillary force: $2\pi R(\gamma_{SV} - \gamma_{SL}) = 2\pi R\gamma\cos\theta$.

For long times, the inertial term disappears and the equation is exactly the one solved by Washburn. For short times, the second derivative term and the viscous one can be neglected and the solution is a constant velocity

$$V_B = \sqrt{\frac{2\gamma \cos\theta}{\rho R}}, \tag{24}$$

called Bosanquet velocity. Times short enough to check this behavior were not accessible to the experimental capabilities at that time, so, Bosanquet's result was seen as a mathematical curiosity. Its importance was only recently considered.[76–82]

On the other hand, Szekely[83] and Levine and coworkers[84] made modifications to the original Washburn approach by introducing the appropriate momentum balance equations and considering the deviation from Poiseuille flow, taking into account entrance effects. However, their equations considered the existence of a single and invariable equilibrium contact angle during the entire rate of capillary rise and made no account of inertial forces.

Siegel, in 1961,[85] was the first to conduct experiments in micro-gravitational conditions. He considered both inertial and dynamic contact angle effects. Many other works were conducted in that condition since this seminal paper. A review is given in Ref. 86.

The first experimental data on normal gravitational conditions where inertial effects are shown is probably the work by Jeje.[87] He did his experiments with thin capillary tubes and recorded them with high-speed camera (1000 fps, an amazing rate for the moment).

Many other authors showed results on capillary rise in tubes were Washburn equations did not match their results. An example is given in Fig. 10. Explanations about anomalous behavior are based on the dynamic contact angle[88–91] and/or on inertial terms.[76–82]

A very simple equation was proposed by Quere[77] following the same approach as Bosanquet

$$\frac{d}{dt}\left(R^2 \rho (x + cR)\frac{dx}{dt}\right) + 8\eta x \frac{dx}{dt} = 2\gamma R \cos\theta. \tag{25}$$

Through the coefficient c, he included the effect of the acceleration of a portion of the liquid to enter into the capillary. θ is not the static, but the dynamic contact angle. In that way, this term is not the driving force. It is a kind of "effective" driving force which includes the dissipation in the vicinity of the contact line.

Hamraoui and coworkers[90] maintained the static contact angle in the right side of the equation (preserving the meaning of that term: capillary driving force)

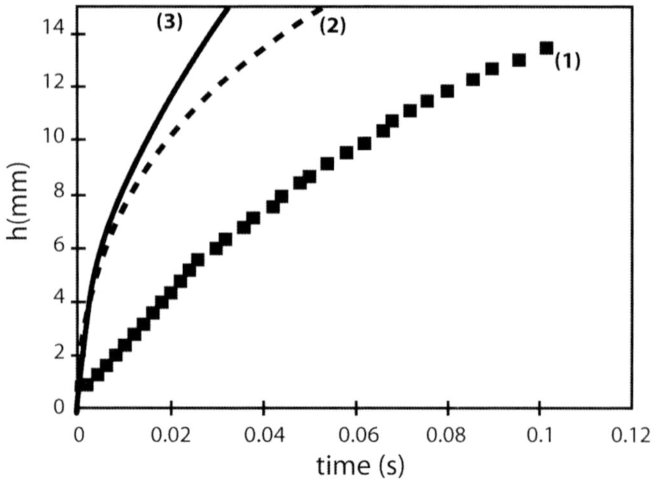

Fig. 10. Variation of the liquid height with time for pentane in a capillary tube: (1) experimental points; (2) Lucas–Washburn equation taking into account the hydrostatic pressure; (3) Lucas–Washburn equation neglecting the hydrostatic pressure. From Ref. 91.

and introduced a friction term which takes into account this dissipation in the contact line.

Finally, Bazilevsky and coworkers[92] modified Quere's equation to take into account fluid elasticity caused by stretching of polymer molecules by velocity gradients when using polymeric solutions of polymers with high molecular mass. The stretching of polymers in the channel gives rise to an extra stress due to Weissenberg effect,[93] which was introduced into the equation by calculation within the Upper Convected Maxwell Model. They found that the constant velocity in the inertial-elastic regime depends nonmonotonally on the radius of the capillary tube (as in Newtonian fluids) displaying a maximum value for a given capillary radius which depends on the elastic and viscous properties of the liquid. This model was partially corroborated by later experiments[81,82] in tubes of different diameters and DNA solutions. The authors claim that this simple method can be used to characterize rheologically DNA solutions and differentiate between single and double stranded DNA, with great importance in biochemical genetics. Some of their results are shown in Fig. 11.

3.1. *Dynamic Contact Angle*

As discussed in the last section, the problem of an interface in motion is not trivial, it is still a matter of huge debate and so, it deserves special attention. The first point to

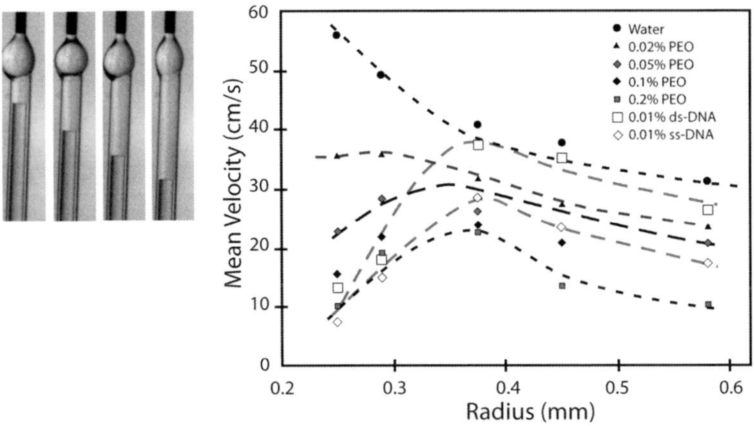

Fig. 11. Left: Sequence of images showing the penetration of 0.01% TE Buffer solution (8 pH) of λ-DNA ($M = 61 \times 10^6$ Da) into a glass capillary, ID = 0.58 mm. The time among pictures is 10 ms. Right: Mean velocity as a function of the capillary radius for different polymeric solutions.[81,82]

note is that when a TCL begins to move on the solid surface, the contact angle does not maintain its static value, neither the advancing nor the receding one, but it adopts a value called dynamic contact angle, which depends strongly on the velocity. To avoid confusion, from now on the advancing and receding static values of the contact angle will be named θ_a^s and θ_r^s, respectively, and θ will be only used for the dynamic one.

The first systematic experimental study on the dynamic contact angle was carried out by Ablett[94] in 1923. His conclusion was that the variation of θ with the TCL velocity is monotonic. Innumerable works have been performed after Ablett's in all different kind of systems and setups: spontaneous spreading of liquid droplets on solid surfaces[95–98]; displacing of a meniscus inside capillary tubes either in forced flows at constant velocity[99–102] or spontaneously at varying velocity[77,78]; immersion in (or withdrawn from) a liquid bath of solid plates or fibers.[103–114] An example of how the dynamic contact angle changes with velocity is shown in Fig. 12: the spherical meniscus between air (on the left of each image) and glycerol is forced to move at different velocities inside a capillary tube.[102]

Theories from two extremely different points of view have been proposed to determine the relation between θ, θ^s, and velocity U. They differ in the way the friction is calculated. In hydrodynamic theories,[112,115–119] the dissipation is assumed to take place very close to the TCL mainly through viscous forces; on the other hand, in the molecular model proposed by Blake[109,120] the dissipation is mostly associated with molecules hopping at the molecular edge of the contact line.

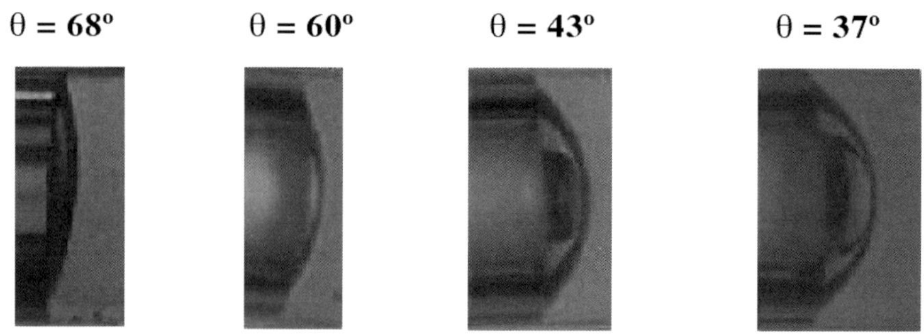

Fig. 12. Air (on the left of each image) displacing glycerol inside a capillary tube. Velocity increases from left to right: $U = 4.3 \times 10^{-6}$; 1.6×10^{-5}; 1.6×10^{-4}, and 3×10^{-4} cm/s.[102]

Although those theories were born as antagonists, some authors[121–123] prefer to give a chance for reconciliation and understand the whole problem of the dynamic contact line as a combination of factors. Within this modern picture, both contributions are complementary and one does not exclude the other. Moreover, Petrov and Petrov[123] understood that they can be simply added and the resultant model is a combination of the molecular and hydrodynamic dissipation. This will be discussed in the final part of the present subsection.

3.1.1. *Hydrodynamic model*

It has been more than 35 years since Huh and Scriven[115] established the classical hydrodynamic assumptions for Newtonian and incompressible liquids drives to nonintegrable singularities in the shear stress tensor evaluated on the TCL[112,115] when considering the nonslip condition. Several ideas were proposed to overcome the failure of the classical assumptions. Most of them consider that within a distance Li of the contact line (called the inner region), the liquid slips at the solid surface[115,116,124] in order to remove the singularity. The most successful and general attempt was proposed by Cox,[119] his strategy takes into account the dominant forces acting in three different length scales which characterize the flow in those three regions (see Fig. 13).

He solved the equations and matched the asymptotic expansions for each of the regions at velocities of the TCL small enough to neglect inertial and viscous effects in comparison with capillary ones. This last condition is insured if the nondimensional capillary number defined as $Ca = \mu U/\gamma$, is much smaller than 1.

The description of the regions is as follows: in the outer region, the viscous effects are negligible in comparison with capillary and gravitational ones; the shape of the

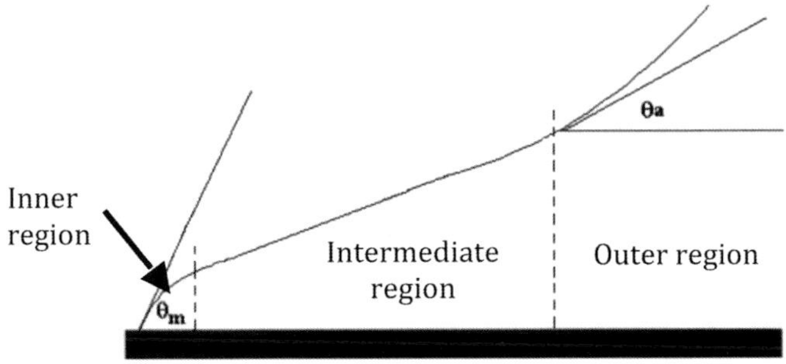

Fig. 13. Moving triple contact line, three different regions characterize the flow.

interface depends on the geometry and it is the same as that in the static situation. It was experimentally shown by Ngan and Dussan[125] that the macroscopic characteristic length is associated with capillarity and gravity. They explored the forced displacement of silicon oil against air in glass slots and measured the variation of the contact angle with velocity and with the spacing between slides (100–1200 μm). They showed that the dynamic contact angle measured as a macroscopic entity (θ_a in Fig. 13) varies with geometry. Nevertheless, Petrov and Sedev[126] based on their own and on other author experiments, stated that geometry is more important for the critical velocity at advancing rather than at receding interfaces.

In the intermediate region, the viscous forces which arise primarily from the local flow field near the moving contact line, become the dominant balance for the surface tension forces. Thus, the flow field and interface shape are independent of the overall system geometry, which has been experimentally corroborated by several works (127–129).

Finally, in the inner region characterized by a microscopic length, the classic hydrodynamics collapses and a new physics governs fluid behaviors. A microscopic dynamic angle is defined (θ_m in Fig. 13).

The physical mechanism ruling this inner region is still an unsolved problem. The theoretical treatments after Huh and Scriven agree on relaxing the no-slip boundary condition,[54,84,112,116,118,119,130–132] but several different conditions have been essayed and the physical basis are still in debate.[133–136]

From the pioneering works in molecular dynamics simulations by Koplik *et al.*[137] and Thompson and Robbins,[138] numerous studies showed results compatible with the existence of a region of a few atomic radii in length where the no-slip condition is violated. Another model was proposed by Dussan in 1976[139] with a caterpillar like movement of the portion of fluid in the region very close to the TCL, in a way that a material spot will advance up to reach the TCL and is finally deposited on the solid surface. In this model, it is not necessary to impose slipping on the TCL because no singularity appears. Six years later, Ngan and Dussan[125] found a procedure to obtain solutions where the exact identification of the physical laws governing the fluid dynamics in the inner region is not required. This method was improved by Dussan, Rame, Garoff, and coworkers in several papers.[125,127–129]

Cox's model[119] relates the dynamic contact angle θ with the static contact angle θ^s, the capillary number Ca, and the fluids viscosity ratio $\lambda = \frac{\mu_{\text{displaced fluid}}}{\mu_{\text{displacing fluid}}}$, in the following way:

$$F(\theta, \lambda) = F(\theta^s, \lambda) + Ca[Ln(K/ls) + Qe], \tag{26}$$

where K is a macroscopic distance (the outer region length scale), ls is the microscopic length up to which the slipping process is allowed, Qe is the dissipation in the inner region and

$$F(\theta, \lambda) = \int_0^\theta \frac{\lambda(x^2 - \text{sen}^2 x)((\pi - x) + \text{sen}x \cos x) + ((\pi - x)^2 + \text{sen}^2 x)(x - \text{sen}x \cos x)}{2\text{sen}x[\lambda^2(x^2 - \text{sen}^2 x) + 2\lambda(x(\pi - x) + \text{sen}^2 x) + (\pi - x)^2 - \text{sen}^2 x)]} dx. \tag{27}$$

For small values of θ and when the motion of the TCL is such that a liquid displaces a gas ($\lambda = 0$), $F(\theta, \lambda)$ tends to θ^3 and the well-known Tanner's law is obtained[95]

$$\theta \alpha Ca^{1/3}. \tag{28}$$

On the other hand, Hoffman[99] experimentally studied the dynamic contact angle — interface velocity relation using different liquids and a large range of velocities. He obtained results compatible with Cox model, seen in Fig. 14.[101]

This agreement between Cox model and experimental data obtained with systems with zero or close to zero static contact angle, is due to the fact that viscous effects are well determined in that case and the factor Qe (molecular dissipation in the very vicinity of the contact line) is negligible compared to the factor $\text{Ln}(K/ls) \sim 14$.

However, the comparison with experimental data for partial or nonwetting situations often gives unreasonable values for ls.[101,102,110,140–142] As an example, Fig. 15 shows data from Ref. 141 where a glycerol/water–air interface moves on a PET surface: It shows good agreement with the cox model (dotted line in figure) in the intermediate region which deviates at the extremes. The authors reported that the value of ls obtained in the fitting was smaller than molecular dimensions, result that has no physical sense.

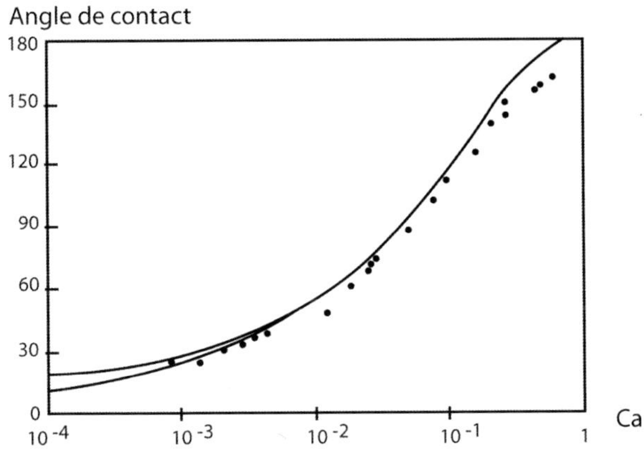

Fig. 14. Variation of the contact angle with the capillary number for silicon oil displacing air. Experimental data were obtained by Hoffman[99] and the solid lines correspond to Cox model.[119] From Ref. 101.

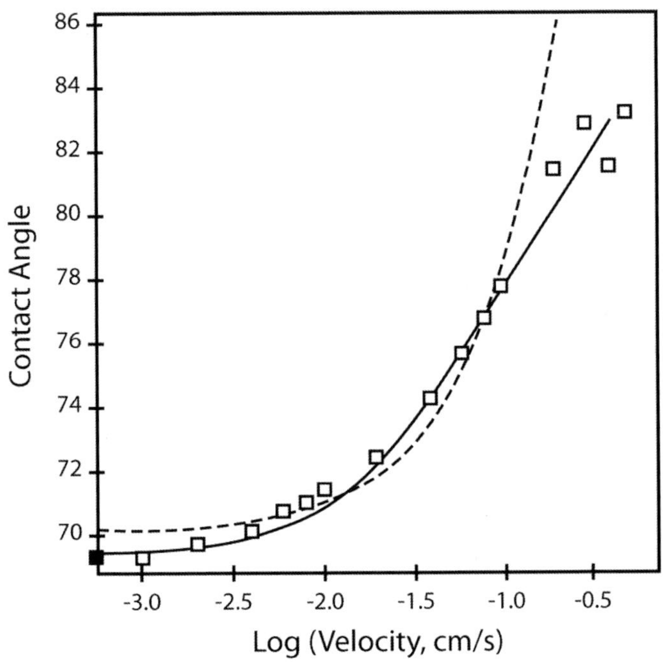

Fig. 15. Variation of the contact angle with the velocity for glycerol/water displacing air on a PET surface. Solid line corresponds to molecular-kinetic theory and dotted line to Cox model.[141]

Cazabat and coworkers considered that the explanation[143] is again given by the *Qe* factor calculated with molecular dynamic simulations by Thompson and Robbins.[138] Using Lennard-Jones potential and $\theta^s = 90°$, they got values as large as 6, meaning that for polar liquids (stronger interactions) *Qe* can be even much larger, so comparable (or larger) than the factor $\mathrm{Ln}(K/ls)$.

Pointing in a different direction but accommodating dissipation through the standard hydrodynamic channels, Shikhmurzaev[144–147] proposed a model based on the analysis of the relaxation of the interfacial tensions while approaching the TCL. To his understanding, the same flux causes a gradient in the interfacial tensions which drastically influence the region in the vicinity of the TCL and also removes the singularity of the moving TCL without any ad hoc assumption.

An important point of the model is that it proposes that as a liquid–gas interface advances on a solid surface, some liquid elements at the liquid–gas interface become transferred to the solid–liquid interface. This flux has been experimentally confirmed[112] and was not taken into account in the conventional model which has, instead, a stagnation point.

In this motion, some diffusive reorganization of the molecules in the interfacial region is required which needs a finite time to be completed. So, an interfacial tension gradient is built up extending a distance $U\tau$ from the contact line (where τ is a characteristic relaxation time). The reorganization also provides a disturbance in the surface tensions from their equilibrium values and the balance of tensions is also modified leading to a nonequilibrium value for the local contact angle.

Even when some authors[110,148] concluded that their experimental results suggest a support for Shikmurzaev model, a major difficulty in its validation is that it introduces two additional phenomenological coefficients linking the flow to the surface tension gradient and the shear stress at the interface. It is not easy to see how these coefficients can be determined *a priori*.

For example, Blake and coworkers[110] experimentally demonstrated that at least for high velocity regimes, the dynamic contact angle does not depend only on the velocity of the interface but on the entire flow. By changing the impinging velocity of a jet towards a plate being drawn from a liquid bath, they found different dynamic contact angles for the same velocity. Later on, by exploring the shape of a dynamic meniscus in the proximity of the TCL (up to 20 mm), Rame and coworkers[148] found that θ and/or *ls* should depend on the velocity and not through the capillary number. This mean that another timescale should be introduced in the system which is compatible with the apparition of a relaxation time as described by Shikmurzaev. However, the new timescale may be also associated with the hoping of molecules in the interface as described by molecular models.

3.2. Molecular Approach

In the molecular dissipation model,[109,149] an ideal solid surface is considered with randomly distributed mono-energetic adsorption sites at a mean distance λ. At equilibrium, the molecules at the contact line oscillate between the sites at the solid–gas and the solid–liquid interfaces with frequency

$$k = \frac{k_B T}{h} e^{-(\Delta G_w / k_B T)}, \tag{29}$$

determined by the activation free energy ΔG_w and the absolute temperature T, k_B, and h are the Boltzmann and Planck constants, respectively.

Under dynamic conditions, when the TCL moves, the net frequency is $k = k^+ - k^-$, where k^+ and k^- are the number of jumps per second in the direction of motion and in the opposite direction

$$k^{\pm} = \frac{k_B T}{h} e^{-\left(\frac{\Delta G_w \mp 0.5 F \lambda^2}{k_B T}\right)}, \tag{30}$$

where F is the unbalanced capillary force per unit line driving the motion: $\gamma(\cos\theta^s - \cos\theta)$ and $F\lambda^2/2$ is the energy associated to the jump. The mean velocity of a unit length of the contact line $U = k\lambda$ is then given by

$$U(\theta) = 2\lambda \frac{k_B T}{h} e^{(-\Delta G_w / N k_B T)} \sinh\left(\frac{\gamma(\cos\theta^s - \cos\theta)\lambda^2}{2 k_B T}\right), \tag{31}$$

where N is the Avogadro number. Some years later, Blake improved his model[149] considering that the hopping frequency is the result of two processes: first the molecules have to reach the contact line and then jump to adjacent places. So, ΔG_w can be written as

$$\Delta G_w = \Delta G_v + \Delta G_s, \tag{32}$$

where ΔG_s takes into account solid–liquid molecular interactions and ΔG_v corresponds to fluid–fluid molecules interactions. According to Eyring's model of viscosity, the first term can be related to the liquid viscosity through

$$\mu = \frac{h}{v_L} e^{-\left(\frac{\Delta G_v}{N k_B T}\right)}, \tag{33}$$

where v_L is the volume of matter performing the jump. For simple molecules, v_L is the volume of the molecule. For polymers, it can be a sub-ensemble, possibly a monomer.

Finally, the equation for the velocity reads

$$U(\theta) = 2\lambda \frac{k_B T}{\mu v_L} e^{-(\Delta G_s / N k_B T)} \sinh\left(\frac{\gamma(\cos\theta_e - \cos\theta)\lambda^2}{2k_B T}\right), \quad (34)$$

which can be simplified to

$$U(\theta) = A \sinh B(C - \cos\theta), \quad (35)$$

with

$$A = 2\lambda \frac{k_B T}{\mu v_L} e^{-(\Delta G_s / N k_B T)},$$

$$B = \frac{\gamma\lambda^2}{2k_B T}, \quad C = \cos\theta^s.$$

More recently,[150] it has been suggested that there is an approximate link between ΔG_s and the liquid–solid work of adhesion $W_a = \gamma(1 + \cos\theta^s)$: $\Delta G_s = W_a\lambda^2 N$. This leads to

$$A = 2\lambda \frac{k_B T}{\mu v_L} e^{-\frac{\lambda^2\gamma(1+\cos\theta^s)}{k_B T}}. \quad (36)$$

Note that when θ^s decreases, that is, when the wetting is better, A decreases while C increases, so, this approximation implies that there may be an optimum wettability to maximize the rate of wetting. Experimental evidence of this fact as well of the relation between the work of adhesion and ΔG_S can be found in the literature.[150]

Contrary to the hydrodynamic approach, the molecular model describes the experimental results in partially wetting (and liquid–liquid) systems (see as example, the fitting in Fig. 15) with physically reasonable adjustable parameters. However, it fails to predict Tanner's law for complete wetting situation.

The other drawback of the model is that there is no independent way to measure the quantities λ and ΔG for a given solid–liquid system and are thus treated as adjustable parameters which can be only qualitatively compared with the parameters obtained for other systems. Of course, the order of magnitude of at least λ is well defined and should be in the molecular size level.

To summarize this short review on dynamic contact angle, it is worth to note that there is a fourth point of view which cannot be considered as a new model because it was born as a combination of the two most popular models. Petrov and Petrov[140] formulated the idea that both wetting line frictions (molecular and hydrodynamic) can be taken into account simply by combining the equations for $U(\theta)$ resulting from both models. In this way, an equation with three adjustable parameters, λ, A, and ln (K/ls) is obtained. Of course, curve-fitting with this equation proved very successful, usually giving better agreement with experimental data than each of the equations alone, especially for receding wetting lines and small contact angles. In addition, the values of the parameters obtained from the analyses appeared reasonable and the problems of each model were overcome. The only remaining concern is that the parameters were increased and it is still not possible to measure them independently.

4. Liquid Films

The technological processes in which it is desired to cover a solid surface with liquid films of very different thicknesses are numerous: impressions and lithographies; painting or covering of metallic plates; production of photographic or X rays films and papers; manufacture of adhesive tape, of optical and magnetic storage media (photography and audio compact discs or tapes), of fabrics, of fibers and cables, among others. It is then understood the huge effort that has been dedicated to the study of the formation, stability, and dynamics of such films.

4.1. *Film Formation*

Coating a solid surface with a liquid film presents difficulties when $S < 0$ (partial and nonwetting conditions) and a thin film is desired, it will be necessary to force the liquid to spread and eventually, to avoid the subsequent retraction of the film. The precise meaning of "thin film" will be explained in Sec. 3.2.

With the first aim: to spread the liquid on the solid, different technological processes use the viscous force exerted when the liquid moves with respect to the solid. The geometrical configuration used to produce the flow determines the name of the process: dip, slot, slide, curtain, knife or blade, and roll coating are some examples, see Ref. 151 for a complete review. Other forces can also be used like centrifugal (spin coating) and electrical (electro-coating). The geometry of the solid often plays a role in the selection of the covering technique, for example, when the surface to be coated is the external surface of a fiber or a rod or the internal surface of a tube, the most commonly used techniques are dip and slug coating for external and internal covering, respectively.

In any case, for partially wetting systems, the difference between the liquid and solid velocities should exceed a critical value (V_c)[152] in order to obtain a liquid film deposited on the solid surface. That critical velocity V_c defines a critical capillary number $Ca_c = \mu V_c/\gamma$ which depends on the receding static contact angle of the system (θ_r^s).

Besides, when $Ca < 10^{-3}$ and $Ca < Ca_c$, the macroscopic shape of the interface is not distorted by viscous effects[153] and keeps the static configuration but with different

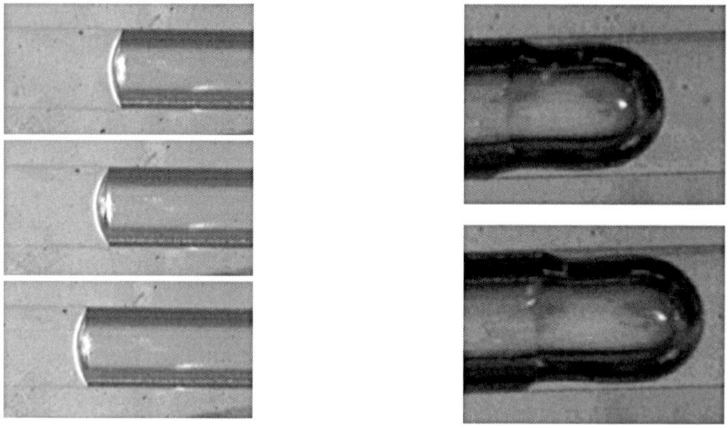

(a) (b)

Fig. 16. A water–glycerol solution (A, in the left and B in the right) is pushed by air inside a PVC capillary tube: (a) $Ca = 2 \times 10^{-4}$, (b) $Ca = 1.6 \times 10^{-3}$. $Ca_c \approx 2 \times 10^{-3}$.[153]

apparent contact angle that was previously discussed. In Fig. 16(a), an example of this situation is shown: an aqueous solution of glycerol (at the left side of the image) is pushed by air inside a PVC capillary tube. On the contrary, if $10^{-3} < Ca < Ca_c$ the macroscopic meniscus has an inflexion point as seen in Fig. 16(b).

In this situation, called "bride's tail",[153] the TCL moves with almost the same velocity as the center of the meniscus. So, the meniscus profile, although highly distorted with respect to the spherical static shape, does not vary with time.

If the relative velocity between the liquid and the solid is increased and $Ca > Ca_c$, the TCL cannot hold the meniscus velocity any more and a macroscopic film of constant thickness h, develops behind the meniscus. Several works have been devoted to study the dependence of h with the velocity in multiple different geometries and in diverse systems. When inertial and gravitational effects are negligible ($Ca < 10^{-2}$), the results can be summarized in a simple equation

$$\frac{h}{a} = \alpha Ca^{\beta}, \tag{37}$$

where the parameter a depends on the geometry: it is the liquid capillary length $\kappa^{-1} = \sqrt{\frac{\gamma}{\rho g}}$ in the planar case and the tube/fiber radius in the cylindrical one, and α and β are fitting parameters, in experimental works;[154–159] or resulting parameters, in theoretical works.[160–162]

Except for the works of Fairbrother and Stubbs and Taylor[154,156] (which give $\alpha = 0.5$, $\beta = 0.5$), most of the works in the area give $\alpha = 0.9 - 1.34$

and $\beta = 0.6 - 0.67$.[155,157–159,161] This general result was also obtained in systems where both fluids are liquids[153,158,162] and for values of the capillary number up to 10^{-1}.[157,158,161,162]

Another feature of cylindrical geometry is the observation of a limiting maximum coating thickness equal to $h/a \cong 0.35$[153,154] theoretically explained by Reinelt and Saffman.[163]

In the past 15 years, many works[164–166] have studied the variation of the thickness with the velocity for larger values of *Ca* when inertial and/or gravitational effects are non-negligible. Experiments were performed in nonconfined systems like plates and fibers being extracted from a liquid bath. Results show that h continues increasing with velocity, but at a slower pace than when viscous effects are dominant and that it tends to an asymptotic value in which cylindrical geometry was found both experimental[153,154] and theoretically,[163] to be $h/a \cong 0.35$.

4.2. *Stability Criteria*

Once a liquid film is formed, the most important issue is whether the film will stay as a film or will broke into droplets. Consider a solid surface of total area A_T, part of which (A) is covered by a liquid film of uniform thickness, h. The free Helmholtz energy of the system is[22]

$$F(h) = A(\gamma_{\mathrm{sl}} + \gamma + P_m(h) + \frac{1}{2}\rho g h^2) + (A_T - A)\gamma_{\mathrm{sf}}, \tag{38}$$

where $P_m(h)$ takes into account molecular interactions and, $\rho g h^2/2$ gravitational effects. If the thickness tends to zero $F(h)$ should tend to $A_T\gamma_{\mathrm{sf}}$, so

$$P_m(0) = \gamma_{\mathrm{sf}} - (\gamma_{\mathrm{sl}} + \gamma) = S. \tag{39}$$

If $0 < h < 10$ nm, gravitational effects are negligible and the short and long range molecular interactions are dominant. While short range interactions (as was previously discussed there is no general description for these forces) are only important for few molecular layers, long range, dipole–dipole van der Waals interactions are dominant in any other case (electrostatic interaction should be added when ions are present in the system) and $P_m(h)$ may be written as

$$P_m(h) = -\frac{A_H}{12\pi h^2}, \tag{40}$$

where A_H is the Hamaker's constant of the system and can be negative or positive.

If the thickness of the film is between 10 nm and 1 μm, both gravitational and molecular effects can be neglected

$$F(h) = A(\gamma_{\mathrm{sl}} + \gamma) + (A_T - A)\gamma_{\mathrm{sf}}. \tag{41}$$

Finally, molecular effects are negligible if $h > 1\ \mu$m

$$F(h) = A\left(\gamma_{\rm sl} + \gamma\frac{1}{2}\rho g h^2\right) + (A_T - A)\gamma_{\rm sf}. \tag{42}$$

4.2.1. *Nanoscopic films ($h < 10$ nm) and spinodal dewetting*

For nanoscopic films ($h < 10$ nm) two independent constants appear in the equilibrium equation: S and A_H and both can be negative or positive. Depending on their signs, the equation has solution or not and the behavior of the liquid film will be very different.

When both constants are positive, the equation has solutions, so, an equilibrium thickness h_c exists and the film spreads on the solid up to that value.[22] If the film is forced to a thickness below h_c, it will brake spontaneously[167,168] into micro-droplets, this process is known as spinodal decomposition.

On the contrary, if they are both negative, there is no solution for the equilibrium equation, so a film cannot coexist with dry solid zones; it retracts increasing its thickness until it becomes macroscopic.

In the other two cases: $(S > 0, A < 0)$ and $(S < 0, A > 0)$ the equation has solution; so, an equilibrium thickness h_c exists.

Besides, if $S > 0$ and $A < 0$ the curve $F(h)$ has a minimum in a value $h_m > h_c$ which means that a film of thickness $h > h_m$ will spread on the solid up to h_m only if it covers all the surface (with no coexistence of film and dry zones). If not enough liquid is available, the film will continue spreading up to the equilibrium value h_c. This condition is called pseudo-partial wetting.

Finally, for $S < 0$ and $A > 0$ $h_c \approx 10$ nm, that is, in the upper limit of the molecular interactions validity. So, if $h < h_c$ spinodal dewetting described above will occur. When $h > h_c$, the film will also brake and retract (because $S < 0$) but as now both molecular interactions and gravitational effects are negligible, the free energy is independent of the thickness.

4.2.2. *Macroscopic films*

For thickness larger than 1 μm the equilibrium equation leads to

$$h_c = \sqrt{-\frac{2S}{\rho g}}. \tag{43}$$

For the total wetting case $(S > 0)$, there is no equilibrium thickness for the film, that is, it cannot coexist with dry solid surface regions, so, it will spread covering the whole area or up to the point in which the molecular forces start to play a role.

On the contrary for partial and nonwetting conditions ($S < 0$)

$$h_c = \sqrt{\frac{2|S|}{\rho g}} = \sqrt{\frac{2\gamma|\cos\theta^s - 1|}{\rho g}} = 2\kappa^{-1}\sin\frac{\theta^s}{2}. \tag{44}$$

A film with $h > h_c$ is stable or, in other words, it can coexist with dry zones of the solid surface. But if the film is forced to adopt a thickness smaller than h_c, it will retract or dewet until its thickness becomes equal or larger than h_c.[169,170]

4.3. *Dewetting of Planar Films*

Dewetting of macroscopic films is a process very frequently found in everyday life; for example, the film that covers a duck when it leaves a water pond, immediately dewets forming big droplets which fall by gravity. In that way, the duck looks dry and clean.

The dynamics in this regime is known as nucleation dewetting and has been studied in different systems and configurations for relatively thick liquid films (from micrometers to hundreds of micrometers).[57,153,159,167–174]

Redon and coworkers[170] were the first to study the dewetting process of macroscopic films deposited on planar and horizontal surfaces. They used different alkanes and poly (dimethylsiloxane) (PDMS) on silanized silicon wafers. In all their studies, inertial effects were negligible against viscous ones. The film is extended on a circular surface and fixed in a perimetral wetting ring. Dewetting is initiated by producing a hole at the center of the film with a jet of air.

They observed that while the central dry zone increases, a bump is built up between the receding contact line and the liquid film; the latter remains static and the receding contact line moves at constant dewetting velocity, V_d. They investigate the dependence of this velocity with the different parameters of the system. The main results are that V_d does not depend on the film thickness (for $h \ll h_c$) and that for viscous and non-polar liquids and small static contact angles (up to 50°) the dewetting capillary number $Ca_d = \mu V_d/\gamma$ scales as the cube of θ^s while the prefactor varies weakly with the studied system. This result, that resembles Tanner's law, was explained[170] by means of a simple hydrodynamic model that assumes a circular cross section for the bump and symmetrical dissipation at both of it ends. Following later observations of asymmetries in the bump's profile[171] this last assumption was modified.

Recent experimental results were obtained by Callegari and coworkers[169] using the same geometry and experimental procedure, but with highly polar liquids (water and glycerol) and PVC solid surfaces which display static contact angles up to 80°

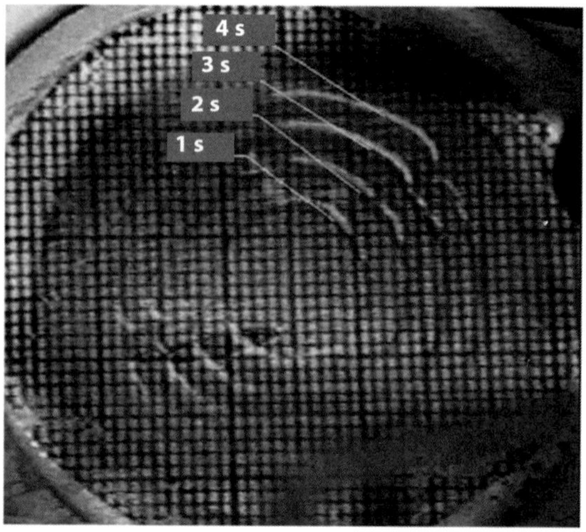

Fig. 17. Dewetting of a water–glycerol solution planar film on a PVC surface. Four successive pictures of an experiment are superimposed. The rim along the receding contact line is clearly visible.[169]

(much higher than in the other works). Four successive pictures of an experiment are superimposed in Fig. 17.

These results confirm two of the previous ones: the presence of a growing bump in the border of the dewetting film and that dewetting velocity does not vary with time nor with the film thickness. However, they disagree with the prediction $Ca_d \propto (\theta^s)^3$ as shown in Fig. 18 were both sets of results are displayed.

While data from Ref. 170 (black symbols) display a linear dependence, that is not the case for results from Ref. 169 (open markers). Callegari *et al.*[102,159] showed that this different behavior may be due to the different role of hydrodynamical and molecular effects in the solid–liquid–fluid systems used in both set of experiments: the first are preponderant in the systems used in Ref. 170 while the latter are more important in those of Ref. 169. As stated previously, this explanation agrees with the picture mostly accepted nowadays in the field.

All the results obtained to this point correspond to liquid films with thickness $h \ll h_c$; as stated, for these films, the dewetting capillary number does not depend on h. But when h becomes comparable to the critical value h_c, gravitational effects decrease the dewetting velocity in the following way

$$\frac{Ca_d}{Ca_{d0}} = 1 - \left(\frac{h}{h_c}\right)^2, \tag{45}$$

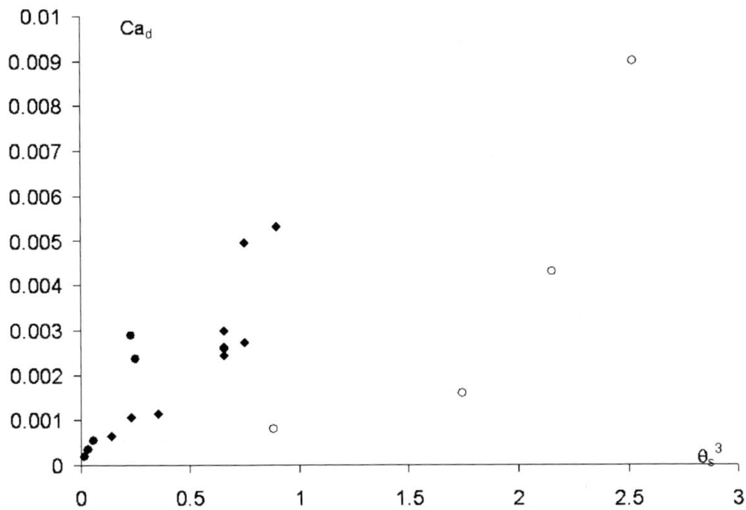

Fig. 18. Variations of Ca_d with $(\theta^s)^3$ for alkanes (black circles) and PDMS with different molecular masses on silanized silicon wafers (black diamonds) from Ref. 170; glycerol–water solutions on PVC surfaces (open symbols) from Ref. 169.

where Ca_{d0} is the constant value corresponding to $h \ll h_c$. This relationship has been plotted as a continuous line in Fig. 19 and provides a very good fit to the experimental results found in Ref. 169 except for those obtained with water for which inertial effects cannot be neglected.

When dewetting occurs at velocities so high that inertial effects cannot be neglected, viscous effects are negligible and a simple analysis on the balance of capillary and inertial forces gives the relation between the dewetting velocity and the film thickness[57,169]

$$V_d = \sqrt{\frac{\rho g(h_c^2 - h^2)}{2\rho h}}. \tag{46}$$

For $h \ll h_c V_d$ tends to $\sqrt{-S/\rho h}$ which makes a big difference with the viscous regime: the dewetting velocity do not have an asymptotic value for small thicknesses but diverges as $\sqrt{1/h}$. Andrieu *et al.*[57] were the first to observe this phenomenon with water dewetting on PET foils. They found that the dewetting velocity is independent of time (as in the viscous/molecular regime) and that superficial waves travel ahead of the bump. Their experimental variation of V_d with h as well as that shown in Fig. 19 with black symbols, agree with the above theoretical expression, but with prefactors smaller than 1.

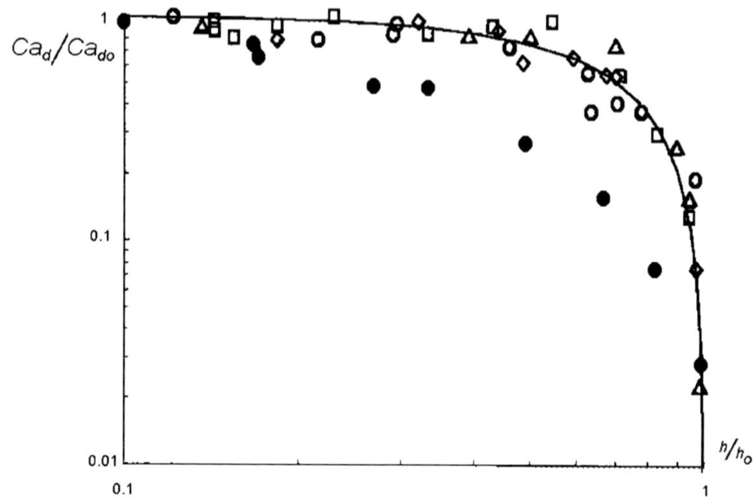

Fig. 19. Variation of the normalized dewetting capillary number with the film thickness normalized by the experimental critical thickness. Open markers: different glycerol–water solutions; black circles: pure water; continuous line: theoretical prediction. From Ref. 169.

A phase diagram allows to characterize the dewetting process[57,169] as a function of two nondimensional numbers defined as

$$RE = \frac{hV_d}{\nu\frac{L_n}{\theta}}, \qquad RL = \frac{L}{h\frac{L_n}{\theta}},$$

where $\nu = \mu/\rho$ is the liquid kinematic viscosity, L is the bump's length, and $L_n = \ln(K/ls)$, being K and ls the macroscopic and microscopic lengths used in the hydrodynamical model describing a moving contact line. Figure 20 shows the phase diagram together with the experimental results from Ref. 169. As observed, almost all data corresponding to pure water experiments (squares) fall in the inertial region.

4.4. *Cylindrical Films*

Films covering cylindrical substrates, either a fiber or the internal surface of a tube, is frequent in many practical domains such as lubrication, surface coating, oil recovery, chemical engineering, analytical or biochemistry, and biology. In addition, from the fundamental point of view, one may expect differences with respect to the planar geometry as will be seen below.

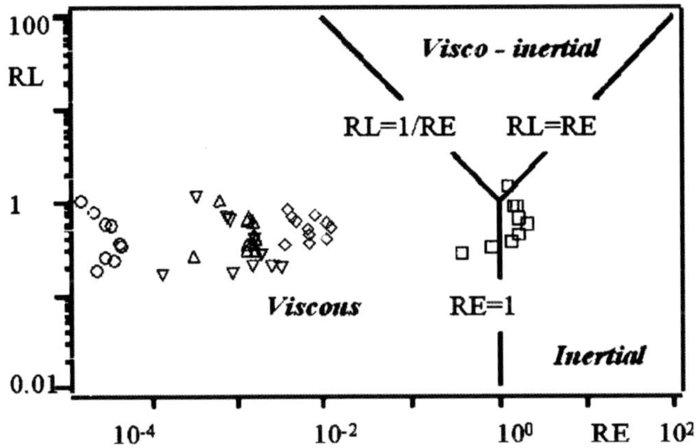

Fig. 20. Phase diagram for dewetting processes, the equations defining the different regions are indicated in the figure.[169]

4.4.1. *Rayleigh instability*

When a fiber is covered with a liquid film of constant thickness, *h*, the area of the liquid–air interface is larger than if the same amount of liquid is distributed in a film of sinusoidal profile with a wave length larger than the radius of the fiber.[175,176] So, a constant thickness film is unstable and spontaneously evolves towards a minimum energy configuration. In Fig. 21, an example is shown: a spider covers the threads

Fig. 21. Rayleigh instability in a spider web.[177] Reprint with permission of the author.

of its web with a thin film of adhesive liquid which naturally evolves to an array of almost equi-spaced droplets separated a distance larger than the radius of the fiber. If the liquid partially wets the solid, the droplets remain isolated; if it is totally wetting, a thin film will link adjacent droplets.[22] The same instability appears if the interior of a capillary tube is covered by an annular film.

Analyzing the Laplace pressure difference in a film sinusoidally perturbed (given by the two principal radii of curvature) it results that the pressure in a valley is larger than in a crest if the wavelength of the perturbation (λ) is larger than $2\pi a$, where a is the fiber radius. In this case, the liquid flows from the valley to the crest and the perturbation is increased. A first order instability analysis shows that the fastest growing perturbation has a wavelength

$$\lambda = 2\pi\sqrt{2}a. \tag{47}$$

Rayleigh instability has been studied in great details for wetting systems: Goren (1962),[178] Goldsmith and Mason (1963),[179] Hammond (1983),[180] Gaulitz and Radke (1987),[181] Aul and Olbrich (1990).[182] Much less efforts have been put in studying partially wetting systems. In particular, only recently the competition between Rayleigh instability and dewetting has been analyzed.[173]

When the film is covering the interior wall of a capillary tube and its thickness is large enough,[182] the crests collapse producing droplets. If the liquid partially wets the solid, the thin film linking the droplets breaks and dewets (see Fig. 22).

On the other hand, when at least one of the extremes of the annular film is free, it can retract before Rayleigh instability appears. Which of the processes occur first depend on its characteristic times, for Rayleigh instability it is[173]:

$$\tau_R = \frac{12\mu a^4}{\gamma h^3}, \tag{48}$$

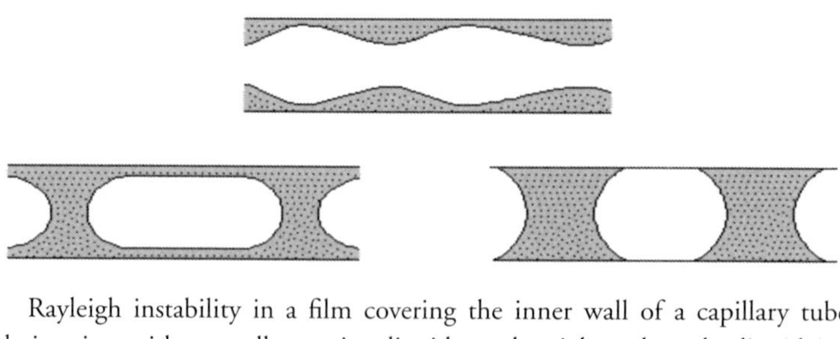

Fig. 22. Rayleigh instability in a film covering the inner wall of a capillary tube; at the left: final situation with a totally wetting liquid; at the right: when the liquid is partially wetting.

and for dewetting process

$$\tau_d = \frac{L}{V_d} \approx \frac{L\mu}{\gamma\theta_s^3}, \tag{49}$$

where L is the film length and it is supposed that any other than viscous dissipation in the TCL can be neglected. Therefore, it is possible to define a characteristic maximal length

$$L_M \approx a\left(\frac{a\theta_s}{h}\right)^3, \tag{50}$$

below which dewetting may be observed because Rayleigh instability is not yet developed. For films larger than L_M, droplets build up all along the capillary. However, some portion of the liquid will normally remain in the interior of the tube. This distinction is essential when intending to dry the interior of a capillary tube.

4.5. *Annular Films Dewetting*

Another novel feature displayed in cylindrical geometry is that the interface curvature gives rise to a driving force varying with the film thickness.[173] This implies that the dewetting velocity should increase with the thickness of the film in the interior of a capillary tube. Only one work studying this process has been published[159]; the authors studied the variation of the dewetting velocity with the film thickness for a wide range of liquid viscosities.

Figure 23 displays a sequence of pictures of an annular film of glycerol–water solution dewetting inside a PVC capillary tube[153] and in Fig. 24 the experimental variation of Ca_d with h/a is shown with triangles. As observed, the data corroborate the theoretical increasing trend, but the experimental enhancement of the dewetting velocity is much larger than the predicted one.

In Ref. 159, the authors explain that this strong increase may be linked to the distribution of the energy dissipation inside the bump located between the contact line and the static constant thickness film. As seen in Fig. 23, these bumps have a strongly asymmetrical shape with a higher slope near the contact line than at the junction with the static film.

Very recent experimental results are superimposed in Fig. 24 with blue markers, those data correspond to the dewetting capillary number measured when films of various water–glycerol solutions dewet on PVC fibers of two different radii.[183] It is observed that they follow the same trend as the values for capillary tubes. To date, there is no complete explanation for this behavior yet.

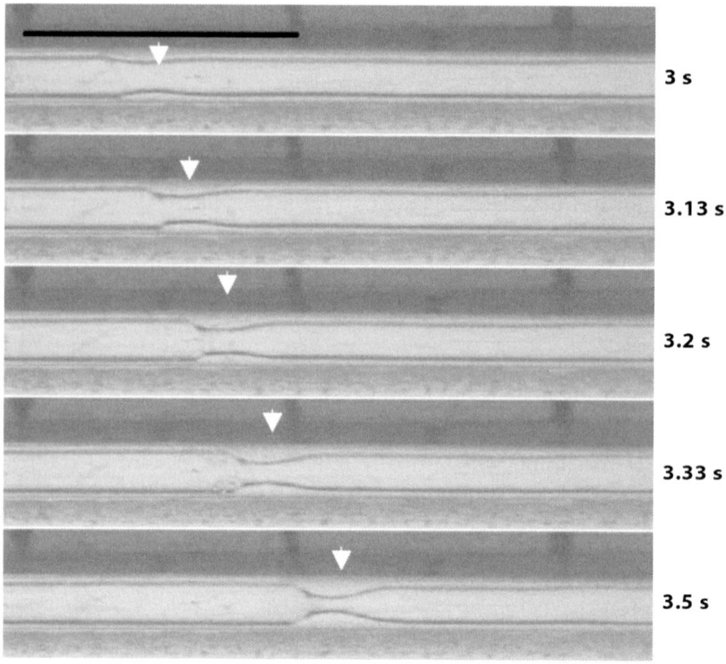

Fig. 23. A water–glycerol solution film dewets inside a capillary tube from left to right. A bump (marked with a white arrow) grows up close to the TCL collecting the liquid. The black line is 1 cm.[153]

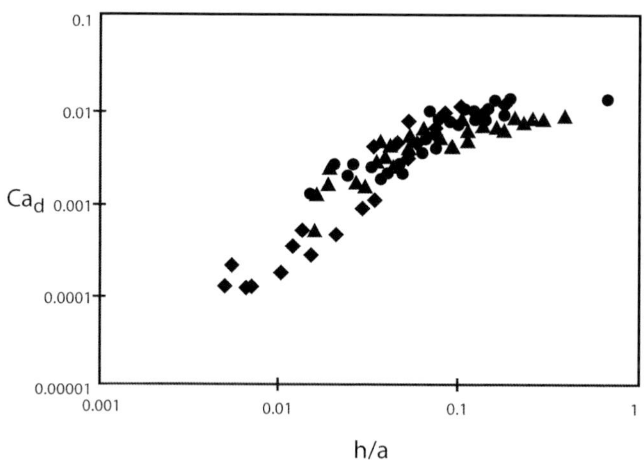

Fig. 24. Dewetting capillary number vs the film thickness normalized with the tube/fiber radius. Triangles: capillary tubes; circles and diamonds: fibers.[183]

References

[1] Eisner T, Aneshansley D. (2000) Defense by foot adhesion in a beetle (*Hemisphaerota cyanea*). *PNAS* **97**: 6568–6573.

[2] Jin H, Kaplan D. (2003) Mechanism of silk processing in insects and spiders. *Nature* **424**: 1057–1061.

[3] Eberhard W, Barrantes G, Ju-Lin W. (2006) Tie them up tight: Wrapping by *Philoponella vicina* spiders breaks, compresses and sometimes kills their prey. *Naturwissenschaften* **93**: 251–254.

[4] Gorb S. (2007) Smooth attachment devices in insects: Functional morphology and biomechanics. *Adv Insect Physiol* **34**: 81–115.

[5] Bush J, Hu D, Prakash M. (2007) The integument of water-walking arthropods: Form and function. *Adv Insect Physiol* **34**: 117–192.

[6] Autumn K, Sitti M, Liang Y, Peattie A, Hansen W, Sponberg S, Kenny T, Fearing R, Israelachvili J, Full R. (2002) Evidence for van der Waals adhesion in gecko setae. *PNAS* **99**: 12252–12256.

[7] Rijke A. (1987) The water repellency of water-bird feathers. *Auk* **104**: 140–142.

[8] Elowson A. (1984) Spread-wing postures and the water repellency of feathers — a test of Rijke hypothesis. *Auk* **101**: 371–383.

[9] Bakken G, Banta M, Higginbotham C, Lynott A. (2006) It's just ducky to be clean: The water repellency and water penetration resistance of swimming mallard Anas platyrhynchos ducklings. *J Avian Biol* **37**: 561–571.

[10] Edwards J. (1983) The biochemistry of cell-adhesion. *Prog Surf Sci* **13**: 125–196.

[11] Bongrand P, Capo C, Depieds R. (1982) Physics of cell-adhesion. *Prog Surf Sci* **12**: 217–285.

[12] Thevenot P, Hu W, Tang L. (2008) Surface chemistry influences implant biocompatibility. *Curr Top Med Chem* **8**: 270–280.

[13] Dubin R, Callegari G, Kohn J, Neimark A. (2008) Carbon nanotube fibers are compatible with mammalian cells and neurons. *IEEE Trans Nanobiosci* 7: 11–14.

[14] London F. (1937) The general theory of molecular forces. *Trans Faraday Soc* **33**: 8–26.

[15] Lennard-Jones. (1924) On the determination of molecular fields. II. From the equation of a state of gas. *Proc Royal Society of London Series A-Mathematical and Physical Sciences* **106**: 463–477.

[16] Keesom W. (1921) On Waal's cohesion forces. *Physikalische Zeitschrift* **22**: 129–141.

[17] Debye P. (1920) The van der Waals cohesion forces. *Physikalische Zeitschrift* **21**: 178–187.

[18] Debye P. (1921) Molecular forces and their electric explanation. *Physikalische Zeitschrift* **22**: 302–308.

[19] Good R, Girifalco L. (1960) A theory for estimation of surface and interfacial energies. 3. Estimation of surface energies of solids from contact angle data. *J Phys Chem* **64**: 561–565.

[20] Churaev N, Sobolev V. (2007) Wetting of low-energy surfaces. *Adv Colloid Interface Sci* **134–135**: 15–23.

[21] Derjaguin B, Churaev N. (1974) Structural component of disjoining pressure. *J Colloid Interface Sci* **49**: 249–255.

[22] De Gennes P, Brochard F, Quere D. (2004) *Capillarity and Wetting Phenomena: Drops, Bubbles, Pearls, Waves*. Springer, New York.

[23] Young T. (1805) An essay on the cohesion of fluids. *Phil Trans R Soc Lond* **95**: 65–87.

[24] Gibbs J. (1928) *Collected Works. Thermodynamics*, Vol. 1. Longmans, New York.

[25] Amirfazli A, Kwok D, Gaydos J, Neumann A. (1998) Line tension measurements through drop size dependence of contact angle. *J Colloid Interface Sci* **205**: 1–11.

[26] Tate T. (1864) On the magnitude of a drop of liquid formed under different circumstances. *Phil Mag* **27**: 176–180.

[27] Harkins W, Brown F. (1919) The determination of surface tension (free surface energy), and the weight of falling drops — The surface tension of water and benzene by the capillary height method. *J Amer Chem Soc* **41**: 499–524.

[28] Strenge K. (1969) Concerning paper tabulated correction factors for drop-weight-volume determination of surface and interface tensions by Lando, Jl and Oakley, Ht. *J Colloid Interface Sci* **29**: 732.

[29] Zhang Z, Mori Y. (1993) Formulation of the Harkins-Brown correction factor for drop-volume description. *Ind Eng Chem Res* **32**: 2950–2952.

[30] Gunde R, Hartland S, Mader R. (1995) Sphere tensiometry: A new approach to simultaneous and independent determination of surface tension and contact angle. *J of Colloid Interface Sci* **176**: 17–30.

[31] Rotenberg Y, Boruvka L, Neumann A. (1984) The shape of nonaxisymmetric drops on inclined planar surfaces. *J of Colloid Interface Sci* **102**: 424–434.

[32] Lecomte du Nouy P. (1919) A new apparatus for measuring surface tension. *J Gen Physiol* **1**: 521–524.

[33] Wilhelmy L. (1863) Ueber die Abhängigkeit der Capillaritäts-Constanten des Alkohols von Substanz und Gestalt des benetzten festen Körpers. *Ann Phys* **195**(6): 177–217.

[34] Tabor D, Winterton R. (1969) Direct measurement of normal and retarded van der Waals forces. *Proc R Soc London Series A-Mathematical and Physical Sciences* **312**: 425–450.

[35] Israelachvili J, Tabor D. (1972) Measurement of van der Waals dispersion forces in range 1.4 to 130 Nm. *Nature-Phys Sci* **236**: 106.

[36] Mangipudi V, Tirrell M, Pocius A. (1995) *Direct Measurement of Molecular Level Adhesion between Poly(ethylene terephthalate) and Polyethylene Films: Determination of Surface and Interfacial Energies*, VSP, Utrecht, The Netherlands, 1254.

[37] Fowkes F. (1964) Attractive forces at interfaces. *Industrial and Engineering Chem* **56**: 40–52.

[38] Fowkes F, Padday J, Uffindel N. (1968) Comments on calculation of cohesive and adhesive energies by Padday, Jf and Uffindell, Nd. *J Phys Chem* **72**: 3700.

[39] Berthelot D. (1898) Sur le Mélange des Gaz. *CR Hebd Acad Sci* **126**: 1703–1706.

[40] Van Oss C, Good R, Chaudhury M. (1988) Additive and nonadditive surface-tension components and the interpretation of contact angles. *Langmuir* **4**: 884–891.

[41] Kwok D. (1999) The usefulness of the Lifshitz–van der Waals/acid-base approach for surface tension components and interfacial tensions. *Colloids Surf a-Physicochem Eng Asp* **156**: 191–200.

[42] Ward C, Neumann A. (1974) On the surface thermodynamics of a two-component liquid–vapor-ideal solid system. *J Colloid Interface Sci* **49**: 286–290.

[43] Morrison ID. (1991) Does the phase rule for capillary systems really justify an equation of state for interfacial-tensions? *Langmuir* 7(8): 1833–1836.

[44] Della Volpe C, Maniglio D, Brugnara M, Siboni S, Morra M. (2004) The solid surface free energy calculation — I. In defense of the multicomponent approach. *J Colloid Interface Sci* **271**: 434–453.

[45] Berg J. (1993) In *Wettability*, J.C Berg, editor, pp. 75. Marcel Deker, New York.

[46] Lee L. (1998) Adhesion and surface-hydrogen-bond components for polymers and biomaterials. *J Adhesion* **67**: 1–18.

[47] Wenzel R. (1936) Resistance of solid surfaces to wetting by water. *Ind Eng Chem* **28**: 988–994.

[48] Cassie A, Baxter S. (1944) Wettability of porous surfaces. *Trans Faraday Soc* **40**: 0546–0550.

[49] Lafuma A, Quere D. (2003) Superhydrophobic states. *Nat Mater* **2**: 457–460.

[50] Johnson R Jr, Dettre R. (1969) Wettability and contact angles. In E Matejivic (ed), *Surface and Colloid Science*, Vol. 2, pp. 85–153. John Wiley and Sons, New York.

[51] Neumann A, Good R. (1972) Thermodynamics of contact angles, I. Heterogeneous solid surfaces. *J Colloid and Interface Sci* **38**: 341–358.

[52] Schwartz L, Garoff S. (1985) Contact angle hysteresis and the shape of the three phase line. *J Colloid Interface Sci* **106**: 422–437.

[53] Marmur A. (1994) Contact angle hysteresis on heterogeneous smooth surfaces. *J Colloid Interface Sci* **168**: 40–46.

[54] Huh C, Mason S. (1977) Steady movement of a liquid meniscus in a capillary tube. *J Fluid Mech* **81**: 401–419.

[55] Johnson R, Dettre R. (1964) Contact angle hysteresis II. Contact angle measurements on rough surfaces. *Adv Chem Ser* **43**: 112–135.

[56] Stokes J, Higgins M, Kushnick A, Bhattacharya S, Robbins M. (1990) Harmonic-generation as a probe of dissipation at a moving contact line. *Phys Rev Lett* **65**: 1885–1888.

[57] Andrieu C, Sykes C, Brochard F. (1996) Dynamics of fast dewetting on model solid substrates. *J Adhesion* **58**: 15–24.

[58] Decker E, Garoff S. (1996) Using vibrational noise to probe energy barriers producing contact angle hysteresis. *Langmuir* **12**: 2100–2110.

[59] Meiron T, Marmur A, Saguy I. (2004) Contact angle measurement on rough surfaces. *J Colloid Interface Sci* **274**: 637–644.

[60] Marmur A. (2000) Wetting on real surfaces. *J Imaging Sci Technol* **44**: 406–409.

[61] Wolansky G, Marmur A. (1999) Apparent contact angles on rough surfaces: the Wenzel equation revisited. *Colloids Surf. A: Physicochem Eng Asp* **156**: 381–388.

[62] Hartmann P, Collet A, Viguier M. (2006) Acrylic copolymers with perfluoroalkylated biphenyl side groups, correlation structure-surface properties. *Macromolecules* **39**: 6975–6982.

[63] Lam C, Wu R, Hair M, Neumann A. (2002) Study on advancing and receding contact angles: Liquid disporting as a cause of contact angle hysteresis. *Adv Colloid Interface Sci* **96**: 169–191.

[64] Chen Y, Helm C, Israelachvili J. (1991) Molecular mechanisms associated with adhesion and contact angle hysteresis of monolayer surfaces. *J Phys Chem* **95**: 10736–10747.

[65] Yasuda T, Miyama M, Yasuda H. (1994) Effect of water immersion on surface configuration of an ethylene-vinyl alcohol copolymer. *Langmuir* **10**: 583–585.

[66] Rangwalla H, Schwab A, Yurdumakan B, Yablon D, Yeganeh M, Dhinojwala A. (2004) Molecular structure of an alkyl-side-chain polymer–water interface: Origins of contact angle hysteresis. *Langmuir* **20**: 8625–8633.

[67] Shanahan M. (1987) Equilibrium of liquid-drops on thin plates — plate rigidity and stability considerations. *J Adhesion* **20**: 261–274.

[68] Princen H. (1969) Capillary phenomena in assemblies of parallel cylinders. 2. Capillary rise in systems with more than 2 cylinders. *J Colloid Interface Sci* **30**: 359–371.

[69] Minor F, Schwartz A, Wulkov E, Buckion L. (1959) Migration of liquids in textile assemblies. *Text Res J* **29**: 931–939.

[70] Dyba R, Miller B. (1969) Evaluation of wettability from capillary rise between filaments. *Text Res J* **39**: 962–970.

[71] Kissa E. (1981) Capillary sorption in fibrous assemblies. *J Colloid Interface Sci* **83**: 265–272.

[72] Lucas R. (1918) Ueber das Zeitgesetz des kapillaren Aufstiegs von Flussigkeiten. *Kolloid Zeitschrift* **23**: 15–22.

[73] Bell J, Cameron F. (1906) The flow of liquids through capillary spaces. *J Phys Chem* **10**: 658–674.

[74] Washburn E. (1921) The dynamics of capillary flow. *Phys Rev* **17**: 273–283.

[75] Bosanquet C. (1923) On the flow of liquids into capillary tubes. *Philos Mag* **45**: 525–531.

[76] Barraza H, Kunapuli S, O'Rear E. (2002) Advancing contact angles of Newtonian fluids during "high" velocity, transient, capillary-driven flow in a parallel plate geometry. *J Phys Chem B* **106**: 4979–4987.

[77] Quere D. (1997) Inertial capillarity. *Europhys Lett* **39**: 533–538.

[78] Zhmud B, Tiberg F, Hallstensson K. 2000 Dynamics of capillary rise. *J Colloid Interface Sci* **228**: 263–269.

[79] Dreyer M, Delgado A, Rath H. (1994) Capillary rise of liquid between parallel plates under microgravity. *J Colloid Interface Sci* **163**: 158–168.

[80] Duarte A, Strier D, Zanette D. (1996) The rise of liquid in a capillary tube revisited: A hydrodynamical approach. *Am J Phys* **64**: 413–418.

[81] Kornev K, Callegari G, Amosova O, Neimark A. (2004) Capillary microfluidics for recognition of DNA hybrids. In *228th ACS National Meeting*. American Chemical Society, Philadelphia.

[82] Kornev K, Callegari G, Neimark A. (2004) Capillary microfluidics for viscoelastic fluids. In W Gutkowski & TA Kowalewski (eds), *ICTAM XXI*, Warsaw, Poland.

[83] Szekely J, Neumann A, Chuang Y. (1971) The rate of capillary penetration and the applicability of the Washburn equation. *J Colloid Interface Sci* **35**: 273–278.

[84] Levine S, Lowndes J, Watson E, Neale G. (1980) Theory of capillary rise of a liquid in a vertical cylindrical tube and in a parallel-plate channel — Washburn equation modified to account for the meniscus with slippage at the contact line. *J Colloid Interface Sci* **73**: 136–151.

[85] Siegel R. (1961) Transient capillary rise in reduced and zero gravity fields. *J Appl Mech* **28**(2): 165–170.

[86] Stange M, Dreyer M, Rath J. (2003) Capillary driven flow in circular cylindrical tubes. *Phys Fluids* **15**: 2587–2601.

[87] Jeje A. (1979) Rates of spontaneous movement of water in capillary tubes. *J Colloid Interface Sci* **69**: 420–429.

[88] Berezkin V, Churaev N. (1982) Changes in contact-angle during the process of capillary rise. *Colloid J USSR* **44**: 376–382.

[89] Ichikawa N, Satoda Y. (1994) Interface dynamics of capillary-flow in a tube under negligible gravity condition. *J Colloid Interface Sci* **162**: 350–355.

[90] Hamraoui A, Thuresson K, Nylander T, Yaminsky V. (2000) Can a dynamic contact angle be understood in terms of a friction coefficient? *J Colloid Interface Sci* **226**: 199–204.

[91] Siebold A, Nardin M, Schultz J, Walliser A, Oppliger M. (2000) Effect of dynamic contact angle on capillary rise phenomena. *Colloids Surf a-Physicochem Eng Asp* **161**: 81–87.

[92] Bazilevsky A, Kornev K, Rozhkov A, Neimark A. (2003) Spontaneous absorption of viscous and viscoelastic fluids by capillaries and porous substrates. *J Colloid Interface Sci* **262**: 16–24.

[93] Weissenberg K. (1947) A continuum theory of rheological phenomena. *Nature* **159**: 310–311.

[94] Ablett R. (1923) An investigation of the angle of contact between paraffin wax and water. *Philos Mag* **46**: 244–256.

[95] Tanner L. (1979) Spreading of silicone oil drops on horizontal surfaces. *J Phys D-Appl Phys* **12**: 1473–1485.

[96] Lelah M, Marmur A. (1981) Spreading kinetics of drops on glass. *J Colloid Interface Sci* **82**: 518–525.

[97] Cazabat A. (1987) How does a droplet spread. *Contemp Phys* **28**: 347–364.

[98] Chen J. (1988) Experiments on a spreading drop and its contact-angle on a solid. *J Colloid Interface Sci* **122**: 60–72.

[99] Hoffman R. (1975) Study of advancing interface .1. Interface shape in liquid-gas systems. *J Colloid Interface Sci* **50**: 228–241.

[100] Calvo A, Paterson A, Chertcoff R, Rosen M, Hulin J. (1991) Dynamic capillary pressure variations in diphasic flows through glass capillaries. *J Colloid Interface Sci* **141**: 384–394.

[101] Fermigier M, Jenffer P. (1991) An experimental investigation of the dynamic contact angle in liquid-liquid systems. *J Colloid Interface Sci* **146**: 226.

[102] Callegari G, Calvo A, Hulin J. (2006) Contact line motion: Hydrodynamical or molecular process? In KL Mittal (ed), *Contact Angle, Wettability and Adhesion*, Vol. 4. VSP, Leiden.

[103] Inverarity G. (1969) Dynamic wetting of glass fibre and polymer fibre. *Brit Polym J* **1**: 245–251.

[104] Schwartz A, Tejada S. (1972) Studies of dynamic contact angles on solids. *J Colloid Interface Sci* **38**: 359–375.

[105] Burley R, Kennedy B. (1976) Study of dynamic wetting behavior of polyester tapes. *Brit Polym J* **8**: 140–143.

[106] Burley R, Kennedy B. (1976) Experimental-study of air entrainment at a solid–liquid–gas interface. *Chem Eng Sci* **31**: 901–911.

[107] Strom G, Fredriksson M, Stenius P, Radoev B. (1990) Kinetics of steady-state wetting. *J Colloid Interface Sci* **134**: 107–116.

[108] Blake T, Clarke A, Ruschak K. (1994) Hydrodynamic assist of dynamic wetting. *Aiche J* **40**: 229–242.

[109] Blake T, Haynes J. (1969) Kinetics of liquid/liquid displacement. *J Colloid Interface Sci* **30**: 421–423.

[110] Blake T, Bracke M, Shikhmurzaev Y. (1999) Experimental evidence of nonlocal hydrodynamic influence on the dynamic contact angle. *Phys Fluids* **11**: 1995–2007.

[111] Elliott G, Riddiford A. (1967) Dynamic contact angles. 1. Effect of impressed motion. *J Colloid Interface Sci* **23**: 389–398.

[112] Dussan E, Davis S. (1974) Motion of a fluid–fluid interface along a solid-surface. *J Fluid Mech* **65**: 71–95.

[113] Chen Q, Rame E, Garoff S. (1996) Experimental studies on the parametrization of liquid spreading and dynamic contact angles. *Colloids Surf a-Physicochem Eng Asp* **116**: 115–124.

[114] Shen C, Ruth D. (1998) Experimental and numerical investigations of the interface profile close to a moving contact line. *Phys Fluids* **10**: 789–799.

[115] Huh C, Scriven L. (1971) Hydrodynamic model of steady movement of a solid/liquid/fluid contact line. *J Colloid Interface Sci* **35**: 85–101.

[116] Greenspan H. (1978) Motion of a small viscous droplet that wets a surface. *J Fluid Mech* **84**: 125–143.

[117] Voinov O. (1978) Asymptotics of the viscous-liquid free-surface at the creeping motion and the dependence of wetting angle on the velocity. *Doklady Akademii Nauk Sssr* **243**: 1422–1425.

[118] Hocking L, Rivers A. (1982) The spreading of a drop by capillary action. *J Fluid Mech* **121**: 425–442.

[119] Cox R. (1986) The dynamics of the spreading of liquids on a solid-surface. 1. Viscous-flow. *J Fluid Mech* **168**: 169–194.

[120] Blake T. (2006) The physics of moving wetting lines. *J Colloid Interface Sci* **299**: 1–13.
[121] Degennes P. (1985) Wetting — statics and dynamics. *Rev Modern Phys* **57**: 827–863.
[122] Cazabat A. (1992) Wetting — from macroscopic to microscopic scale. *Adv Colloid Interface Sci* **42**: 65–87.
[123] Petrov P, Petrov J. (1992) A combined molecular-hydrodynamic approach to wetting kinetics. *Langmuir* **8**: 1762–1767.
[124] Hocking L. (1977) Moving fluid interface. 2. Removal of force singularity by a slip-flow. *J Fluid Mech* **79**: 209–229.
[125] Ngan C, Dussan E. (1989) On the dynamics of liquid spreading on solid-surfaces. *J Fluid Mech* **209**: 191–226.
[126] Petrov J, Sedev R, Petrov P. (1992) Effect of geometry on steady wetting kinetics and critical velocity of film entrainment. *Adv Colloid Interface Sci* **38**: 229–269.
[127] Dussan E, Rame E, Garoff S. (1991) On identifying the appropriate boundary-conditions at a moving contact line — an experimental investigation. *J Fluid Mech* **230**: 97–116.
[128] Marsh J, Garoff S, Dussan V. (1993) Dynamic contact angles and the fluid–fluid interface shape near a moving contact line. *Phys Review Lett* **70**: 2778–2781.
[129] Chen Q, Ramé E, Garoff S. (1995) The breakdown of asymptotic hydrodynamic models of liquid spreading at increasing capillary number. *Phys Fluids* 7: 2631–2639.
[130] Lowndes J. (1980) The numerical-simulation of the steady movement of a fluid meniscus in a capillary-tube. *J Fluid Mech* **101**: 631–646.
[131] Neogi P, Miller C. (1983) Spreading kinetics of a drop on a rough solid-surface. *J Colloid Interface Sci* **92**: 338–349.
[132] Tilton J. (1988) The steady motion of an interface between 2 viscous-liquids in a capillary-tube. *Chem Eng Sci* **43**: 1371–1384.
[133] Dussan E. (1979) Spreading of liquids on solid-surfaces — static and dynamic contact lines. *Annu Rev Fluid Mech* **11**: 371–400.
[134] Ruckenstein E, Dunn C. (1977) Slip velocity during wetting of solids. *J Colloid Interface Sci* **59**: 135–138.
[135] Tolstoi DM. (1952) Molecular theory for slippage of liquids over solid surfaces. *Dokl Acad Nauk SSSR* **85**: 1089–1092 (in Russian).
[136] Ruckenstein E. (1995) The moving contact line of a droplet on a smooth solid. *J Colloid Interface Sci* **170**: 284–286.
[137] Koplik J, Banavar J, Willemsen J. (1988) Molecular-dynamics of Poiseuille flow and moving contact lines. *Phy Rev Lett* **60**: 1282–1285.
[138] Thompson P, Robbins M. (1989) Simulations of contact-line motion — slip and the dynamic contact-angle. *Phys Rev Lett* **63**: 766–769.
[139] Dussan V. (1976) The moving contact line: the slip boundary condition. *J Fluid Mech* 77: 665–684.
[140] Petrov J, Petrov P. (1992) Forced advancement and retraction of polar liquids on a low-energy surface. *Colloids Surf* **64**: 143–149.

[141] Hayes R, Ralston J. (1993) Forced liquid movement on low-energy surfaces. *J Colloid Interface Sci* **159**: 429–438.

[142] Blake T, De Coninck J. (2002) The influence of solid–liquid interactions on dynamic wetting. *Adv Colloid Interface Sci* **96**: 21–36.

[143] Cazabat A, Gerdes S, Valignat M, Villette S. (1997) Dynamics of wetting: From theory to experiment. *Interface Sci* **5**: 129–139.

[144] Shikhmurzaev Y. (1993) The moving contact line on a smooth solid-surface. *Int J Multiphase Flow* **19**: 589–610.

[145] Shikhmurzaev Y. (1996) Dynamic contact angles and flow in vicinity of moving contact line. *Aiche J* **42**: 601–612.

[146] Shikhmurzaev Y. (1997) Moving contact lines in liquid/liquid/solid systems. *J Fluid Mech* **334**: 211–249.

[147] Shikhmurzaev Y. (1997) Spreading of drops on solid surfaces in a quasi-static regime. *Phys Fluids* **9**: 266–275.

[148] Rame E, Garoff S, Willson K. (2004) Characterizing the microscopic physics near moving contact lines using dynamic contact angle data. *Phys Rev E Stat Nonlin Soft Matter Phys* **70**: 031608.

[149] Blake T. (1993) Dynamic contact angles and wetting kinetics. In JC Berg (ed), *Wettability*, pp. 251–302. Marcel Dekker, New York.

[150] Vega M, Gouttiere C, Seveno D, Blake T, Voue M, De Coninck J, Clarke A. (2007) Experimental investigation of the link between static and dynamic wetting by forced wetting of nylon filament. *Langmuir* **23**: 10628–10634.

[151] Ruschak K. (1985) Coating flows. *Annu Rev Fluid Mech* **17**: 65–89.

[152] Blake T, Ruschak K. (1979) Maximum speed of wetting. *Nature* **282**: 489–491.

[153] Callegari G. (2003) Demojado en geometrías simples. PhD Thesis, Departamento de Física. Buenos Aires University, Buenos Aires.

[154] Fairbrother F, Stubb J. (1935) Studies in electro-endosmosis. Part VI. The 'bubble-tube' method of measurement. *J Chem Soc* **1**: 527–529.

[155] Bretherton F. (1961) The motion of long bubbles in tubes. *J Fluid Mech* **10**: 166–188.

[156] Taylor G. (1961) Deposition of a viscous fluid on the wall of a tube. *J Fluid Mech* **10**: 161–165.

[157] Chen J. (1986) Measuring the film thickness surrounding a bubble inside a capillary. *J Colloid Interface Sci* **109**: 341–349.

[158] Cachile M, Chertcoff R, Calvo A, Rosen M, Hulin J, Cazabat A. (1996) Residual film dynamics in glass capillaries. *J Colloid Interface Sci* **182**: 483–491.

[159] Callegari G, Calvo A, Hulin J. (2005) Dewetting processes in a cylindrical geometry. *Eur Phys J E* **16**: 283–290.

[160] Landau L, Levich B. (1942) Dragging of a liquid by a moving plate. *Acta Physicochim (USSR)* **17**: 42–54.

[161] Park C, Homsy G. (1984) 2-phase displacement in Hele Shaw cells — theory. *J Fluid Mech* **139**: 291–308.

[162] Teletzke G, Davis H, Scriven L. (1988) Wetting hydrodynamics. *Revue de Physique Appliquee* **23**: 989–1007.
[163] Reinelt D, Saffman P. (1985) The penetration of a finger into a viscous fluid in a channel and tube. *SIAM J Sci Stat Comput* **6**: 542–561.
[164] DeRyck A, Quere D. (1996) Inertial coating of a fibre. *J Fluid Mech* **311**: 219–237.
[165] Quere D. (1999) Fluid coating on a fiber. *Annu Rev Fluid Mech* **31**: 347–384.
[166] Rebouillat S, Steffenino B, Salvador B. (2002) Hydrodynamics of high speed fibre impregnation: The fluid layer formation from the meniscus region. *Chem Eng Sci* **57**: 3953–3966.
[167] Reiter G. (1992) Dewetting of thin polymer films. *Phys Rev Lett* **68**: 75–78.
[168] Reiter G. (1993) Unstable thin polymer films: Rupture and dewetting processes. *Langmuir* **9**: 1344–1351.
[169] Callegari G, Calvo A, Hulin J. (2002) Experimental results of dewetting in the visco-gravitational regime. *Colloids Surf a-Physicochem Eng Asp* **206**: 167–177.
[170] Redon C, Brochard-Wyart F, Rondelez F. (1991) Dynamics of dewetting. *Phys Rev Lett* **66**: 715–718.
[171] Andrieu C, Sykes C, Brochard-Wyart F. (1994) Average spreading parameter on heterogeneous surfaces. *Langmuir* **10**: 2077–2080.
[172] Debregeas G, Martin P, Brochard-Wyart F. (1995) Viscous bursting of suspended films. *Phys Rev Lett* **75**: 3886–3889.
[173] Callegari G, Calvo A, Hulin J, Brochard-Wyart F. (2002) Dewetting versus Rayleigh instability inside capillaries. *Langmuir* **18**: 4795–4798.
[174] Herminghaus S, Seeman R, Jacobs K. (2002) Generic morphologies of viscoelastic dewetting fronts. *Phys Rev Lett* **89**(5): 056101.
[175] Plateau J. (1873) *Statique experimentale et theorique des liquides soumis aux seules forces moleculaires*. Gauthiers-Villars, Paris.
[176] Rayleigh J. (1899) *Scientifique Papers*. Cambridge University Press, Cambridge.
[177] Ackley S. (2002) Soul of the healer, Spider web. *The Permanent J* **6**(4): 64.
[178] Goren S. (1962) The instability of an annular thread of fluid. *J Fluid Mech* **12**: 309–319.
[179] Goldsmith H, Mason S. (1963) Flow of suspensions through tubes. 2. Single large bubbles. *J Colloid Sci* **18**: 237–261.
[180] Hammond P. (1983) Nonlinear adjustment of a thin annular film of viscous fluid surrounding a thread of another within a circular cylinder pip. *J Fluid Mech* **137**: 363–384.
[181] Gauglitz P, Radke C. (1988) An extended evolution equation for liquid film breakup in cylindrical capillaries. *Chem Eng Sci* **43**: 1457–1465.
[182] Aul R, Olbrich W. (1990) Stability of a thin annular film in pressure-driven low-Reynolds-number flow through a capillary. *J Fluid Mech* **215**: 585–599.
[183] Sánchez D, Chertcoff R, Calvo A, Callegari G. (2007) Dewetting in fibers. Poster #31 in *Pan-American Advanced Studies Institute on Interfacial Fluid Dynamics: From Theory to Applications*, Mar del Plata, Argentina, 6–17th August 2007.

Chapter 9

Chemical Characterization of Biological and Technological Surfaces

Peter Kruse

As many examples in this book illustrate, surfaces and interfaces in biological systems are vastly diverse in form and function. Inevitably, their function depends on their chemistry. Their chemical characterization is therefore crucial to the understanding of biological interfaces and for the design of new biological materials. The task of chemical characterization can be subdivided to obtain two types of information: (1) the chemical identity and relative abundance of atoms, i.e. the chemical composition of the material, and (2) the chemical environment of these atoms, i.e. the chemical structure of the material. Accordingly, a wide range of techniques are available, including electron and optical spectroscopies, mass spectrometric techniques, ion beams, scattering and diffraction techniques. All these techniques have their own unique challenges and strengths. Often a complete picture of the chemistry of a particular system can only be obtained by combining information from a number of techniques. In this chapter, sample preparation and requirements, the type and analysis of data obtained as well as the kind of information that can be extracted will be discussed for a range of techniques, with references given for the interested reader. It has furthermore been realized in recent years that in most cases not the average chemical composition of biological surfaces and interfaces is relevant to their function, but rather the distinct heterogeneity of interfacial chemistry on the micrometer and nanometer scale determines function. Therefore, a discussion of spectromicroscopic techniques is also included. The repertoire of surface-analytical techniques is vast and extremely powerful, ranging from simple bench-top routine instruments to highly specialized and expensive facilities and synchrotron-based techniques. That power has to be wielded with caution in order to avoid embarrassing mis- or over-interpretation of data sets. Fortunately, experts and specialized facilities are widely available and the study of biological samples is becoming more and more prevalent and maturing quickly.

1. Introduction

This chapter deals with analytical methods that assist in establishing the chemical identity of species at biological and technological surfaces. In order to fully characterize the chemistry of a surface, it is necessary to establish identities and quantities of the atoms present, as well as their bonding environment. Some techniques, such as nuclear reaction analysis or Auger electron spectroscopy (AES), excel at giving us the chemical composition, while other techniques, such as vibrational spectroscopy, primarily yield information about bonding. This is, however, not a practical way of classifying chemical characterization techniques since some of the most popular techniques (e.g. X-ray photoelectron spectroscopy (XPS)) can yield some information about both, but rarely can a problem be solved by application of a single technique. Almost always it will be necessary to build the chemical model of a surface as a composite of information from different techniques.

Instead, this chapter follows a different route to classification. The techniques are divided by the degree to which they can access interfaces within the sample, for even a surface exposed to regular atmosphere will be covered in a thin film of water. Here, the different ways of probing a surface come into play. In principle, we can distinguish between photon-based techniques, particle beam-based techniques, and proximity probes. There are combinations, of course, such as XPS, where incoming photons trigger the emission of electrons from the surface, which are then analyzed and detected. The mean free path of the emitted electrons will only be a few nanometers, not enough to study cells in their natural environment, for example. For XPS analysis — as for any particle beam-based technique — samples have to be placed inside a vacuum chamber. Proximity probes may overcome this limitation by placing a probe (a scanned tip or nanoparticles) near the object of study. This same trick is also used to overcome the lack of spatial resolution that photon-based techniques commonly suffer. After some brief comments on common sample preparation procedures, this chapter first discusses the application of purely photon-based techniques to open surfaces, buried interfaces, and *in situ* systems. Then, electron-based techniques are discussed, which are extremely important in traditional surface analysis of hard materials, but are now being increasingly adapted to biological systems. This is followed by a brief foray into other particle-based techniques, which require the use of dedicated facilities and the collaboration with specialists, but frequently reward the effort with precious morsels of information which were otherwise elusive. Mass spectrometric techniques are discussed in a separate section, with special emphasis on secondary ion mass spectrometry (SIMS), which has developed into a veritable workhorse in the study of biological systems and biomaterials despite the drawback of requiring

Fig. 1. Use of proximity probes for chemical analysis of heterogeneous surfaces. (Cartoon from the "lab initio" strip by Nick D. Kim at http://www.nearingzero.com/. Reproduced with permission.)

ex situ analysis in a vacuum chamber. Finally, the chapter is rounded out with a discussion of a small selection of proximity probe-based techniques that can yield chemical information.

Surface analysis is never as straightforward as the cartoon in Fig. 1 might suggest; measurements are usually much more indirect and there are many possible artifacts. Surface sensitivity of a technique implies that surface contamination is disproportionately important, such as adventitious carbon which can accumulate as a sample is transported through ambient conditions. Variations in humidity may also impact surface structure and composition. This is why sample preparation is of highest importance for surface analysis, as will be discussed in the next section. Furthermore, each technique has its own set of challenges, which is why it can be beneficial to compare the performance of different techniques for the same system.[1] The reader is referred to a number of books,[2–4] reviews,[5–9] and original papers for further technical details of each method and examples of how particular problems can be tackled by combining data from different techniques.

2. Sample Preparation

Surface sensitivity is both a blessing and a curse, in that even though it allows us to measure selectively what we are interested in (the surface), it also makes it possible for contaminants to dominate the measurement. In particular, any kind of particle-based technique (electron spectroscopy, Sec. 4; positron, neutron, or ion beam technique, Sec. 5; mass spectrometry, Sec. 5.7) requires the surface to be loaded into a vacuum chamber for analysis. This not only precludes *in situ* measurements of living systems, but it also requires to either remove any volatile species (water, solvents, a wide range of small molecules, and certain elements) from the sample prior to measurement, or to cool the sample to a temperature where the vapor pressure of its constituents has become negligible (less than about 10^{-8} Torr, depending on the technique). We will not discuss any drying or purification techniques here, since they are highly sample-specific. Instead we will discuss sample freezing, which for example allows XPS measurements of hydrated species, polishing, and sputtering, which are common methods to remove contamination. The two latter techniques remove material from the surface. They are therefore best suited for preparation of cross-section samples or buried interfaces for analysis or for the pre-treatment of surfaces before the actual sample is prepared.

A particular challenge arises when the sample is either probed with a charged particle beam, or charged particles are ejected during the measurement (even if they themselves are not used for detection). Any nonconducting sample will charge up under these circumstances. As a consequence, incoming charged particles might be deflected from the sample, resulting in a mysterious drop in signal; the sample might decompose as the result of charge accumulation, resulting in the creation of new species that are not representative of the original sample; or outgoing charged particles will leave with a different kinetic energy or even be recaptured, leading to artifacts in the spectrum. Principally, there are two approaches to deal with this challenge. The first approach is to compensate charge *in situ* during the measurement using a flood gun or other method of charge compensation, essentially supplying low energy charged particles to the sample to compensate any lost charge. This approach is particularly common in XPS and SIMS, but may be problematic in cases where these low energy charged particles may interfere with measurements based on the detection of low energy charged particles, such as low energy Auger electrons. The second approach is to modify the sample to make it conducting, most commonly by applying a thin conducting coat. If this modification is done in a well-controlled manner, it can be calculated back out of the data and valid conclusions about the original sample can be drawn. Most popular are carbon, gold, platinum, and osmium coatings. Platinum and osmium coatings can be applied thinner and more uniform, but at additional expense.

The two big challenges with coated samples are to keep the sample true to its original state during coating and to keep the sample accessible in the case of limited probe depth of particle beams, in particular low energy electrons which may have mean free paths of just a few nanometers.[10]

2.1. *Freezing*

The purpose in freezing samples prior to analysis in vacuum is to reduce the vapor pressure of all relevant species. In the case of biological samples, water is usually the limiting factor. Either the water content is part of the sample composition of interest, or water plays a structural role. In order to obtain reproducible results, clear protocols have to be devised, such as the protocol for XPS analysis of hydrated samples in ultra-high vaccum (UHV) described by Castner and Ratner from the University of Washington in Seattle.[6] As part of this protocol, the sample is hydrated by exposure to water and cooled below $-100°C$ in a dry nitrogen atmosphere. By warming the sample to about $-90°C$ in UHV, ice will sublime off the sample. The sublimation process is stopped via quick cooling below $-120°C$ when only about a 2 nm layer of ice remains atop the sample. This layer is thin enough to perform XPS analysis through it, done at $-160°C$. By acquiring the XPS signal at multiple take-off angles (see Sec. 4.2), the exact layer thickness can be determined. Typically, a spectrum of a dehydrated sample is taken afterwards by warming the sample to room temperature in vacuum and re-cooling it to $-160°C$ for analysis. Each analysis facility has established its own protocol depending on the characteristics of the samples and the capabilities of the equipment.

2.2. *Polishing*

Polishing is understood as removal of material from the sample surface in a way that the sample surface overall becomes flatter, smoother, and more homogeneous. While this very efficiently removes contamination, it typically also removes the surface layer and is therefore inappropriate to apply to a sample surface just before measurement. It can be an appropriate way of preparing a sample of which the bulk composition shall be studied via a surface sensitive technique, for thinning of cross-sectional samples or for preparation of a suitable specimen geometry even before the relevant surface chemistry is to take place. We can distinguish between mechanical polishing, chemical polishing, or electropolishing, as well as mixed forms such as chemical mechanical polishing, which is very common in the semiconductor industry. For biological specimen, mostly mechanical polishing is relevant of frozen samples, although all techniques are popular and common for semiconductors and metals, which might have technological applications in contact with biological systems. Polishing recipes are very sample-specific and will have to be determined on a case-by-case basis after a good literature search.

2.3. *Sputtering*

Sputtering is the process of ablating material from the sample with a particle beam, usually positively charged ions such as Ar^+ or Ga^+. It is not to be confused with sputter deposition, in which case debris from the sputtering of a sacrificial specimen (target) of desired composition is intentionally re-deposited on another sample, for example as a conductive coating. Ion milling and focused ion beam (FIB) are also forms of sputtering. In cases where the impacting ions are reacting with surface species before leavening the sample we also talk about reactive ion etching (RIE). Plasma cleaning is a related technique. Sputtering is a destructive technique and therefore most commonly used in surface analysis to establish a depth profile of the sample by repeated cycles of measurement and sputtering. Dynamic SIMS is exceptional in that the measurement itself is based on sputtering the sample and detecting the ablated fragments. Here, depth profiling occurs naturally. Sputtering — similar polishing — can also be used to thin cross-sectional specimen, prepare samples prior to surface chemistry taking place or remove contaminated surface layers for bulk analysis. Sputtering results in severe damage to the sample, such as implantation of incident ions, changes to the chemical bonding arrangements within the sample, and layer mixing. More recently, gentle methods with large ions like Bi_3^+ or C_{60}^+ are being developed for applications to soft materials and biological samples. Sputter yields may vary for different species, leading to differential sputtering. The incident ions have to be selected accordingly. The best sputter yield is achieved at a characteristic incident angle, typically around 45° and for mass matched incident ions and atoms in the sample. Effectively, it is necessary to maximize the momentum transfer according to classical mechanics just like in a game of billiards. Simulations have been conducted to understand this process. Another potential challenge can be the redeposition of sputtered material. In all cases, one needs to make sure not to sputter with an ion of the same element that one wants to detect (e.g. Ga). While *ex situ* sputter cleaners are available in many analysis facilities, instruments for very surface sensitive techniques like electron spectroscopic methods (XPS, AES) and SIMS have *in vacuo* facilities for surface cleaning and sample preparation (often in conjunction with annealing).

3. Optical Spectroscopies (Photon Based)

Standard optical spectroscopy is ill-adapted to study surfaces mainly for two reasons: (a) The standard geometry for optical measurements is in transmission. Even if the specimen is transparent in the desired wavelength range, a single monolayer at a surface is far too little sample to achieve good signal-to-noise in standard instruments in this geometry, even though in some instances meaningful data has been obtained. (b) In the case of simple reflection or scattering off a surface, one needs to be concerned about the average travel distance of photons in the sample. This is not typically a concern for metal surfaces, but can be an issue for oxides, polymers, or biological specimen, where the probed sample depth can be significantly larger (microns) than just the surface layer. As a result, the signal from the bulk of the sample will overpower the signal from surface or interface species, unless either optically active species are limited to the interface region or a technique is employed that guarantees selectivity to interfacial signals, such as nonlinear effects that occur only at sites with broken symmetry (see Sec. 3.5). On the other hand, the deeper reach of optical spectroscopy may allow us to access buried interfaces which may not otherwise be accessible to particle beam-based techniques. Single molecule spectroscopy[11] will not be discussed here since it rarely serves the analysis of an interface. Nuclear magnetic resonance (NMR, MRI), which works with photons in the radio frequency range, has been used to study some systems with very high surface areas, but suffers from an extraordinarily low sensitivity (due to the low energy transitions), which precludes interfacial studies on the vast majority of samples. It is therefore also not included here. Another omission here is UV–vis spectroscopy, which can be applied to surfaces but typically gives little information and is not used too commonly in surface science.

3.1. *Photon Sources*

The power of an optical spectroscopic technique significantly depends on the photon source available. Lab-based photon sources are not always sufficiently bright to achieve the desired signal-to-noise ratio. Also, efforts to improve monochromaticity typically require to give up brightness. In cases where only one specific wavelength is required (e.g. Raman, see Sec. 3.4; second harmonic generation (SHG), Sec. 3.5), a laser can be utilized as a source. The coherent, intense, monochromatic laser light can only be tuned to a different wavelength with great effort (dye laser) and is therefore unsuitable in cases where a larger range of wavelengths needs to be scanned (e.g. infrared (IR) spectroscopy, Sec. 3.2.). In the X-ray region, no widely tunable lab sources are available at all. Therefore, synchrotron facilities are a powerful alternative.[12] Synchrotrons are expensive facilities which can sometimes be hard to get access to. Nevertheless, they have serious advantages over lab-based techniques in many circumstances. Not only do they provide higher brightness by several orders of magnitude and therefore a superior signal-to-noise ratio, but also they allow for tunable photon energies. They can deliver bright beams of monochromatic photons of tunable wavelengths from the IR range continuously all the way to hard X-rays. While the entire range is not available at any single beamline, each synchrotron typically has a large number of beamlines that together cover the most relevant spectral regions. The higher brightness can sometimes become a challenge on light-sensitive soft or biological samples due to the possibility of radiation damage, but much recent effort has gone into understanding and addressing these issues and biological samples are now routinely studied at synchrotrons around the world.[13,14]

3.2. *Infrared Spectroscopy*

Due to its particular energy range, IR spectroscopy detects molecular vibrations and hence information about chemical bonds.[15] This can be utilized to detect particular functional groups (amines, alcohols, aldehydes, carboxylic acids, etc.) to indicate the presence or absence of certain molecules in the analysis area.[16–18] Alternatively, if the species present are known, IR can follow the modification of different bond strengths due to changes in the chemical environment, e.g. the transition from a triple bond to lower bond orders during chemisorption of CO onto various sites, or the dissociation of chemisorbed species.[19] Crystalline phases (including their surfaces and interfaces) also experience collective vibrational modes, so-called phonons. These can be detected and studied using vibrational spectroscopy (IR, Raman, high-resolution electron energy loss spectroscopy (HREELS)). Depending on the sample, IR can be performed in transmission geometry, external reflection, or internal reflection. In particular, single external reflection IR spectroscopy is widely known as Reflection-Absorption InfraRed Spectroscopy (RAIRS). If "true" surface problems are studied (i.e. monolayer coverages or less), the signal becomes very weak. For that reason, multiple internal and external reflection geometries have been devised.[15] Internal reflection is often referred to as Attenuated Total Reflection (ATR) since the incident angle has to fulfill the total reflection condition known from classical optics which depends on the ratio of the refractive indices of the two media. The big advantage of IR spectroscopy over many other techniques is that it can be applied under a variety of occasionaly extreme conditions,[20] and therefore used to study a large diversity of systems. IR microscopy is also of interest for heterogeneous samples, although the lateral resolution is limited to a few microns, unless the experiment is conducted in the near field (see Sec. 6.5 at the end of this chapter). Apart from lab-based systems, IR spectroscopy and microscopy also greatly benefit from the higher brightness of synchrotron-based sources.

In all these studies, IR spectroscopy is limited by certain selection rules. Most generally, there needs to be a change in dipole moment as a result of the particular vibrational mode. Ar, N_2, and O_2, for example, cannot be detected by IR in the gas phase, whereas CO_2 and H_2O can. (This is the reason the former are not considered to be greenhouse gases, whereas the latter two are.) At a surface, the situation can get a lot more complex. In the case of (sub)monolayer adsorption on metal surfaces, for example, it has to be considered that any charge or dipole sitting above the surface in the form of a chemisorbate will cause a rearrangement of the mobile electron cloud in the metal below, giving rise to what is known as an image charge or an image dipole, respectively. For the purpose of IR, the whole system needs to be considered in order to decide whether the overall dipole moment is changing or not. Only the component of a dipole perpendicular to the surface will be detectable, since dipoles parallel to the surface are canceled out by their image.

3.3. *Surface Plasmon Resonance (SPR)*

SPR refers to an optical technique which is a very popular detection scheme in biosensors. It is most commonly not used to study the surface itself, but rather to study the interaction of a solute with a reaction partner that has been immobilized at the surface.[21–24] The principle of SPR is illustrated in Fig. 2. A laser beam is coupled into a high refractive index glass prism and totally internally reflected off its base. The reflected beam is then detected outside the third side of the prism. The base is coated with a thin metal layer (usually gold) and SPR exploits the peculiar physics of total reflection. Instead of being strictly reflected at the edge of the glass prism,

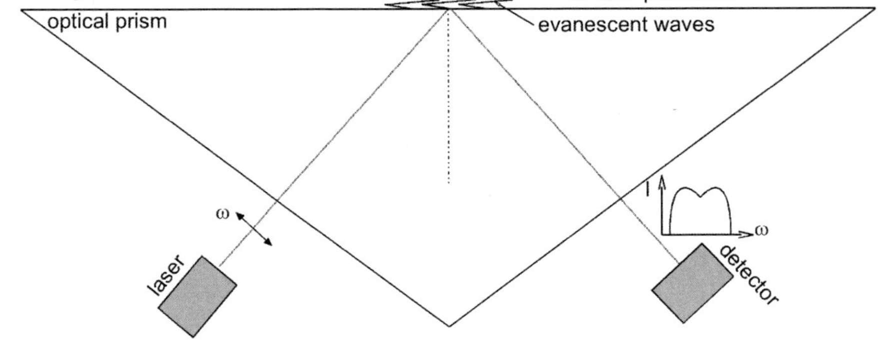

Fig. 2. Principle and experimental set-up for surface plasmon resonance.

quantum mechanics dictates that the photons have a certain probability of tunneling outside, creating an evanescent field that drops off exponentially over the range of a few nanometers. This is the same principle as employed in the attenuated total reflection geometry of IR. In the case of SPR, however, the photons are in the visible range and capable of exciting plasmons (collective electronic excitations) in the thin gold layer. This resonance with the plasmons at the surface of the gold layer leads to an evanescent field exponentially dropping off into the solution and interacting with molecules up to about 150 nm above the sensor surface (hence the name of the technique). Energy from the reflected light beam inside the prism will be lost due to these processes, leading to a dip in reflected beam intensity as a function of the properties (dielectric constant and hence refractive index) of the film above the sensor surface. This dip can be observed by scanning the incident angle of the light beam into the prism around the critical angle required for total internal reflection. The dip will occur only in a very narrow angular range around this critical angle. An alternative detection scheme would pick up surface plasmon field-enhanced fluorescence in the layer above the gold film from the sample side of the configuration.[24]

Most commonly, SPR is not employed as a spectroscopic method to identify chemical properties, but in order to monitor changes to the sensor surface over time. Chemical information can be gained by tuning the selectivity of the surface coating to certain chemicals or biochemical binding events. The assumption is then that the change in signal must be due to the particular analyte that the surface has been sensitized to. It is not difficult to see how false positive readings are possible if the system was not characterized carefully enough w.r.t. interactions with other species present. SPR is a popular method not only for the detection of particular chemical or biological species in a sample, but also for the study of binding and reaction kinetics.

3.4. *Raman Spectroscopy*

The Raman effect refers to the energy loss (or gain) a photon may experience upon being scattered from a molecule. Usually, this interaction is elastic in nature, unless the photon can be completely absorbed by the molecule due to the match in energy with a transition between two quantum mechanical states of the molecule (absorption spectroscopy). In rare cases, inelastic scattering takes place. Photons in the UV, visible, or near IR region can loose just enough energy to a molecule to excite a vibrational transition. By detecting the resulting lower-energy photons, a vibrational spectrum of the sample can be obtained, with information not dissimilar to that obtained in IR (see Sec. 3.2). There are several important differences, however, which have recently favored the application of Raman spectroscopy. First of all, the selection rules in Raman spectroscopy differ from IR. A change in the polarizability of the molecule is now required rather than a change in dipole moment, making it possible to study species that are not IR active. Often, the IR and Raman spectra of a molecule are complementary in which modes can be detected. The second very important aspect is sensitivity. With the advent of powerful, compact, and affordable lasers in the desirable ranges (shorter wavelengths such as in the UV favor the probability of inelastic scattering while longer wavelengths such as in the near IR reduce the fluorescence background), the signal-to-noise ratio in Raman spectroscopy has greatly improved along with its reliability making it fit for the mainstream alongside IR. There are phenomena in Raman spectroscopy, however, which enhance its sensitivity to certain species by several orders of magnitudes, such as resonant Raman (important e.g. in the case of carbon nanotubes[25,26]) and surface enhanced Raman spectroscopy (SERS) — with enhancements reported that exceed a billion times! Furthermore, since Raman can operate in the visible range and the precise wavelength to some extent freely chosen, the choice of optical materials in a Raman instrument is much simplified. Due to the shorter wavelength utilized, Raman microscopes have a better spatial resolution than IR microscopes to begin with, allowing a facile choice of the sample area to be spectroscopically analyzed or mapped. Confocal Raman microscopes bring further enhancement to the spatial resolution,[26] but the ultimate spatial resolution in Raman spectroscopy is possible due to local applications of the SERS, as discussed in more detail in Sec. 6.4.

Raman microscopy can be applied to determine binding arrangements in a variety of samples. Often, it is combined with elemental analysis to fully characterize the sample. In the case of titanium dioxide coatings as a base layer for biomimetic apatite deposition, for example, EDX (see Sec. 4.4) and XPS (see Sec. 4.2) were used to identify the species present in the sample, but only Raman spectra could distinguish between the different phases of titanium oxide.[27] In the case of single-walled carbon nanotubes (SWCNTs) Raman spectroscopy can yield a wealth of information, as illustrated in Fig. 3.[25] The features appearing between 100 and 300 cm^{-1} are radial breathing modes which are fingerprint signals indicating the presence of SWCNTs with different diameters and chiralities. The D peak observed at $\sim$1350 cm^{-1} is commonly associated with carbon atoms with four bonds, i.e. defects in the sidewalls of the nanotubes. The G band observed in the region of 1400–1700 cm^{-1} is characteristic of any kind of graphitic structure, such as graphite or single- or multi-walled carbon nanotubes. Electron–phonon coupling in bundles of metallic SWCNTs causes the shoulder in the G band. In the range of 2500–2800 cm^{-1} an additional mode is observed, which is believed to be an overtone of the D mode and is hence labeled as D*. The shape of the G band as well as the G band and D* mode positions depend on whether the nanotubes are doped, which was utilized in Ref. 25 to discuss inadvertent doping of SWCNT's during sonication in a number of solvents. While Raman spectroscopy was able to determine which samples exhibited doping,

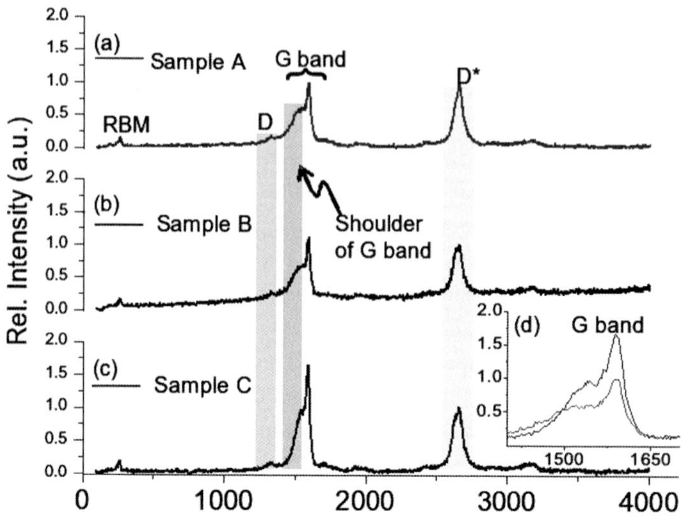

Fig. 3. Raman spectra of annealed SWCNTs. (a) Sample A pressed onto a PTFE membrane, (b) sample B in a sealed capillary tube, and (c) as-is unannealed SWCNTs, Sample C, also pressed onto a PTFE membrane. The inset (d) corresponds to the G band of samples A and C, respectively. (Reproduced with permission from Ref. 25. Copyright 2008 American Chemical Society.)

the identity of the dopant ultimately had to be determined by XPS (see Sec. 4.2). These examples show the power of Raman spectroscopy in determining sample structure and properties.

3.5. *Second Harmonic and Sum Frequency Generation (SHG and SFG)*

The phenomena underlying these nonlinear optical techniques only occur at locations with broken symmetry, such as surfaces and buried interfaces. Even though a large bulk volume is probed by the incident photons, and photons are collected again from a large sample volume, the frequency doubled or added photons all originate from surfaces and interfaces, imparting an exquisite selectivity onto these techniques[28,29] (Certain chiral crystals allow for bulk nonlinear optical effects, but they shall not be of concern here). In SHG, a single monochromatic beam impinges onto the surface, and photons are collected specifically at double the frequency (i.e. half the wavelength). In SFG, multiple monochromatic incoming laser beams combine, most commonly one beam in the visible range and one beam in the IR. Scanning the IR laser over a range of frequencies allows for the collection of vibrational spectra of the samples in a technique referred to as sum frequency spectroscopy due to the resonant enhancement of the SFG signal near vibrational modes. In order to be detected, these vibrational modes of the sample have to be both IR and Raman active. Difference frequency generation happens always at same time as SFG, but SFG is the more common phenomenon to take advantage of for data collection and interpretation. SHG and SFG can yield information about changes in concentration, electron density, or molecular orientation.[30,31] Phase transitions can also be observed and SFG can give structural information similar to IR or Raman spectroscopy, but with superior surface sensitivity. Both techniques are best at monitoring changes in dynamic systems *in situ*, while other complementary techniques are used for detailed chemical analysis.[8,20]

3.6. *X-ray Absorption Spectroscopy (XAS, NEXAFS, STXM, PEEM)*

X-ray absorption spectroscopy (XAS) is probably the most challenging of all absorption spectroscopies. First of all, a tunable X-ray source is required, restricting XAS to synchrotron facilities. Secondly, X-ray optics (deflection, focusing, mono-chromatization) is a lot more challenging to accomplish than optics for the visible or IR range. As a results, X-ray beamlines for XAS are significantly larger, more expensive and more cumbersome than, for example, a commercial IR spectrometer. The spectral range of soft and hard X-ray covers photons that have enough energy to cause the removal of core electrons from the atoms in the sample, i.e. core ionization. The required ionization energies are characteristic of each element and tabulated for reference.[32] XAS in its basic form is suitable for determining elemental composition of a sample, although determination of percentage composition depends on careful calibration. There is a lot more information in an X-ray absorption spectrum, however. Closer inspection of the absorption edge reveals both a pre-edge Near-Edge X-ray Absorption Fine Structure (NEXAFS) due to incomplete ionization events where the excited electron is captured in higher, previously unoccupied orbitals and an post-edge Extended X-ray Absorption Fine Structure (EXAFS) due to interference between scattered photoelectrons. Furthermore, the interaction between photons and electrons depends on the polarization direction of the photons relative to the orientation of the orbital occupied by the electron. Therefore, the information in an X-ray absorption spectrum is very rich, albeit unfortunately at times too complex to interpret unambiguously. Joachim Stöhr has written a monograph on NEXAFS which is widely considered the standard reference in the field.[33]

Core shell ionization of an atom can have a variety of consequences. As it will be discussed in Sec. 4, the energy liberated when the hole that was left behind is filled by another electron from a higher energy level can be dispersed either by ejection of an Auger electron or in the form of an X-ray photon (fluorescence). In addition to the classical transmission experiment where the sample is positioned between the photon source and a detector, two other detection schemes become possible: electron yield (total or partial) and fluorescence yield. Electron yield detection effectively renders NEXAFS to be a surface sensitive technique, because Auger electrons can only escape from the top few nanometers of the sample. The difference between total and partial yield detection is whether all electrons are counted regardless of kinetic energy or whether a filter or cut-off is applied only to select electrons within a certain range of energies. Fluorescence yield allows for the study of bulk samples without too much concern for surface contamination. Transmission mode requires very thin samples and is rarely used, except in the case of scanning transmission X-ray microscopy (STXM) as discussed below. There are other schemes to detect X-ray induced processes in the sample, such as X-ray exited optical luminescence, which has been successfully used to sort out bonding arrangements in thin films and nanostructures.[34] A particular challenge is posed by the tendency of samples, in particular organic compounds, polymers and biological specimen to degrade under intense X-ray irradiation. Nevertheless, a very wide range of samples have been made accessible to XAS.[13,35,36]

It is not surprising that significant interest has gone towards obtaining spatially resolved spectroscopic information ("Spectromicroscopy"). For the purpose of vacuum-based surface studies in particular, Photo-Electron Emission Microscopy (PEEM) has been adopted to work with tunable X-ray lightsources to perform spectromicroscopy (X-PEEM).[13,14,37,38] A PEEM instrument consists of a system of electrostatic and magnetic lenses mounted directly in front of the sample (which is irradiated from the side at a shallow angle), which transmit all electrons emitted from the sample in a way that the map of their emission location from the sample is magnified and projected onto a CCD camera. Typical fields of view range from 1 to 100 μm, with resolutions down to a few nanometers under ideal conditions. Contrast in an image formed under illumination with a mercury arc lamp (classical PEEM) results from the convolution of a number of effects, such as sample workfunction and topography. In X-PEEM effectively the total electron yield is recorded, with topographic and workfunction effects only playing a minor role. PEEM instruments can also be fitted with an electron energy analyzer, in which case only electrons within a particular range of energies ("band pass," partial electron yield) are detected. Figures 4 and 6 show data obtained with such an instrument connected, respectively, to the spectromicroscopy (SM, 11ID-2, X-ray range 200–2000 eV) and variable line spacing plane-grating monochromator (VLS-PGM, 10ID-1, X-ray range 5.5–250 eV) beamlines at the Canadian Light Source (CLS) in Saskatoon, Saskatchewan. The data in Fig. 6 specifically makes use of the analysis of the photoelectron energy and is discussed

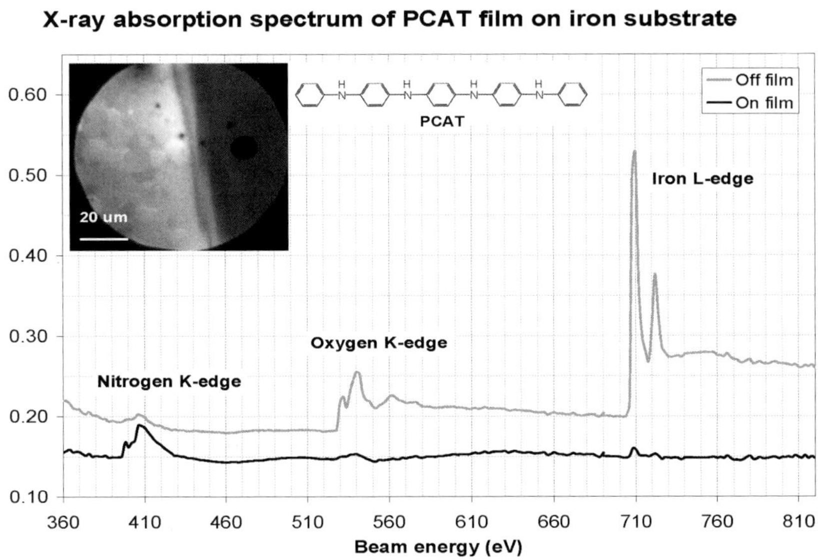

Fig. 4. X-ray adsorption spectromicroscopy of the PCAT film edge. (Reproduced with permission from Ref. 38. Copyright 2008 American Chemical Society.)

further in the sections dealing with photoelectron spectroscopy. The data in Fig. 4 is straightforward XAS data, even though it was recorded in partial electron yield mode. Visible in mercury arc lamp illumination mode (inset top left) is the edge of an organic film on a steel sample. The phenyl-capped aniline tetramer (PCAT) molecule contains nitrogen atoms and its presence is therefore easily detected at the nitrogen edge: Compare the two spectra taken on and off the PCAT film. Since the film is fairly thick (10 nm), the underlying iron sample is not visible through the film. As discussed earlier, photoelectrons do not have a large escape depth, making X-PEEM a fairly surface sensitive technique. We can also detect the native oxide film on the steel when we take a spectrum away from the film. Note that the oxide film has the same thickness below the film, but the signal got attenuated. The detailed edge structure could in principle be modeled to give more detailed information about bonding arrangements within the sample, but for such a complex system as presented here such calculations are far from trivial.

The desire to obtain X-ray adsorption spectra from small features on chemically heterogeneous samples has also lead to the development of STXM.[13,39–41] In this synchrotron-based technique, a Fresnel zone plate is used to collimate the photon beam from the monochromator into a small spot (as small as 30 nm). Similar to the geometry used in transmission electron microscopy (TEM), the sample has to be thin enough (a few hundred nanometers or less) for some of the X-ray beam to be transmitted. The transmitted portion of the beam is detected behind the sample in a traditional absorption spectroscopy arrangement. In places where the sample is optically denser at a particular wavelength due to the presence of a particular species in high concentration, the detector will register a drop in X-ray intensity. The stage with the sample is scanned in front of the beam (since moving the beam is a challenging proposition) to provide concentration maps of different species while the photon energy is stepped after each scan to eventually collect a full XAS spectrum for each pixel. Such a complete data set ("stack") can take hours to acquire. (You will not be surprised to find the analysis software to be full of features to compensate for drift and stitch data pieces together.) In this configuration, samples can be studied in ambient and controlled conditions as opposed to being limited to clean surfaces in vacuum as was the case with X-PEEM. X-ray fluorescence or electron yield detectors can also be added to the STXM instrument, although the most common way to study surfaces or interfaces in STXM is in cross section using the transmission geometry much like in TEM.

4. Electron Spectroscopies

The most common forms of electron spectroscopy are ultra-violet photoelectron spectroscopy (UPS) and XPS, but this chapter will also deal with several other related techniques. AES and X-ray fluorescence detect processes that take place after the photoelectron has been emitted and the resulting hole is being filled with electrons from outer shells. The term electron spectroscopy (an external beam of electrons is involved in probing the sample) has to be distinguished from the term electronic spectroscopy (involving electronic excitations, i.e. changes in the electron configuration of the sample itself), although the majority of electron spectroscopies discussed here also involve electronic excitations (so does XAS in the previous section). HREELS on surfaces is the exception, since here vibrational modes are probed in the sample in analogy to Raman spectroscopy.

Before discussing each technique in detail, let us first consider what happens at the atomic scale that gives rise to the different phenomena and spectroscopies. An incident beam of reasonably energetic electrons or X-ray photons (a few keV) can easily knock out electrons from the sample, whether they are core electrons or valence electrons. This process is illustrated in Fig. 5. If the incident photon energy is only in the range of a few electron volts (i.e. in the ultra-violet spectral region), the binding energy of core electrons cannot be overcome and the photoelectron signal will consist of former valence electrons with a characteristic kinetic energy. UPS, therefore, gives information about binding situations and valence band structures in sample, whereas in the case of XPS the photon energy is sufficient to remove core electrons, and the

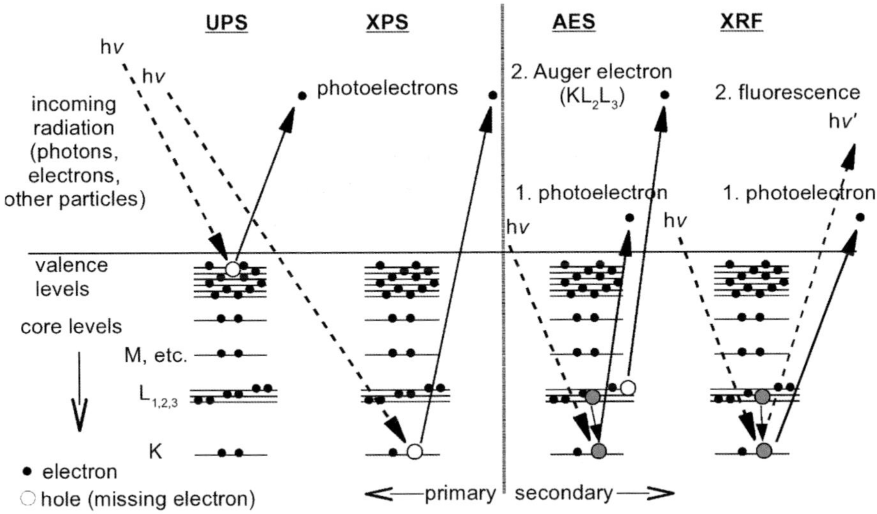

Fig. 5. Knocking out electrons. Illustration of the principles of electron-based techniques.

characteristic kinetic energies of electrons emitted from these discreet levels are useful in establishing chemical identities of atoms present in the sample. These kinetic energies can be measured in an electron energy analyzer and allow us to draw conclusions about the original binding energy of the electron according to:

$$E_{\text{bind}} = h\nu - E_{\text{kin}}$$

after compensating for the workfunction. In the case of isolated atoms or molecules the binding energy is known as ionization energy, which according to Koopman's theorem corresponds to the absolute value of the negative quantity known as orbital energy, which in turn is of importance in quantum mechanical calculations. Once a photoelectron has been emitted from a core shell, the resulting hole needs to be filled because it is energetically favorable to have this low energy spot filled with one of the many electrons occupying higher orbitals. This process liberates energy, which must be dissipated in one of two ways, giving rise to the next two spectroscopic methods. It can either be given to another electron that is then ejected from the atom as an Auger electron or it can be emitted as a photon, typically in the X-ray spectral range as X-ray fluorescence. The techniques detecting these species are known as Anger electron spectroscopy (AES) and X-ray fluorescence (XRF, this one goes by many other acronyms, see Sec. 4.4). In addition to some general references,[42–44] more specific pointers will be given for each technique throughout the sections.

4.1. *UV Photoelectron Spectroscopy (UPS)*

UV photons (typically from a helium lamp with lines at 21.2 eV and 40.8 eV or from a tunable synchrotron beamline) are used in UPS to remove outer shell ("valence") electrons from atoms, molecules, or aggregates. In the case of extended solids, valence orbitals are typically delocalized over so many atoms that they become bands with a range of possible energies rather than crisply defined orbitals. As illustrated in Fig. 6 (left part of the spectra, below about 15 eV on the binding energy scale), the resulting features are broad, smeared out, and difficult to model. Additionally, low energy electrons are easily deflected by electric or magnetic stray fields, such as charges on the sample or residual magnetization in steel parts of the experimental set-up. These are the reasons why UPS is not amongst the work horses of routine surface analysis, even though it can provide extremely valuable information.[42,44] It has found particular applications in recent studies of organic–inorganic interfaces, which are important in molecular electronics, organic electronics, light emitting or light harvesting devices, and corrosion inhibition. In the example of the data shown in Fig. 6, differences between UPS spectra of a clean metallic iron surface and a clean and a contaminated semiconducting iron oxide surface can be observed. The difference in onset of the initial peak (left side of figure) is due to the difference in band structure between a

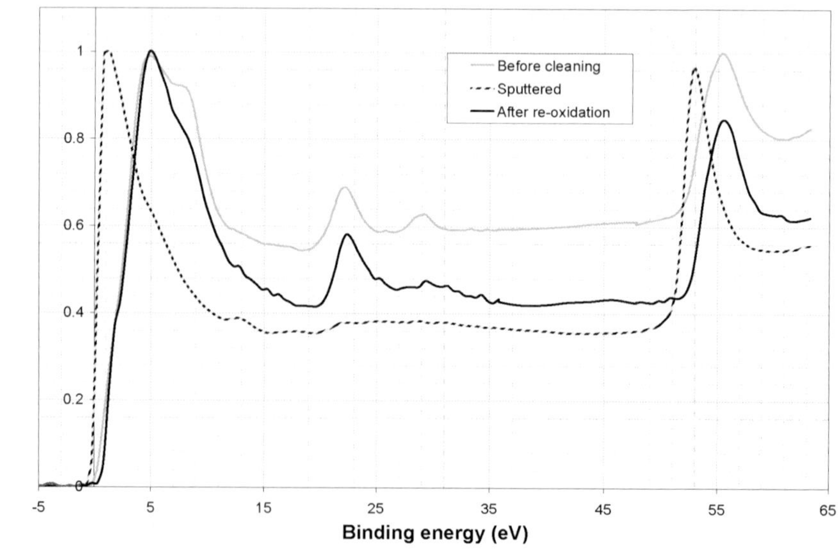

Fig. 6. Photoemission spectra ($h\nu = 160$ eV) of iron substrate before and after sputter cleaning and after re-oxidation. (Reproduced with permission from Ref. 38. Copyright 2008 American Chemical Society.)

metal and a semiconductor. In data not reproduced here, the study goes on to consider the effect of depositing corrosion-inhibiting organic molecules onto this surface on the electronic structure of the oxide.[38] In sufficiently well-studied systems, detailed information about the electronic structure of the interface can be derived.[45] Not surprisingly, interest for application in biological systems has been moderate until now, although corrosion also plays an important role in biomaterials and biosensors, which crucially depend on the interfacial design of the electronic structure.

4.2. *X-ray Photoelectron Spectroscopy (XPS, ESCA)*

XPS is historically also known as electron spectroscopy for chemical analysis (ESCA), because it can be used to map out the chemical composition of samples. The energy of incident X-ray photons is sufficient to knock out core electrons from inner shell orbitals. These orbitals are near the atomic nucleus and therefore quite contracted. They have only very limited spatial overlap with the corresponding (in energy) orbitals of neighboring atoms, meaning that these orbitals do not participate in the formation of bonds (molecules) or bands (solids) to any significant extent. As a result, their energies are still as discreet and well defined as they were in an isolated atom. One small difference is that the valence electron density might have changed once the atom became part of a molecule or solid. Different atoms have different electron affinities, resulting in a redistribution of electron density, with ionic bonding forming the extreme case. This has a (small) effect on the energetic locations of the core orbitals, known as chemical shifts in XPS. The same basic equation is still valid for the calculation of the electron binding energies. In addition to the valence band structure of the sample, Fig. 6 includes also core level peaks, in particular oxygen 2s at 22 eV and iron 3p at 54 eV. The chemical shift in the iron peak due to oxidation is very clearly visible. This data was taken at a synchrotron lightsource, meaning that the incident photon energy is tunable. As a result, it is possible to tune the depth resolution of the signal, because for a given binding energy the kinetic energy of the photoelectron is directly correlated with the energy of the incoming photon. The mean free path of an electron in a solid between inelastic collisions is dependent on its kinetic energy (with a minimum around 50 eV).

A more typical set of XPS data taken with a laboratory-based instrument is shown in Fig. 7. Since there are only a limited number of suitable sources (the line width is important since it impacts the resolution achievable in the photoelectron spectrum and whether a particular chemical shift can be detected) in a laboratory setting, the depth resolution cannot be tuned continuously. The most common sources derive themselves from the Al Kα line at 1486.7 eV and the Mg Kα line at 1253.6 eV. Often, a combination of both of these anodes is found in a laboratory X-ray source

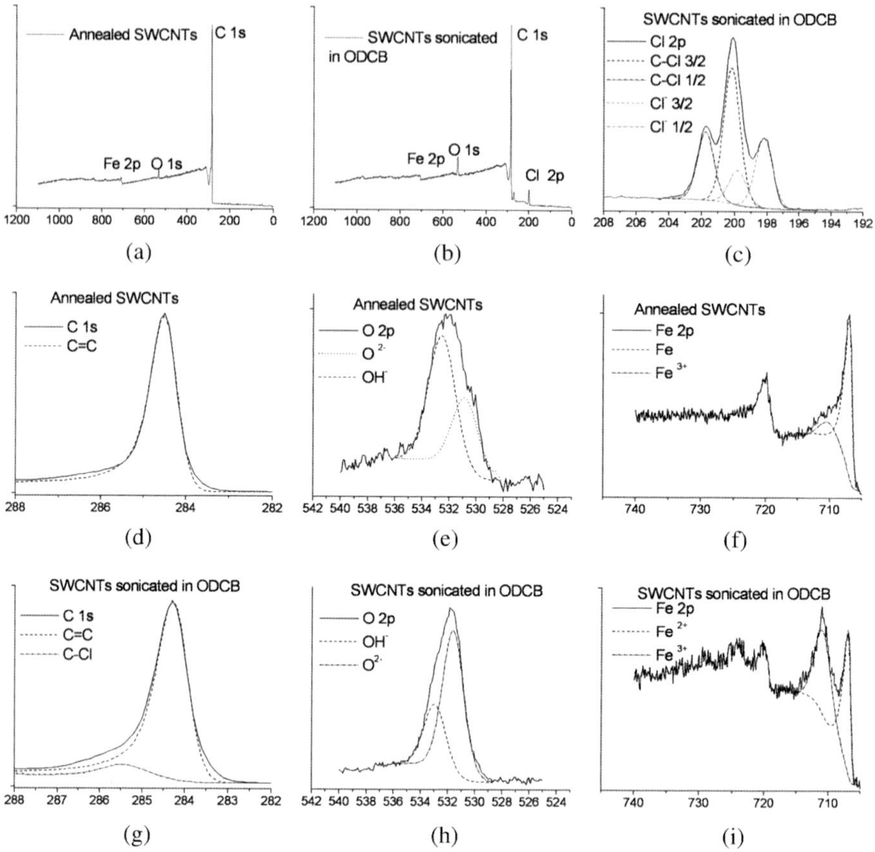

Fig. 7. XPS core level full spectra of (a) annealed SWCNT and (b) SWCNTs sonicated in ODCB; chlorine peaks are observed. XPS high-resolution spectra of the (c) Cl 2p peaks of the sonicated sample, (d) C 1s peak, (e) O 1s peak, (f) Fe 2p peaks for the annealed SWCNTs, (g) C 1s peak, (h) O 1s peak, (i) Fe 2p peaks for SWCNTs sonicated in ODCB. (Adapted from Ref. 25 with permission. Copyright 2008 American Chemical Society.)

in order to allow the operator to avoid interference of Auger electron signals with photoelectron signals of interest by shifting the kinetic energy of the latter. As we will see, the kinetic energy of Auger electrons does not depend on the energy of the incident photon or electron. While the combination of the two available photon sources is not usually particularly beneficial to depth profiling experiments, there are techniques available that allow for the extraction of depth information even without the use of a synchrotron. It is possible to look at the distribution of different species in different peaks of the same element, such as the 2p and 3d peaks of gallium and arsenic with mean free paths of 0.472 nm (As 2p), 0.762 nm (Ga 2p), 2.161 nm (As 3d), and 2.189 nm (Ga 3d) to determine their relative abundance at the top vs the bottom of a thin surface oxide film.[46] A more generally applicable method, however, is the use of angular resolved photoelectron spectroscopy (ARPES, AR-XPS).[47] The sample is rotated ("rocked") in front of the detector. While the penetration depth of the incoming photons is never the limiting factor, the emitted photoelectrons will be detected at different angles, which means that they have to transverse different thicknesses of overlying sample in order to reach the detector positioned at a particular angle. Thus, electrons from different depths of the sample will be detected. This technique has some technical challenges because the spot on the sample from which signal is detected will not stay the same for measurements at different angles. Apart from possible offsets of the rotation axis from the plane of the sample surface, the shape of the spot will also change from a circle for normal incidence detection to an increasingly stretched ellipse for shallower angles. This is not a problem for very homogeneous samples, as long as the impact of changing sample population on the signal intensity is taken into account. For samples with inhomogeneities along the surface, however, lateral variations (e.g. rust spots) may be mistaken for features of the depth profile.

The data set shown in Fig. 7 intentionally goes well beyond the simple identification of present and absent chemical elements that casual XPS users are too often satisfied with. A survey spectrum such as shown in panels (a) and (b) accompanied by a calibrated table of percent composition is undoubtedly valuable, but a lot more information about the sample can usually be obtained with high-resolution scans of relevant peaks. In this case, we identify that the starting material of annealed carbon nanotubes contains iron and oxygen in addition to carbon as the obvious main constituent of the nanotubes. The presence of oxygen as well as adventitious carbon is not surprising for a sample that has been exposed to an ambient environment. The presence of iron is due to residual catalyst particles in the material, a remnant from the production process and in this case the true culprit behind a peculiar behavior of the sample. Namely, after sonication in a chlorinated organic solvent the nanotubes show doping (in their Raman spectra, see Sec. 3.4) and the presence of chlorine (see survey spectrum in Fig. 7(b)). The two survey spectra in Figs. 7(a) and 7(b) are rather unhelpful in understanding the origin of this behavior, since the presence of chlorine could simply be due to entrapped solvent. A good look at the high-resolution chlorine, carbon, iron, and oxygen XPS spectra (Figs. 7(c)–7(i)) tells us the details of the story (together with other analytical methods outlined in the original paper[25]). In the Cl 2p peak, we notice the common phenomenon of doublet splitting due to spin-orbit coupling of the electrons. Two different chlorine species are identified, each exhibiting two peaks labeled as 1/2 and 3/2. In practice, spectra will not always be fitted using doublets when the splitting is too small to have a visible impact (it is surprising what one can get away with), but that is a shortcut that does not apply in this case. Both organic chlorine (covalently bonded to a carbon atom, due to trapped solvent) and chloride ions are present, so we can conclude that a new species is present in our samples in addition to some trapped solvent. Incidentally, the doping observed in the Raman spectra also manifests itself in a small but distinct shift (about 0.2 eV) of the carbon 1s peak towards lower binding energies (Figs. 7(d) and 7(g)). The presence of a chloride salt in the sonicated sampled as indicated by the chlorine spectrum is reflected in a change of the structure of the iron 2p peak, where a notably higher percentage of iron in the +2 and +3 oxidation states (higher binding energy peaks) is now detected. The iron catalyst nanoparticles were attacked by chlorine radicals from the sonochemically decomposing solvent and iron chlorides are known dopants for carbon nanotubes. We also observe an increase of oxide relative to hydroxyl and water species (O 1s spectra, Figs. 7(e) and 7(h)) after sonication, likely because iron nanoparticles underwent oxidation. A detailed look at high-resolution XPS spectra is therefore often worthwhile.

Since it is impossible to cover every aspect of this versatile technique in detail, the interested reader is referred to a number of useful reference books and websites.[32,42–44,48–51] I do want to point out a number of important developments in XPS instrumentation that are relevant to the study of biological samples. The difficulty of working with insulating samples was already discussed in Sec. 2, but modern XPS instruments come with excellent charge compensation. However, a small shift may be introduced into the spectra by the use of such equipment. The operator will try to calculate that back out of the spectra by setting a well-known peak to a fixed value, e.g. the carbon 1s peak, thus removing information such as the doping induced shift shown in Fig. 7(g). For samples in powder form (such as the carbon nanotubes used here) another pathway is possible, which is the pressing of the powdered sample into indium foil for analysis in the hope that charge compensation becomes unnecessary and the absolute position of the peaks can once again be obtained. But the study of biological and other insulating samples is now routinely possible in XPS, without the application of a damaging coat. Another issue is sample inhomogeneity. Here, several approaches are possible. XPS mapping with resolutions from a few micrometers to a few hundred micrometers is possible, but time consuming and therefore expensive, although data acquisition is vastly sped up in modern instruments. The spot size is limited to the micron range. Some instruments also offer "point and shoot" style data taking from particular sample areas by combining the XPS with a scanning electron microscope (SEM) rather than the more common coarse positioning camera. The ultimate spatial resolution (a few dozen nanometers; at the expense of limiting energy resolution to 0.5 eV or worse, making high-resolution work impossible) is achieved in an X-PEEM instrument, as discussed in Sec. 3.6. Figures 4 and 6 give an illustration of what is possible with such an instrument. As discussed earlier, synchrotron-based XPS[13,52] offers many advantages, such as better signal-to-noise ratio, higher energy resolution, and facile depth profiling, but more recent work also includes XPS work on samples held in ambient conditions[53,54] which avoids arduous freezing protocols[6] for analysis in vacuo. Even though XPS by itself is a veritable workhorse technique, I want to conclude this section by once again pointing out the utility of combining several techniques in order to construct a complete story.[1,8,25,55]

4.3. *Auger Electron Spectroscopy (AES, SAM, PAES)*

Auger electrons are common byproducts of the irradiation of matter with high-energy electrons, photons, or other particles. Apart from forming the basis for the dedicated technique of AES, Auger peaks are also always present in XPS spectra and will be labeled in their unique nomenclature as indicated in Fig. 5 by listing the location of the original core hole, the original location of the electron dropping into that hole and the original location of the electron that was ejected in the Auger process. The main difference between AES and XPS is that AES typically uses an electron gun (energies between 2 and 10 keV, sometimes more) as excitation source, which is significantly cheaper and safer to operate than an X-ray source, but is not very monochromatic. This is unproblematic because the Auger peak energy is independent from the energy of the incoming electron or photon. Hence, it is also possible to distinguish them from photoelectrons by changing the incident energy, which is easily accomplished with most electron guns or by switching between different lab-based photon sources (e.g. Mg vs Al cathodes for X-ray) or tuning synchrotron radiation with a monochromator. The peak energy depends on several orbital energies of a core-ionized atom. This makes the theoretical prediction of peak locations extremely difficult. Since AES is a very useful analytical tool, its practitioners have adopted the pragmatic approach of referring to a standard book[56] with all possible reference spectra of elements, against which data from actual samples are compared. This book also contains useful information for quantifying Auger spectra and is an indispensible tool for any AES operator.[56] There are also other good introductory books available.[43] AES has experienced a recent renaissance due to the advent of field emission scanning Auger microscopes (FE-SAM) which are indispensible tools in the analysis of nanostructures due to their high spatial resolution. Scanning Auger microscopes have been around for a long time, but the introduction of FE guns for high spatial resolution (a few nanometers) and hemispherical analyzers with high energy resolution has boosted their utility tremendously.[57–59] Auger electrons can be emitted as the result of a variety of processes, including positron annihilation, as will be discussed in Sec. 5.2.[60,61] The spectra are essentially identical, save for the elimination of the continuous secondary electron background. This huge background is also the reason that Auger spectra are commonly recorded using an oscillating signal and a lock-in amplifier and are displayed in derivative mode.[56]

4.4. *X-ray Fluorescence (EDS, EDX, WDX, XRF)*

XRF probes the photons emitted while a core hole is being filled with an electron from a higher shell. (see Fig. 5) This technique is most commonly found as a accessory in electron microscopes, in particular SEM[27,62,63] and TEM.[64] Here, the core holes are byproducts of the incident probe electron beam knocking out secondary electrons from sample atoms. The energy (and therefore wavelength) of the emitted X-rays is characteristic of each element. The difference between energy-dispersive X-ray spectroscopy (EDS, EDX) and wavelength-dispersive X-ray spectroscopy (WDX, WDS) is the use of either a compact and economical solid state energy detector or a larger and more expensive monochromator to resolve the wavelengths of the emitted photons with much better resolution. Despite of what it might appear like, when used in an SEM this is not actually a surface-analysis technique, but rather a bulk technique, since it probes about a micrometer deep. Nevertheless, it is frequently used to get a first impression of surface or near-surface chemical composition, as illustrated in Fig. 8 which shows a typical EDX spectrum obtained in a FE-SEM from a sample of a specially processed biomedical alloy. The spectrum at the bottom was taken as the averaged signal from the rectangular area of the sample marked towards the right of the crude SEM image shown. Commercial EDS systems come with software that will wager a guess at the quantitative composition of the sample, also shown here at the top right. Usually these guesses have to be treated with caution (background, miss-assigned peaks, heterogeneous samples, etc.), but they can serve as a guide for the qualitative comparison of samples. In particular, lower beam energies in SEM lower the fluorescence yield and decrease accuracy of quantitative determinations. It is possible, however, to develop this into a quantitative technique with proper calibration, well-prepared samples, and the use of standards. The example shown in Fig. 8 did not use a standard. Since the electron beam energy was set at 10 keV, essentially no peaks above 5 keV can be detected (rule of thumb, useful data up to about half the beam energy). Given the expected composition of the sample, this is sufficient in this case, but may have to be adjusted for other samples. The trade-off of higher beam energies can sometimes be sample damage, leading to desorption of species and false determinations of the chemical composition. Another side-effect of electron beam irradiation of a sample in imperfect vacuum (as in virtually all electron microscopes operated by multiple users) is the build-up of carbon on the surface. A carbon peak at the far left of the spectrum in Fig. 8 has not been labeled to exclude from the sample composition, since no carbon was expected to be part of the sample. It can be seen that the alloy consists chiefly of titanium, with some vanadium and aluminum added. The iron peak can be largely ignored since it is at too high of an energy to be included in quantitative considerations. Oxygen is expected to be present due to a thin native

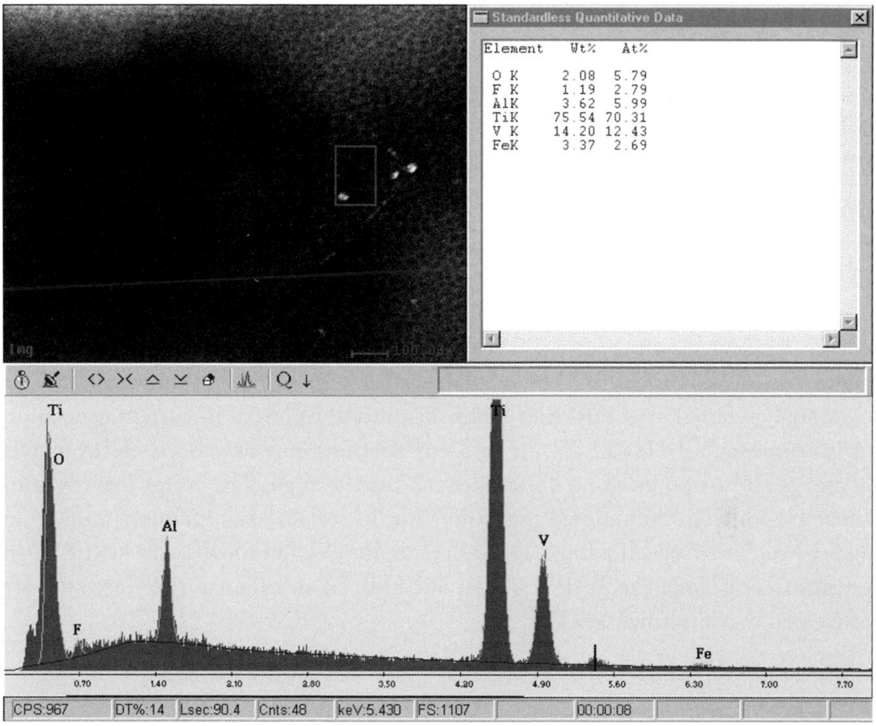

Fig. 8. EDX spectrum and elemental quantification (without standard) of a treated sample of a biomedical alloy.

oxide at the sample surface. Two lessons can be learned from oxygen in this sample. Firstly, our assignment of oxygen to the surface oxide rather than a bulk species is not based on EDX data but rather on our knowledge of the sample. Secondly, the oxygen peak is far from well defined in the spectrum. It is overlayed by a much stronger titanium peak and the spectrum would appear perfectly satisfactorily interpreted without adding oxygen to the mix. It is us who knew what sample composition we expected and what peaks we wanted to select. The software will propose a range of possible elements for each peak, and we have to decide which one makes sense for our sample. For example, we did not expect significant amounts of dysprosium to be present in our sample, a perennial favorite of the software. If a second one of our samples gave a different percentage of oxygen within a factor of 2 or so, no conclusion should be drawn with respect to any differences in sample composition. Fluorine was also detected, a remnant of the chemical treatment. While the numbers might suggest an accurate determination of the fluorine content, a glance at the raw data in the spectrum below will remind us that the fluorine peak is rather tentative and not a thought should be wasted on quantification.

4.5. *Electron Energy Loss Spectroscopy (EELS)*

An electron beam impacting a sample does not have to knock out secondary electrons, it may also just be elastically or inelastically scattered. Elastic interactions lead to crystallographic information, such as a TEM image. Inelastically scattered electrons either loose or gain energy from the sample in the scattering process. There are two very distinct techniques that are both commonly referred to as EELS. The high-energy incident electrons in a TEM can excite electronic transitions in a sample, giving rise to TEM-EELS.[65,66] Here, of relevance is the surface technique EELS, also commonly referred to as HREELS (High Resolution EELS) in its current generation of instruments.[67] In HREELS, the incident electrons only have a few electron volts of energy and are bounced off the surface at a shallow angle. The energy loss (or gain) occurs to (or from) vibrational transitions (or, in certain cases, phonons), much in analogy to Raman spectroscopy. The difference in selection rules makes HREELS an excellent complement to RAIRS, much like bulk IR and Raman spectroscopies are often seen as complementary.[18,68]

5. Particle Beams

The techniques discussed in this section are very powerful, since different types of interactions of the sample with various particles give a large variety of information, making it easier to obtain a more complete picture of the sample. Not all particle beam techniques are discussed here, there are far too many of them. Some of them work best on hard, atomically ordered surfaces in UHV (e.g. Helium ion scattering). Below is a survey of popular techniques for probing the chemistry of surfaces. Many of the techniques require dedicated particle sources (neutrons from a nuclear reactor or a spallation source, positrons derived from a nuclear reactor or a linear accelerator (LINAC), 1 MeV deuterium ions generated by an accelerator, etc.). The most important and widespread technique is SIMS, which does not require large equipment, but is nevertheless typically found only in dedicated analysis facilities due to the complexity of operation and (in particular) data analysis.

Figure 9 illustrates the principles behind some of the techniques that are discussed in these sections. See each subsection for details. A nice set of lecture notes detailing practical aspects and theoretical background of particle beam interactions with solids is also available on the internet courtesy of Willie Lennard at the University of Western Ontario.[69]

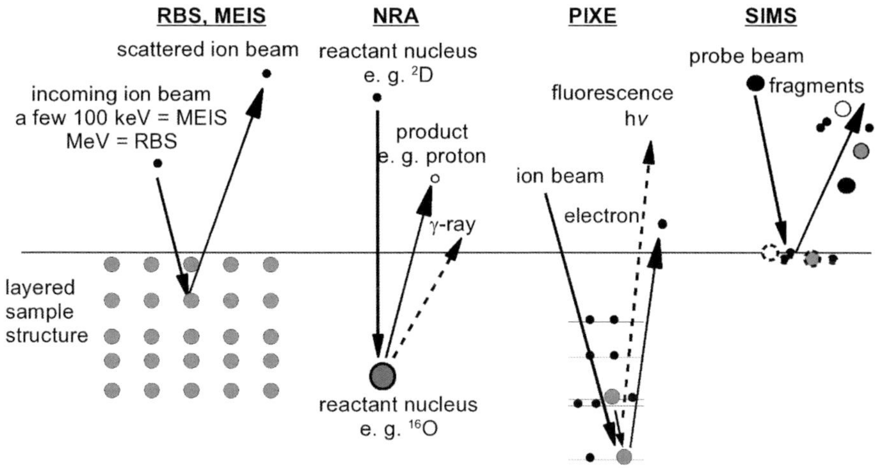

Fig. 9. Particle beam interactions with samples.

5.1. *Small Angle Neutron Scattering (SANS)*

SANS is a diffraction rather than a spectroscopic technique, but it is very valuable for obtaining structural information and complements small angle X-ray scattering due to the different nature of the interaction with the sample. While X-rays preferentially interact with the electron cloud and atomic positions are merely inferred from locations of high electron densities, neutrons directly interact with the nuclei. The interaction cross sections are tabulated and the relative intensity of the interaction of different nuclei with neutrons is often counterintuitive to the outsider, but highly complementary to that of X-rays (e.g. in the case of hydrogen, which is difficult to detect with X-rays due to its low electron density). For this reason and due to the scarcity of facilities working with neutrons (you need either a nuclear reactor or access to a spallation source) collaboration with an expert in the field is imperative. Nevertheless, both neutron and X-ray specular reflection have successfully been applied to both hard materials and biological systems, resulting in a wealth of structural information.[9,70,71]

5.2. *Positron Spectroscopy*

When a positron meets an electron, they form a short-lived species called positronium before they annihilate and emit two gamma rays with characteristic energies in opposite directions. Most important for its application in materials analysis is the fact that the positron first roams around the solid sample before settling down with an electron, because it usually will end up at a defect, meaning that positron annihilation spectroscopy (PAS) is extremely valuable for the study of defects (or interfaces, which by definition are also defects) in crystalline materials (e.g. semiconductors), thin films or nanostructures.[72] The most common detection method measures the positron lifetime (which depends on how quickly it finds the defect and the nature of the defect), although other, more sophisticated techniques are being utilized. Varying the incident positron beam energy will change the penetration depth of the positrons into the sample and allow for nondestructive depth profiling with a resolution in the nanometer to micrometer regime, depending on the positron source. The most important lab-based positron source is derived from the decay of a sodium isotope (^{22}Na), but nuclear reactor or LINAC-based sources are far more intense and offer superior control over the positron beam energy. Slow positron beams, as they can also be achieved by moderation of positrons from sodium sources, are a pre-requisite for making PAS surface sensitive or perform depth profiling of thin film samples. In all these cases, we can no longer rely on the gamma ray emitted during the decay of ^{22}Na in order to time the positron lifetime. Possible ways out are pulsed beams (technically challenging), or the use of different detection schemes, such as angular correlation or Doppler broadening. With a focused positron beam, it is also possible to use PAS as a microprobe, but the resolution will ultimately be limited by the travel distance of positrons inside the sample before annihilation. This distance can be hundreds of nanometers in well-ordered samples such as silicon wafers, but significantly shortens in defect-rich or disordered samples. Nevertheless, technical challenges have so far limited the utility of positron microscopy.

There are a wide range of positron-based techniques. In addition to PAS, certain derivatives of classical surface-analytical techniques are also worth mentioning here. Low energy electron diffraction (LEED) has found its counterpart in low energy positron diffraction (LEPD) and AES (see Sec. 4.3) is related to positron-annihilation induced Auger electron spectroscopy (PAES).[60,61] The advantages of the positron-based techniques are the removal of the secondary electron background signal, higher surface sensitivity, and easier modeling due to simpler interactions with the sample atoms in the case of LEPD. A drawback is the need for higher intensity positron sources.

5.3. *Rutherford Backscattering Spectrometry (RBS)*

Rutherford scattering refers to the deflection of an incoming, moving nucleus (typically a light ion such as H^+, He^+, or He^{2+} with a kinetic energy in the MeV range) due to interaction with an atomic nucleus at rest and more or less firmly held in place in the sample. Ions scattered further into the sample are implanted and evade detection, but ions backscattered in the process can be detected once they leave the sample surface. Hence, the name Rutherford backscattering spectrometry (RBS) for this technique.

It is important to choose the ions and their kinetic energy appropriately in order to avoid nuclear reactions (see Sec. 5.5). While the actual Rutherford scattering process is purely between the nuclei, the ions will undergo inelastic interactions with the electrons on their way through the sample. The energy with which the ions leave the sample has information encoded onto it both about the length of the path traveled and the type of atoms encountered. Hence, an RBS spectrum (count rate vs energy for a given pair of incident and detection angles) will consist of characteristic peaks for each element encountered. Each of these peaks will be elongated towards lower energies depending on the depth range at which it was encountered. Since the energy is lowered both by encounter with lighter elements and by travel deeper into the sample, films of heavy elements on top of films of light elements are best resolved. For example, the surface oxide of a silicon wafer will result in an oxygen peak riding on a silicon background, whereas a gold film on top of a silicon wafer will result in a gold peak which is nicely separated from the silicon peak. An effect known as "channeling" can be used on ordered substrates such as single-crystalline silicon to suppress the signal from the substrate by aligning the lattice of the substrate with the beam directions to minimize interactions. The interested reader is referred to the literature[1,8,9,55,63,73] and local ion beam facilities for more details on the intricacies and power of RBS.

5.4. *Medium Energy Ion Scattering (MEIS)*

In order to achieve higher surface sensitivity and higher resolution in depth profiling, the kinetic energy of the ion beam must be reduced compared with RBS and the kinetic energy resolution must be improved as well. In MEIS, kinetic energies below 300 keV (typically around 100 keV) are used, and the position sensitive detector is preceded with a toroidal electrostatic analyzer, yielding a two-dimensional data set (intensity vs energy and angle) for each analyzer energy. The analyzer energy is stepped through the range of interest for the elements contained in the sample. This way thin films can be analyzed with up to atomic layer resolution. Due to the higher surface sensitivity and atomic resolution of MEIS, it is typically performed in UHV in order to minimize contamination. The underlying principle of MEIS is similar to RBS, except that a much lower background from the bulk substrate can be achieved, especially if a geometry is utilized where both the incident beam and the detected beam are located along a channeling direction of the substrate.[1,8,55]

5.5. *Nuclear Reaction Analysis (NRA)*

Nuclear reaction analysis is exactly that: a beam of nuclei (e.g. deuterium ions) is accelerated towards the sample and undergoes a nuclear reaction with the nuclei in the sample.[62,74–76] An example (illustrated in Fig. 9) would be the quantification of oxygen by counting the protons resulting from

$$^{2}D^{+}(1\ \text{MeV}) + {}^{16}O \rightarrow {}^{17}O + {}^{1}H^{+}.$$

The protons have a kinetic energy which is characteristic of this particular nuclear reaction (as opposed to the reaction $^{2}D^{+}(1\ \text{MeV}) + {}^{12}C \rightarrow {}^{13}C + {}^{1}H^{+}$, for example). By recording the emitted protons and their kinetic energy, spectra can be obtained such as those presented in Fig. 10. Several important features of the NRA experiment are illustrated by the data in Fig. 10. Firstly, NRA is a quantitative technique and very reliably so, as long as it is used in conjunction with a standard. The count rate in NRA is proportional to the concentration of the target atoms in the sample, but the proportionality factor is a function of the experimental set-up and has to be determined. Furthermore, only the total number of atoms is measured. If the thickness of an oxide film is of interest, as in the example given, assumptions have to be made about the chemical composition, density, and compactness of the sample. Only then can the film thicknesses of the samples (Figs. 10(b) and 10(c)) be meaningfully compared to the standard (Fig. 10(a)). If the nuclear reaction takes place deeper inside the sample (up to about half a micrometer), the protons can undergo inelastic scattering inside the sample before being detected. This will manifest itself in a broadening in the peak. NRA can therefore be used to determine concentrations as well as depth profiles of certain elements, provided a suitable nuclear reaction exits for their identification. The data in Fig. 10 show three distinct peaks from the reaction of deuterons with oxygen (leading to two different ^{17}O final states and hence to protons with two distinct energies) and with carbon. Adventitious carbon is an inevitable contamination on sample surfaces that have been processed in ambient. Other nuclear reactions are possible, but the technique is mostly limited to the quantitative determination of light elements such as hydrogen, boron, carbon, nitrogen, and oxygen.[69]

5.6. *Particle-induced X-ray Emission (PIXE)*

Particle-induced X-ray emission (PIXE, this acronym is also sometimes referred to just as proton-induced X-ray emission) is in many ways analogous to XRF (see Sec. 4.4.), but has distinct advantages that justify the greater technical effort.[73,77] A comparison of the principles of XRF (Fig. 5) and PIXE (Fig. 9) reveals that both are essentially based on the creation of holes in the core shells and the emission of the same characteristic X-ray radiation upon filling of that hole with an upper shell electron. The decisive difference between the two techniques is in the way the incident particles (electrons vs light ions) interact with the sample. In the case of an electron beam,

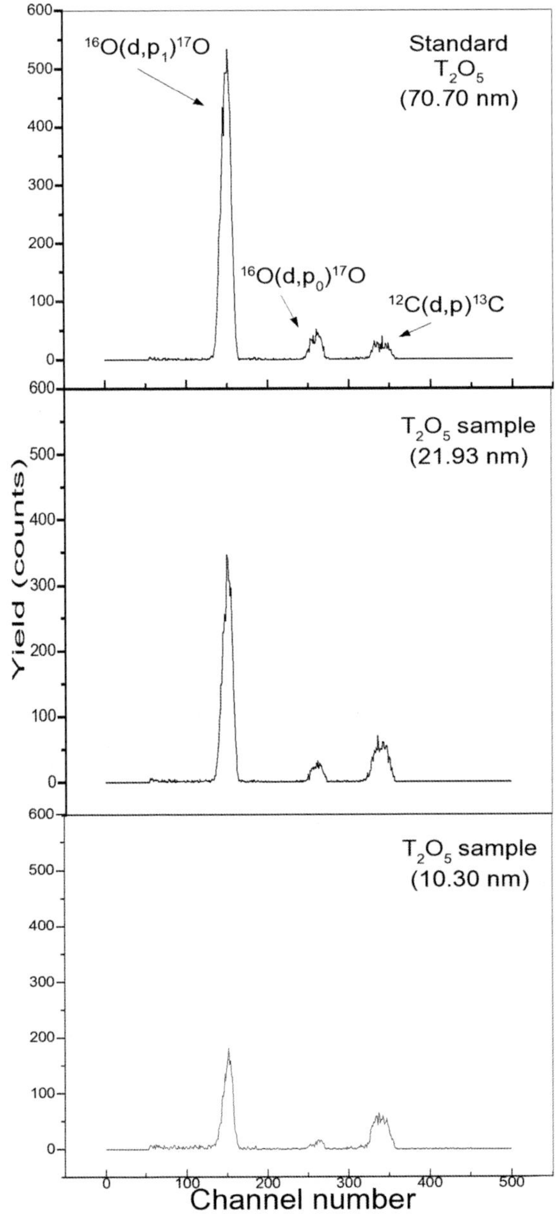

Fig. 10. NRA spectra of (a) the Ta_2O_5 calibrated standard (70.7 nm), (b) electrochemically grown Ta_2O_5 (21.9 nm), and (c) electrochemically grown Ta_2O_5 (10.3 nm). (Reproduced with permission from Ref. 62. Copyright 2007 Elsevier.)

a significant background is noticeable (see Fig. 8), which can obscure small peaks. This background is due to bremsstrahlung (breaking radiation) which is generated when electrons slow down due to inelastic collisions with the sample. Almost the entire energy of an electron can be lost this way, leading to a wide spread of the bremsstrahlung that can underlie the entire spectrum. For heavier particles (protons are 1836 times as heavy as electrons; other light ions are even heavier than that), the situation changes and barely any background impacts the spectrum past about 1 keV. As a trade-off, elements below sodium cannot be detected in most PIXE set-ups since their soft X-rays do not penetrate the detector window, in addition to their commonly small fluorescence yields. This means that carbon, nitrogen, oxygen, and fluorine are among the elements not commonly detectable by this technique. Another advantage of PIXE however, is the way that light ions are getting scattered inside the sample. Whereas electrons generate signal from a large plume (about 1 μm in diameter, starting from a nanometer sized entry point at the surface) inside the sample, severely deteriorating spatial resolution, ion beams stay a lot more focused and tend to probe a much narrower sample volume, leading to lateral resolutions of 100 nm or less. Surface sensitive PIXE (SPIX) is possible in principle,[69] but not reported to be in widespread use.

5.7. *Mass Spectrometry*

Mass spectrometry can be a very sensitive means of detecting chemical species coming off a surface. The challenges are two-fold: (a) the species have to be removed from the surface with minimal damage, and (b) the plethora of atomic clusters and fragments coming off the surface have to be identified and interpreted with respect to their relation to the original structure.[6] Mass spectrometric analysis is by definition destructive, the art of it consist in controlling the damage in a way that the detected fragments can be brought into relation to the original specimen. Sections 5.7.1–5.7.3 will discuss different approaches to these challenges, which result in several well-established techniques.

5.7.1. *Dynamic secondary ion mass spectrometry (SIMS)*

Dynamic SIMS has long been an established technique for the detection of low concentration species in bulk samples (e.g. dopant levels in semiconductors). While the sample is being sputtered (see Sec. 2.3), a quadrupole mass analyzer detects the resulting fragments in real time, resulting in a depth profile of the desired species in the sample or, with calibration, an indication of the concentration of the desired species in the bulk of the sample.[1,78,79] The method is fast and simple, but a quadrupole mass analyzer can only tune into one particular mass-to-charge ratio at a time, or maybe at best multiplex between a few different fragments during one run. Data about fragments other than the one currently being monitored is wasted, with a huge penalty in the signal-to-noise ratio as soon as more than one species is of interest. The intense sputtering results in sample damage and mixing of layers in delicate samples, limiting the spatial resolution of the depth profile. For many cases, in particular the analysis of hard samples with known components of unknown concentration or distribution, this is just fine. In many cases, however, including delicate biological specimen or pure surface studies, static SIMS will be the more appropriate technique, especially since modern static SIMS instruments can also do depth profiling (see next section). The differences are in the price and speed, with quite affordable table-top, high turnover and simple to use dynamic SIMS instruments now being widely available.

5.7.2. *Static secondary ion mass spectrometry (SIMS)*

Time-of-flight SIMS (ToF SIMS) or static SIMS (SSIMS) are two names used to refer to a variant of SIMS where all fragments leaving the surface can be analyzed simultaneously. For this to happen, the sputter rate has to be reduced significantly (hence static) and a different type of mass spectrometer needs to be utilized, namely a ToF tube, where fragments are sorted by their mass-to-charge ratio according to their different speeds at identical kinetic energies. This method is more expensive and elaborate than dynamic SIMS, but it is also much more accurate and versatile. It can be applied to a variety of hard and soft samples and combined with a separate sputtering gun for depth profiling.[6,9,76,80] Nevertheless, being a vacuum technique (since it is particle beam based), all samples have to be or be made vacuum compatible. Cryo-stages in the load-lock and the analysis chamber are common accessories for the study of biological samples. There are many ways in SIMS to deal with insulating samples, many more than there are available in electron spectroscopy, since we are not trying to measure the kinetic energy of the charged particles leaving the surface. (The constant kinetic energy during passage of the ToF tube is achieved by passing the fragments through a constant electric field for acceleration.) In an effort to keep sample damage to a minimum (it is a destructive technique to be sure, but we do not want to damage the sample before we get a chance to analyze it), a number of increasingly sophisticated primary ion sources have been developed. While argon, cesium, and molecular oxygen positive ions are most commonly used for sputter depth profiling, popular primary ions for the analysis beam are gallium, indium, bismuth clusters, and (more recently) fullerenes. The goal for a primary beam in ToF SIMS is to knock out surface species with high efficiency while avoiding excessive fragmentation of sensitive species (especially in biological samples) and minimizing implantation, mixing, and other mechanical damage to deeper layers of the sample before they can be analyzed in depth profiling.

SIMS data commonly is presented in one of three ways: (1) mass spectra, (2) composition images or maps, or (3) depth profiles. Mass spectra are particularly useful if the chemical composition of the sample is unknown. Both, the presence or absence of certain elements and the bonding environment (molecular fragments) of the sample atoms can be interfered, although often indirectly. As mentioned earlier, the interpretation of mass spectrometric data can be very complex and always warrants consultation with an expert. The average composition of a surface is undoubtedly a useful first indication of its condition, but many relevant surface properties derive from its heterogeneity. ToF SIMS is extremely well suited to surface chemical mapping, as long as the primary ion beam can be focused and rastered across the surface. Gallium ion beams can be focused down to a few nanometers, allowing for nanostructural characterization. Since the analysis has to take place pixel by pixel, acquiring detailed surface maps can be arduous and time consuming, but it can be done where needed. Mapping can be combined with depth profiling do give a three-dimensional compositional picture of the sample. Because in ToF SIMS always a complete mass spectrum is recorded in parallel at each pixel, it is not hard to see why SIMS data sets commonly measure several gigabytes in size and rarely leave the file servers of the analysis facility. Instead, the user will be provided with analyzed and processed extracts of the data according to the desired pieces of information. Figure 11 shows an example of such an extract. Only a few fragments were selected for display in the depth profiles. SIMS spectra were also obtained of those samples, and some of them were presented as supporting information, but is simply impossible to provide the complete data set, as users of optical spectroscopy would expect.[76] The data presented in Fig. 11 are depth profiles of porous oxide films grown on tantalum surfaces in the presence of fluoride and sulfate ions. The analysis took place using a pulsed bismuth ion source as the primary ion beam with a spot size of about 1.5 μm and Cs^+ ions for sputter depth profiling. The traces were selected to represent the relevant species: $^{18}O^-$ was chosen to represent oxygen, because its presence in the oxide film was so intense that the $^{16}O^-$ channel got saturated. The $^{19}F^-$ channel also got saturated for the fluorine rich layer, which is why $^{19}F_2^-$ was chosen as a representative fragment. Interference with $^{18}OH^-$ would also have been a concern for $^{19}F^-$. Numerous fragments could have been chosen to represent the sulfate, except that $^{32}S^-$ would have not been a good choice because of interference with $^{16}O_2^-$. Most fragments showed the same trend as $^{32}S^{16}O^-$ chosen here for display. Some Ta-containing fragments representing tantalum oxide and tantalum fluoride (and possibly oxyfluoride) species are also displayed. It has to be kept in mind that fragments can also be formed from recombination of species that have been sputtered from the surface.

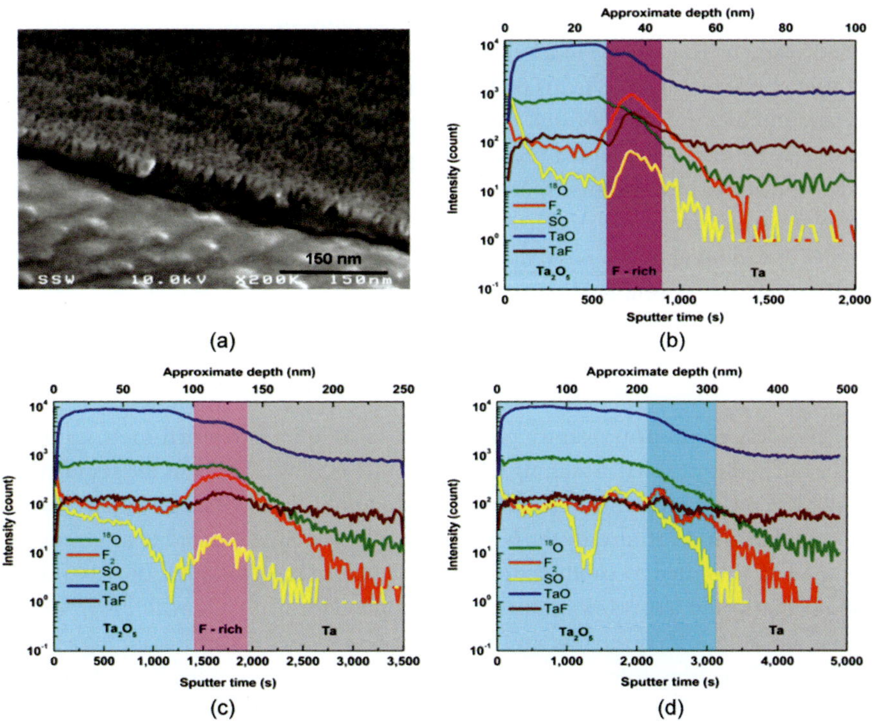

Fig. 11. SEM micrograph of cross-sectional structure and ToF-SIMS depth profiling data of tantalum oxide films grown for different times. (a) Close-up of the edge of a partially detached 5 min oxide film. (b) Depth profile of a film grown for 1 min. (c) Depth profile of a film grown for 10 min. (d) Depth profile of a film grown for 20 min. (Reproduced with permission from Ref. 76. Copyright 2008 American Chemical Society.)

5.7.3. *Self-assembled monolayer desorption ionization mass spectrometry (SAMDI)*

As we have seen, mass spectrometry is crucially dependent on a safe and effective method for removing individual molecules or fragments from a sample. This can be a challenge in the case of fragile but nonvolatile species, such as many of the molecules that make up biological systems or soft biomaterials. SIMS and MALDI (matrix-assisted laser desorption ionization) are examples for techniques that are continuously being refined with this criterion in mind. Milan Mrksich at the University of Chicago has developed a clever alternative called SAMDI.[81] In this technique, molecules of interest are attached to a gold surface via a thiole linker. The sulfur–gold bond is sufficiently strong to allow for the formation of ordered and stable self-assembled monolayers. In fact, such model systems are frequently employed for studying the interactions of biological systems with bioactive surfaces, requiring no further sample preparation in a number of cases. Nevertheless, the sulfur–gold bond can be broken selectively by irradiation with a laser, resulting in the gentle introduction of large and fragile molecules into the mass spectrometer. This method has been applied to a wide range of systems, from small molecule reaction mechanism studies and electrochemical studies to various activity essays, the observation of protein–protein interactions and high-throughput screening.[81]

6. Proximity Probes

The broadest group of proximity probe-based techniques is referred to as scanning probe microscopy (SPM). Scanning tunneling microscopy (STM) was the grandfather of all SPM techniques when it was first presented publicly in 1982. Its impact was so enormous that in 1986 already the Nobel Prize in Physics was awarded to its inventors. Since STM is limited to conducting samples, it was quickly followed (in 1986) by a variant that could also work on insulating samples albeit initially with lower spatial resolution, atomic force microscopy (AFM). From now on, the idea of scanning a probe across a surface using some sort of probe–surface interaction as a feedback parameter to follow surface topography at a constant distance could be (and to a large extend has been) generalized to virtually any kind of interaction. The field has gotten incredibly broad and the interested reader is referred to the microscopy chapter of this book for further detail and several good books on the topic for a thorough treatment.[82,83] Here, we will only discuss a few examples which are pertinent to chemical analysis. Some mentioning will also be made in this section of the idea to introduce nanoparticles as local probes. Especially when coupled with the surface-enhanced Raman effect this method holds significant promise.

6.1. *Elastic and Inelastic Tunneling Spectroscopy*

When the STM was originally constructed, its purpose was not microscopy; it was simply intended as an adjustable tunneling gap for the purpose of tunneling spectroscopy. Tunneling spectroscopy experiments on large area samples had been conducted for decades, but were commonly hampered by the difficulty of producing defect-free films between the electrodes. Elastic tunneling spectroscopy can yield very detailed information about the electronic structure of the two electrodes and to some extent the dielectric properties of the film in between. Inelastic tunneling occurs when the tunneling electrons loose or gain energy in interaction with the film between the electrodes. The modes excited inside the film are typically vibrations, so that a vibrational spectrum of molecules between the electrodes can be obtained in analogy to Raman spectroscopy or EELS, except that in this case the electrons are in the process of tunneling when the energy transfer occurs. If these concepts are now translated to the geometry of an STM, detailed spatially resolved information about the electronic structure (density of states) of the substrate-adsorbate system and vibrational spectra of adsorbate molecules can be obtained, to the point of being able to resolve the location of certain vibrational modes within the molecule. The interested reader is referred to the literature for details[82,83] and should be cautioned at this point that these kind of experiments are far from routine and are best left to a collaboration with an expert in order to avoid embarrassing over-interpretations of artifacts.

6.2. *Force and Chemical Force Spectroscopy*

AFM has already been discussed in the microscopy chapter of this book. The technique has recently been taken to the extreme of chemically identifying single atoms.[84] Numerous textbooks are available to the interested reader for more detailed information.[82,83] I do want to give a brief overview, however, over the possibilities arising from the use of an AFM tip to perform force spectroscopy at defined locations on a sample. While in principle any kind of force (electrostatic, magnetic, chemical, van der Waals, etc.) can be measured as a function of tip position over the sample, the most common measurements involve plastic or elastic deformation (indenting), adhesion (van der Waals), friction (lateral force), and chemical (bond formation) force spectroscopy. Deformation and friction are useful to characterize mechanical properties of substrates either to establish these properties for a new material or in order to distinguish different materials, e.g. a polymer or monolayer coating from a metal substrate, or different polymers from each other. In this case, a hard tip (e.g. silicon nitride or diamond) is preferred. Adhesion and chemical force measurements would typically involve a tip that has been coated with one of the materials (or molecules) of interest. The tip would be brought into contact with the substrate and slowly retracted while the pulling force curve is recorded to the point where the force drops to zero. The unfolding of proteins or unzipping of DNA can also be monitored in this fashion. These techniques are very far from routine and bear many pitfalls. Really the only way to get started is by talking to somebody with experience in the field. Just to name a few points of importance, one needs to insure to measure molecular properties in the appropriate environment (e.g. vacuum vs air vs solution), eliminate any kind of tip (incl. cantilever, tip holder) to sample interactions other than between the species of interest, carefully characterize the tip (radius, shape, coating) and repeat the experiments many times over since our conventional perceptions of thermodynamics and kinetics only apply to very large systems (statistics!).

6.3. *Scanning Electrochemical Microscopy (SECM)*

Allen Bard at the University of Texas Austin has combined the idea of SPM with miniature-size electrodes to construct a local probe of the electrochemical properties of a surface, which is referred to as SECM.[85] This is a very versatile and powerful concept, because it can be used to study the chemistry of both conducting and insulating samples. The Faradaic current between the tip and an electrode at the far end of the electrochemical cell depends on the motion of charged species in solution, which will be partially blocked if the electrode is brought close to the sample, allowing for topographic imaging. More interesting from a chemical analysis perspective is the possibility that a particular species which is formed at the tip electrode will react only with certain features at the sample surface, therefore allowing the mapping of such features. This has been used in the mapping of biological specimen. It is also possible to incorporate a conductive sample as an electrode, permitting us to map electrochemical reactions along a heterogeneous surface.

6.4. *Locally Enhanced Raman Effect*

When Raman spectroscopy was discussed in Sec. 3.4., the possibility of SERS was mentioned. While in the simplest case SERS is used to amplify the signal from a thin film of analyte on a nanostructured metal film, the significant signal enhancements can be utilized to the point of performing Raman spectroscopy with very high (a few nanometers) spatial resolution. This requires positioning a small probe strategically near or in the sample. For accessible surfaces, the use of an AFM or STM tip for tip-enhanced Raman spectroscopy comes to mind, a simple concept the technical challenges of which have only recently been mastered.[86,87] While proof of concept studies had been published much earlier, the method is now finally ripe for applications, such as the spatially resolved study of carbon nanotubes.[86] This technique allows for arbitrary placement of the probe tip on the sample surface, while a much larger area of the sample is irradiated with a laser, and the Raman signal is collected from the entire sample. In the case of inaccessible interfaces, for example in cells inside a mouse, one might have to give up on the idea of arbitrary placement, but spatially resolved information due to a SERS effect from nanoparticles can nevertheless be obtained.[88] Alternatively, nanoparticles can be sparsely embedded into a substrate matrix in order to study Raman enhancement at a few, well-spaced spots. This can be used as a local probe of thin film properties[89] or even taken down to the point of single molecule studies.[90]

6.5. *Nearfield Optical Methods*

Near-field scanning optical microscopy (NSOM; also known as scanning near field optical microscopy — SNOM) is designed to beat the diffraction limit, with the concept first proposed in the 1930s. If an aperture was positioned much closer to the sample than the wavelength of the light ("near field"), then the lateral spatial resolution would be dictated by size of the aperture, not by the wavelength of the light as it would be if the aperture was far from the sample ("far field"). However, the technology to actually build such a microscope only matured in the 1990s, when coated glass fibers were pulled out to very thin tips and piezoelectric scanners made it possible to position these tips with nanometer resolution. IR, UV/vis, and Raman spectroscopies can now be applied with very high spatial resolution (around 50 nm) in a variety of different illumination and detection geometries. Optical spectroscopy was discussed earlier in this chapter. The interested reader is referred to a number of excellent books on the topic of scanning probe microscopy with chapters about NSOM and references therein.[82,83]

7. Summary

Within the spatial restraints of this chapter, only a small selection of surface-analytic techniques could be mentioned. The reader hopefully comes away with a feeling for the possibilities of surface analysis and a few good leads for further reading. The current trend in the field is very strongly towards the development of methods for the study of soft and *in situ* samples outside a well-defined vacuum environment. Given the present focus on adapting the existing powerful techniques to the study of biological systems, the reader can be assured that even if this chapter did not yet hold the solution to his/her problem, somebody is probably working on it right now.

Acknowledgment

The author wishes to thank his past and current students for taking and processing the data presented in several figures in this chapter, in particular Kevin Moonoosawmy, Sherdeep Singh, Stephen Yue Wang, Mark Greiner, and Hany El-Sayed. The data illustrating this chapter were taken at a number of different facilities across Canada, including Surface Science Western, Interface Science Western, the Electrochemical Technology Centre at Guelph, Watlabs at Waterloo, and the Canadian Lightsource in Saskatoon. We thank all the friendly people at these facilities for their kind and patient assistance. The work was financially supported by the Natural Science and Engineering Research Council of Canada.

References

[1] Seah MP, Spencer SJ, Bensebaa F, Vickridge I, Danzebrink H, Krumrey M, Gross T, Oesterle W, Wendler E, Rheinlander B, Azuma Y, Kojima I, Suzuki N, Suzuki M, Tanuma S, Moon DW, Lee HJ, Cho HM, Chen HY, Wee ATS, Osipowicz T, Pan JS, Jordaan WA, Hauert R, Klotz U, van der Marel C, Verheijen M, Tamminga Y, Jeynes C, Bailey P, Biswas S, Falke U, Nguyen NV, Chandler-Horowitz D, Ehrstein JR, Muller D, Dura JA. (2004) Critical review of the current status of thickness measurements for ultrathin SiO_2 on Si Part V: Results of a CCQM pilot study. *Surf Interface Anal* **36**: 1269–1303.

[2] O'Connor DJ, Sexton BA, Smart RStC. (eds) (2003) *Surface Analysis Methods in Materials Science*, 2nd ed. Springer Verlag, Heidelberg, Germany.

[3] Woodruff DP, Delchar TA. (1994) *Modern Techniques of Surface Science*, 2nd ed. Cambridge University Press, Cambridge, UK.

[4] Vickerman JC. (1997) *Surface Analysis: The Principal Techniques*. John Wiley, Chichester, Sussex, UK.

[5] Kasemo B. (2002) Biological surface science. *Surf Sci* **500**: 656–677.

[6] Castner DG, Ratner BD. (2002) Biomedical surface sciences: Foundations to frontiers. *Surf Sci* **500**: 28–60.

[7] Zahn DRT, Gavrila GN, Salvan G. (2007) Electronic and vibrational spectroscopies applied to organic/inorganic interfaces. *Chem Rev* **107**: 1161–1232.

[8] Faupel F, Willecke R, Thran A. (1998) Diffusion of metals in polymers. *Mater Sci Eng* **R22**: 1–55.

[9] Bucknall DG. (2004) Influence of interfaces on thin polymer film behaviour. *Prog Mater Sci* **49**: 713–786.

[10] De Stasio G, Frazer BH, Gilbert B, Richter KL, Valley JW. (2003) Compensation of charging in X-PEEM: A successful test on mineral inclusions in 4.4 Ga old zircon. *Ultramicroscopy* **98**: 57–62.

[11] Nie S, Zare RN. (1997) Optical detection of single molecules. *Annu Rev Biophys Biomol Struct* **26**: 567–596.

[12] Ferrer S, Petroff Y. (2002) Surface science done at third generation synchrotron radiation facilities. *Surf Sci* **500**: 605–627.

[13] Bluhm H, *et al.* (2006) Soft X-ray microscopy and spectroscopy at the molecular environmental science beamline at the advanced light source. *J El Spec Rel Phen* **150**: 86–104.

[14] Morin C, Hitchcock AP, Cornelius RM, Brash JL, Urquhart SG, A. Scholl A, Doran A. (2004) Selective adsorption of protein on polymer surfaces studied by soft X-ray photoemission electron microscopy. *J El Spec Rel Phenom* **137–140**: 785–794.

[15] Hirschmugl CJ. (2002) Frontiers in infrared spectroscopy at surfaces and interfaces. *Surf Sci* **500**: 577–604.

[16] Bertilsson L, Liedberg B. (1993) Infrared study of thiol monolayer assemblies on gold: Preparation, characterization, and functionalization of mixed monolayers. *Langmuir* **9**: 141–149.

[17] Weldon MK, Stefanov BB, Raghavachari K, Chabal YJ. (1997) Initial H_2O-induced oxidation of Si(100)-(2x1). *Phys Rev Lett* **79**: 2851–2854.

[18] Rivillon S, Amy F, Chabal YJ. (2004) Gas phase chlorination of hydrogen-passivated silicon surfaces. *Appl Phys Lett* **85**: 2583–2585.

[19] Siaj M, McBreen PH. (2005) Creating, varying, and growing single-site molecular contacts. *Science* **309**: 588–590.

[20] Rupprechter G, Weilach Ch. (2008) Spectroscopic studies of surface-gas interactions and catalyst restructuring at ambient pressure: Mind the gap! *J Phys: Condens Matter* **20**: 184019.

[21] Schuck P. (1997) Use of surface plasmon resonance to probe the equilibrium and dynamic aspects of interactions between biological macromolecules. *Annu Rev Biophys Biomol Struct* **26**: 541–566.

[22] McDonnell JM. (2001) Surface plasmon resonance: Towards an understanding of the mechanisms of biological molecular recognition. *Curr Opin Chem Bio* **5**: 572–577.

[23] Myszka DG. (1999) Improving biosensor analysis. *J Mol Recog* **12**: 279–284.

[24] Knoll W, Park H, Sinner E-K, Yao D, Yu F. (2004) Supramolecular interfacial architectures for optical biosensing with surface plasmons. *Surf Sci* **570**: 30–42.

[25] Moonoosawmy KR, Kruse P. (2008) To dope or not to dope: The effect of sonicating single-wall carbon nanotubes in common lab solvents on their electronic structure. *J Am Chem Soc* **130**: 13417–13424.

[26] Balasubramanian K, Friedrich M, Jiang C, Fan Y, Mews A, Burghard M, Kern K. (2003) Electrical transport and confocal Raman studies of electrochemically modified individual carbon nanotubes. *Adv Mater* **15**: 1515–1518.

[27] Gua YW, Taya BY, Limb CS, Yong MS. (2005) Biomimetic deposition of apatite coating on surface-modified NiTi alloy. *Biomater* **26**: 6916–6923.

[28] Williams CT, Beattie DA. (2002) Probing buried interfaces with non-linear optical spectroscopy. *Surf Sci* **500**: 545–576.

[29] Eisenthal KB. (1996) Liquid interfaces probed by second-harmonic and sum-frequency spectroscopy. *Chem Rev* **96**: 1343–1360.

[30] Li Q, Hua R, Chou KC. (2008) Electronic and conformational properties of the conjugated polymer MEH-PPV at a buried film/solid interface investigated by two-dimensional IR-visible sum frequency generation. *J Phys Chem B* **112**: 2315–2318.

[31] Li Q, Hua R, Cheah IJ, Chou KC. (2008) Surface structure relaxation of poly(methyl methacrylate). *J Phys Chem B* **112**: 694–697.

[32] http://xdb.lbl.gov/ (Synchrotron parameters, binding energies).

[33] Stöhr J. (1992) *NEXAFS Spectroscopy*, Series in Surface Science Vol. 25. Springer, Berlin.

[34] Zhou X, Sham T-K, Wu Y, Chong Y-M, Bello I, Lee S-T, Heigl F, Regier T, Blyth RIR. (2007) X-ray excited optical luminescence from diamond thin films: The contribution of sp^2- and H-bonded carbon to the luminescence. *J Am Chem Soc* **129**: 1476–1477.

[35] Schoell A, Zou Y, Huebner D, Urquhart SG, Schmidt T, Fink R, Umbach E. (2005) A comparison of fine structures in high-resolution x-ray-absorption spectra of various condensed organic molecules. *J Chem Phys* **123**: 044509.

[36] Johnstone J, Peacock C, Roberts KJ. (1999) Application of polarized NEXAFS spectroscopy to the structural characterization of condensed molecular surfaces and interfaces. *J Cryst Growth* **198/199**: 275–281.

[37] Christensen S, Lanke UD, Haines B, Qaqish SE, Paige MF, Urquhart SG. (2008) Structural and compositional mapping of a phase-separated Langmuir–Blodgett monolayer by X-ray photoelectron emission microscopy. *J El Spec Rel Phenom* **162**: 107–114.

[38] Greiner MT, Festin M, Kruse P. (2008) Investigation of corrosion-inhibiting aniline oligomer thin films on iron using photoelectron spectroscopy. *J Phys Chem C* **112**: 18991–19004.

[39] Ade H, Hitchcock AP. (2008) NEXAFS microscopy and resonant scattering: Composition and orientation probed in real and reciprocal space. *Polymer* **49**: 643–675.

[40] Thibault P, Dierolf M, Menzel A, Bunk O, David C, Pfeiffer F. (2008) High-resolution scanning x-ray diffraction microscopy. *Science* **321**: 379–382.

[41] Felten A, Bittencourt C, Pireaux J-J, Reichelt M, Mayer J, Hernandez-Cruz D, Hitchcock AP. (2007) Individual multiwall carbon nanotubes spectroscopy by scanning transmission X-ray microscopy. *Nano Lett* 7: 1530–6984.

[42] Hüfner S. (1996) *Photoelectron Spectroscopy: Principles and Applications*, 2nd ed. Springer-Verlag, Heidelberg.

[43] Briggs D, Grant JT (eds). (2003) *Surface Analysis by Auger and X-ray Photoelectron Spectroscopy*. Surface Spectra, UK.

[44] http://www.xpsdata.com/ (XPS & UPS resources, background and spectra).

[45] Schlaf R, Merritt CD, Picciolo LC, Kafafi ZH. (2001) Determination of the orbital lineup at reactive organic semiconductor interfaces using photoemission spectroscopy. *J Appl Phys* **90**: 1903–1910.

[46] Ghosh SC, Biesinger MC, LaPierre RR, Kruse P. (2007) X-ray photoelectron spectroscopic study of the formation of catalytic gold nanoparticles on UV-ozone oxidized GaAs(100) substrates. *J Appl Phys* **101**: 114322.

[47] Krishnan S, Ayothi R, Hexemer A, Finlay JA, Sohn KE, Perry R, Ober CK, Kramer EJ, Callow ME, Callow JA, Fischer DA. (2006) Anti-biofouling properties of comblike block copolymers with amphiphilic side chains. *Langmuir* **22**: 5075–5086.

[48] http://srdata.nist.gov/xps (XPS data).

[49] http://www.nist.gov/srd/nist71.htm (IMFP calculator).

[50] Surface Science Spectra database at http://scitation.aip.org/ssso/ (overwhelmingly XPS spectra, but also some SIMS, AES or other techniques).

[51] Castle JE. (2007) Module to guide the expert use of x-ray photoelectron spectroscopy by corrosion scientists. *J Vac Sci Technol A* **25**: 1–27.

[52] Lorusso GF, DeStasio G, Casalbore P, Mercanti D, Ciotti MT, Cricenti A, Generosi R, Perfetti P, Margaritondo G. (1997) Photoemission analysis of chemical differences between the membrane and cytoplasm of neuronal cells. *J Phys D: Appl Phys* **30**: 1794–1799.

[53] Ketteler G, Ashby P, Mun BS, Ratera I, Bluhm H, Kasemo B, Salmeron M. (2008) In situ photoelectron spectroscopy study of water adsorption on model biomaterial surfaces. *J Phys: Condens Matter* **20**: 184024.

[54] Yamamoto S, Bluhm H, Andersson K, Ketteler G, Ogasawara H, Salmeron M, Nilsson A. (2008) In situ x-ray photoelectron spectroscopy study of water adsorption on metals and oxides at ambient conditions. *J Phys: Condens Matter* **20**: 184025.

[55] Harmer SL, Goncharova LV, Kolarova R, Lennard WN, Munoz-Marquez MA, Mitchell IV, Nesbitt HW. (2007) Surface structure of sphalerite studied by medium energy ion scattering and XPS. *Surf Sci* **601**: 352–361.

[56] Davis LE, MacDonald NC, Palmberg PW, Riach GE, Weber RE. (1976) *Handbook of Auger Electron Spectroscopy*, 2nd ed. Physical Electronics Industries, Inc., Eden Prairie, Minnesota.

[57] Teo M, Wong PC, Zhu L, Susac D, Campbell SA, Mitchell KAR, Parsons RR, Bizzotto D. (2006) Characterization of a Co–Se thin film by scanning Auger microscopy and Raman spectroscopy. *Appl Surf Sci* **253**: 1130–1134.

[58] Stenger DA, Georger JH, Dulcey CS, Hickman JJ, Rudolph AS, Nielsen TB, McCort SM, Calved JM. (1992) Coplanar molecular assemblies of amino- and perfluorinated alkylsilanes: Characterization and geometric definition of mammalian cell adhesion and growth. *J Am Chem Soc* **114**: 8435–8442.

[59] Teeter G. (2005) The reaction kinetics with the CdTe(111)-B surface: Formation of metastable Cu_xTe ($x\sim2$). *J Chem Phys* **123**: 184713.

[60] Weiss AH, Yang G, Kim JH, Nangia IA, Fazleev NG. (1996) Application of positron annihilation induced Auger electron spectroscopy to the study of surface chemistry. *J Radioanal Nucl Chem* **210**: 423–433.

[61] Mayer J, Schreckenbach K, Hugenschmidt C. (2007) Surface investigation of Si(1 0 0), Cu, Cu on Si(1 0 0), and Au on Cu with positron annihilation induced Auger-electron spectroscopy. *Surf Sci* **601**: 2459–2466.

[62] El-Sayed H, Greiner MT, Kruse P. (2007) Selective electroplating of copper lines on pre-patterned tantalum oxide thin films. *Appl Surf Sci* **253**: 8962–8968.

[63] Toworfe GK, Composto RJ, Shapiro IM, Ducheyne P. (2006) Nucleation and growth of calcium phosphate on amine-, carboxyl- and hydroxyl-silane self-assembled monolayers. *Biomater* **27**: 631–642.

[64] Ghosh SC, Biesinger MC, LaPierre RR, Kruse P. (2007) The role of proximity caps during the annealing of UV-ozone oxidized GaAs. *J Appl Phys* **101**: 114321.

[65] Hitchcock AP, Dynes JJ, Johansson G, Wang J, Botton G (2008) Comparison of NEXAFS microscopy and TEM-EELS for studies of soft matter. *Micron* **39**: 311–319.

[66] Lazar S, Botton GA, Wu M-Y, Tichelaar FD, Zandbergen HW. (2003) Materials science applications of HREELS in near edge structure analysis and low-energy loss spectroscopy. *Ultramicroscopy* **96**: 535–546.

[67] Ibach H, Mills DL. (1982) *Electron Energy Loss Spectroscopy and Surface Vibrations.* Academic Press, New York.
[68] Eves BJ, Lopinski GP. (2005) Formation and reactivity of high quality halogen terminated Si(111) surfaces. *Surf Sci* **579**: L89–L96.
[69] http://publish.uwo.ca/~wlennard/NotesIBA/headerpage.htm (Lecture notes on particle beam techniques by Dr. Willie Lennard at the University of Western Ontario).
[70] Melnichenkoa YB, Wignall GD. (2007) Small-angle neutron scattering in materials science: Recent practical applications. *J Appl Phys* **102**: 021101.
[71] Krueger S. (2001) Neutron reflection from interfaces with biological and biomimetic materials. *Curr Opin Coll Interface Sci* **6**: 111–117.
[72] Philipose U, Saxena1 A, Ruda HE, Simpson PJ, Wang YQ, Kavanagh KL. (2008) Defect studies of ZnSe nanowires. *Nanotechnology* **19**: 215715.
[73] Lieb K-P. (1999) Thin film analysis with nuclear methods. *Contemp Phys* **40**: 385–413.
[74] Davies JA, Norton PR. (1980) Absolute coverage measurement of adsorbed CO and D_2 on platinum. *Nucl Instrum Methods* **168**: 611–615.
[75] Lennard WN, Tong SY, Mitchell IV, Massoumi GR. (1989) An alternative technique for surface oxygen coverage measurements. *Nucl Instrum Methods Phys Res B* **43**: 187–192.
[76] Singh S, Greiner MT, Kruse P. (2007) Robust inorganic membranes from detachable ultra-thin tantalum oxide films. *Nano Lett* **7**: 2676–2583.
[77] Garmana EF, Grime GW. (2005) Elemental analysis of proteins by microPIXE. *Prog Biophys Molec Biol* **89**: 173–205.
[78] Dickinson M, Heard PJ, Barker JHA, Lewis AC, Mallard D, Allen GC. (2006) Dynamic SIMS analysis of cryo-prepared biological and geological specimens. *Appl Surf Sci* **252**: 6793–6796.
[79] Guerquin-Kern J-L, Wu T-D, Quintana C, Croisy A. (2005) Progress in analytical imaging of the cell by dynamic secondary ion mass spectrometry (SIMS microscopy). *Biochim Biophys Acta* **1724**: 228–238.
[80] Chan ChM, Weng L-T. (2006) SSIMS analysis of organics, polymer blends and interfaces. *Appl Surf Sci* **252**: 6570–6574.
[81] Mrksich M. (2008) Mass spectrometry of self-assembled monolayers: A new tool for molecular surface science. *ACS Nano* **2**: 7–18.
[82] Meyer E, Hug HJ, Bennewitz R. (2004) *Scanning Probe Microscopy: The Lab on a Tip.* Springer Verlag, Berlin.
[83] Bonnell D (ed.) (2001) *Scanning Probe Microscopy and Spectroscopy. Theory, Techniques and Applications*, 2nd ed. Wiley-VCH, New York.
[84] Sugimoto Y, Pou P, Abe M, Jelinek P, Perez R, Morita S, Custance O. (2007) Chemical identification of individual surface atoms by atomic force microscopy. *Nature* **446**: 64–67.
[85] Bard AJ, Denuault G, Lee C, Mandler D, Wipf DO. (1990) Scanning electrochemical microscopy: A new technique for the characterization and modification of surfaces. *Acc Chem Res* **23**: 357–363.
[86] Hartschuh A, Sanchez EJ, Xie XS, Novotny L. (2003) High-resolution near-field Raman microscopy of single-walled carbon nanotubes. *Phys Rev Lett* **90**: 095501.

[87] Hartschuh A, Anderson N, Novotny L. (2003) Near-field Raman spectroscopy using a sharp metal tip. *J Microscopy* **210**: 234–240.

[88] Keren S, Zavaleta C, Cheng Z, de la Zerda A, Gheysens O, Gambhir SS. (2008) Non-invasive molecular imaging of small living subjects using Raman spectroscopy. *Proc Nat Academ Sci* **105**: 5844–5849.

[89] dos Santos Jr. DS, Goulet PJG, Pieczonka NPW, Oliveira Jr. ON, Aroca RF. (2004) Gold nanoparticle embedded, self-sustained chitosan films as substrates for surface-enhanced Raman scattering. *Langmuir* **20**: 10273–10277.

[90] Pieczonka NPW, Aroca RF. (2008) Single molecule analysis by surfaced-enhanced Raman scattering. *Chem Soc Rev* **37**: 946–954.

Part III

Methods for Modifying Man-Made Surfaces

"As they did not know that it was impossible they did it"
Anonymous

Chapter 10

Laser Interference Metallurgy

Frank Mücklich and Andrés Fabián Lasagni

Interfering laser beams of a high-power pulsed laser provides the opportunity of applying a direct lateral treatment of the surface microstructure of metals and ceramics in the micro/nanoscale based on photothermal mechanisms. The "Laser interference metallurgy" allows the creation of periodic patterns with a well-defined long-range order at the scale of typical microstructures (from the sub-micrometer level up to micrometers). This technique is an approach to initiate lateral patterned metallurgical processes such as melting, recrystallization, recovery, and defect and phase formation and with an additional long-range order given due to high-power interference fringe patterns. For polymers the dominating photo thermal and chemical interaction produces rather precise topography pattern. This was found to be a tool to investigate systematically the influence of various surface topography features, e.g. on the contact guidance of cell adhesion.

1. Introduction

The topography and microstructure of material surfaces strongly influence their properties and even their functions. From biomimetic studies, we know that surface topography or structure changes various properties. For example, the investigations of the skin or surface of various fishes, insects, and plants[1] have shown that optimized microstructures render the surface functions, such as anti-reflection of the moth eyes.[2] Composite structures in micro/nanoscale are used in bones.[3] In a more general sense, a well-defined surface structure, e.g. spatially modulated chemical composition, crystallinity, topography, or structure, will give the surface a completely new property. Therefore, the creation of well-defined regular surface structures may lead to advanced functionalization as well as completely new or enhanced properties.[1,2]

Generally, patterning techniques can be basically divided into two groups: (1) methods which require mechanical processes to produce the structure, and (2) methods which utilize beams of electromagnetic waves (e.g. visible light, UV) or particles (e.g. electrons, ions).[4] The first group includes techniques based on printing, molding, and embossing making use of a master to get the final structure.[5] The second group includes optical lithography,[6] E-beam lithography,[7] laser direct writing,[8] holographic patterning,[9] and direct laser interference patterning[10] amongst others.

Direct laser interference patterning[10] uses a single preparation step whereas other holographic methods comprise at least two different steps (irradiation, etching, developing, etc.) In addition, in the case of laser interference patterning, no masks are required. Relatively large areas can be directly structured (of the order of cm^2) in a short time scale. For example, a periodic line-like structure with a feature spacing of 10 μm on a $3 \times 3\,mm^2$ area can be produced in one single laser pulse of 10 ns while direct laser writing requires about 56 h to complete the task (using a beam scanning speed of 30 μms^{-1}).[11]

Direct laser interference patterning has been applied in different materials such as semiconductors, metals, ceramics, and polymers.[12–19] In the case of metals, the interference pattern produced by beams of a high-power pulsed laser permits a direct, local, and periodic heating of the surfaces through the photothermal interaction. In the process of light interaction with metals, the nonthermal energy is absorbed by conducting electrons vibrating with the electrical field vector and thermalizing the lattice by inelastic impacts.[20] Thus, different metallurgical effects such as melting, recrystallization, quenching, recovery, defect or phase formation can be exploited. Therefore, the name Laser Interference Metallurgy (LIMET) was introduced.[10,21]

Laser Interference Lithography has been already used for the fabrication of anti-reflective "moth-eye" (see also Chapter 4) structures and the self-cleaning "lotus effect." In general, the laser interference patterns are applied to a glass pane coated with a photo-sensitive lacquer, and a striped pattern is imprinted on the lacquer through exposure.[22] Periodically structured silicon surfaces 300 nm with depth varying between 35 and 190 nm prepared by this method were examined in the range 200 nm $< \lambda <$ 3000 nm. A decrease in the reflectivity that becomes stronger with increasing structure depth was observed below 1000 nm, being this broad-band reduction caused by diffraction effects at short wavelengths and by the "moth-eye effect" at long wavelengths.[23] Additionally, the enhancement of the wetting properties of smart photochromic surfaces by interferometric patterning employing green light has been reported in Ref. 24. The periodic gratings with 2.5–10.0 μm period were fabricated in a green-curable matrix with spiropyran molecules. The structured surfaces exhibited photocontrolled and reversible wettability, and enhanced hydrophilicity with respect to the native substrates.

In this chapter, different aspects of the "LIMET" method will be introduced. In the first part of the chapter, the interference theory will be described showing the possibility of creating many different periodic arrays depending on the number of laser beams, laser intensity as well as the geometrical configuration of the beams. Additionally, the design of advanced patterns will be verified by calculating the solution of the inverse problem which means that for a desired periodical pattern it is possible to establish a configuration of electromagnetic waves that reproduces the pattern. The second part of this chapter will be dedicated to practical examples showing the possibility of fabricating periodic arrays in polymers and metals. The fabrication of periodic structures in metals will be described in more detail showing the different metallurgical effects that can be achieved including recrystallization, control of texture, formation of intermetallic phases between others. Apart from that, in the case of thin metallic film systems, the changes in the topographic types that can be obtained will be explained in terms of the laser fluence which is required to melt or vaporize one or more of the layers of the film. For a better understanding of the patterning process, some thermal models will be introduced. Finally, the in vitro cell response of micropatterned polymer surface will be described showing the possibility of modifying the cell orientation.

2. Interference Principle

The simplest form of light is a monochromatic, linear polarized plane wave. This is a sufficient approximation of a real laser beam. Under the assumption of plane waves, the total field (E) of the interference pattern can be obtained by adding all the individual "j" beams[25]

$$E = \sum_{j=1}^{n} E_j = \sum_{j=1}^{n} E_{j0} e^{i(kr - \omega t)}, \tag{1}$$

where E_{j0} is the amplitude of the electric field of the j-beam, r is the coordinate along the direction of propagation, ω is the angular frequency, t is the time, and k is the wave number

$$k = \frac{2\pi}{\lambda}, \tag{2}$$

with λ denoting wavelength of the light source.

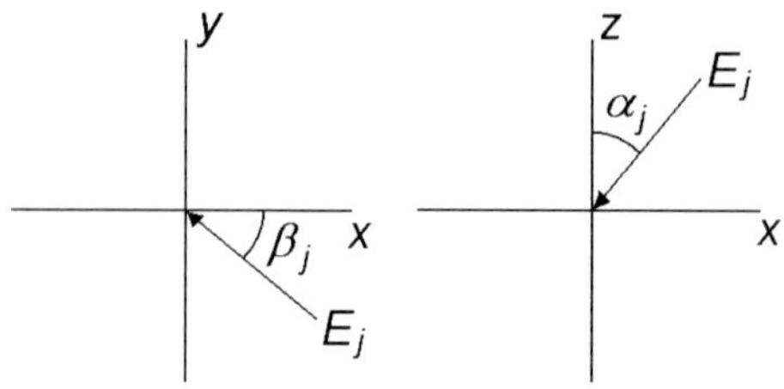

Fig. 1. Representation of an electromagnetic wave in 3D. The sample is located at the position $z = 0$ parallel to the plane x, y (interference plane).

The product between k and r can be written in terms of α_j and β_j which are the angles between the beams with respect to the interference plane (Fig. 1)

$$k \cdot r = \begin{pmatrix} k \cdot \sin\alpha_j \cdot \cos\beta_j \\ k \cdot \sin\alpha_j \cdot \sin\beta_j \\ k \cdot \cos\alpha_j \end{pmatrix} \cdot \begin{pmatrix} x \\ y \\ z \end{pmatrix}. \tag{3}$$

Rewriting (1) in terms of α_j, β_j, x, y and considering $\omega = 0$ we obtain

$$E = \sum_{j=1}^{n} E_j = \sum_{j=1}^{n} E_{j0} e^{-ik \sin\alpha_j (x\cos\beta_j - y\sin\beta_j)}. \tag{4}$$

The intensity distribution of the overlapped beams is given by

$$I = \frac{c\varepsilon_0}{2} |E|^2, \tag{5}$$

where c is the speed of light and ε_0 the permittivity of free space.

The two-beam interference ($n = 2$) produces a linear pattern. Assuming that $E_{01} = E_{02} = E_0$; $\alpha_1 = \alpha_2 = \alpha$; $\beta_1 = 0, \beta_2 = \pi$, the intensity distribution I can be expressed as

$$I = 2c\varepsilon_0 E_0^2 \cos(kx\sin\alpha)^2. \tag{6}$$

The calculated line-type interference pattern is shown in Fig. 2(a). The period of the structure (P) is given by

$$P = \frac{\lambda}{2\sin\alpha}, \tag{7}$$

where 2α is the angle in between the laser beams.[26]

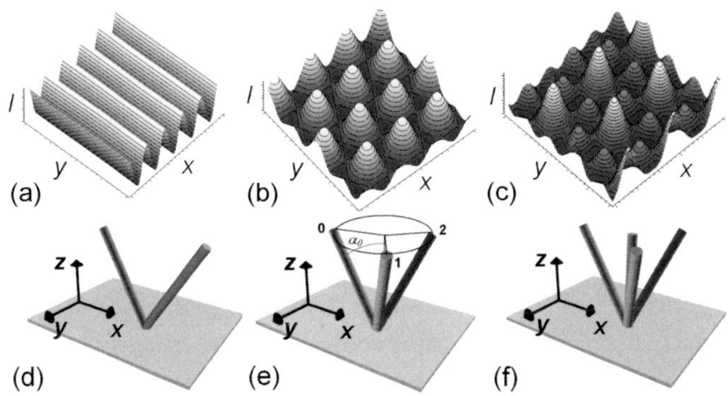

Fig. 2. Calculated intensity distribution for (a) two-beam interference (d), (b) three-beam interference assuming symmetrical configuration (e), and (c) four-beam interference under the conditions shown in (f).[21] Reproduced with permission — Copyright Carl Hanser Verlag GmbH & Co. KG.

Three-beam interference ($n = 3$) produces different 2D arrays depending on the magnitude of the electric field of each beam and the geometric configuration. For symmetric configuration ($E_{01} = E_{02} = E_{03} = E_0$; $\alpha_1 = \alpha_2 = \alpha_3 = \alpha$; $\beta_1 = 0, \beta_2 = 2/3, \pi, \beta_3 = -2/3\pi$), the intensity distribution $I(x, y)$ is given by[25]

$$I = \frac{9}{2} c\varepsilon_0 E_0^2 \left[\begin{array}{l} \left(- \cos(kx \sin\alpha) - \cos\left(k \sin\alpha \left(-\frac{x}{2} - y\frac{\sqrt{3}}{2}\right)\right) \right. \\ \quad \left. - \cos\left(k \sin\alpha \left(-\frac{x}{2} + y\frac{\sqrt{3}}{2}\right)\right)\right)^2 \\ \quad + \left(- \sin(kx \sin\alpha) + \sin\left(k \sin\alpha \left(-\frac{x}{2} - y\frac{\sqrt{3}}{2}\right)\right) \right. \\ \quad \left. - \sin\left(k \sin\alpha \left(-\frac{x}{2} + y\frac{\sqrt{3}}{2}\right)\right)\right)^2 \end{array} \right]. \tag{8}$$

Under these conditions, the periodic dot pattern shown in Fig. 2(b) is obtained.

If a fourth beam is added, and considering ($E_{01} = E_{02} = E_{03} = E_{04} = E_0$; $\alpha_1 = \alpha_2 = \alpha_3 = \alpha$, $\alpha_4 = 0$; $\beta_1 = \beta_4 = 0$, $\beta_2 = 2/3\pi$, $\beta_3 = -2/3\pi$), the periodic pattern shown in Fig. 2(c) is obtained. The intensity distribution shows an intense central peak surrounded by six peaks with a lower intensity arranged in a hexagonal form.

3. Design of Periodical Structures

In order to apply the laser interference structuring method, the configuration of laser beams that produce desired interference pattern and hence energy distribution on the surface of the sample has to be calculated prior to the experiment.[27] Calculating the configuration of electromagnetic waves that reproduce a desired interference pattern is an inverse problem[28] and its solution is not known in general.

Each laser beam "j" with wave vector k_j contributes a new degree of freedom to the interference pattern. With relatively few laser beams, the geometry and symmetry of the interference pattern remain simple. More complicated structures will in general require more generating laser beams for the construction of the interference pattern.[28,29]

Mützel *et al.*[28] have presented a numerical method to solve the inverse problem in the general case that the laser beams are not in phase. For the laser configuration used in the present study, all laser beams are in phase and have the same angular frequency. Therefore, the mathematical procedure of solving the inverse problem can be simplified in the following way.

Since the pattern is described in terms of individual beams with wave vectors k_j, it is convenient to examine the pattern in *k-space*[28] by applying the discrete Fourier transformation (FT) to the desired interference pattern (see Figs. 3(a) and 3(b)). To reproduce the target pattern, the most significant points (vectors k_j) in Fourier space must be used. These vectors contain all the information which is necessary to reproduce the pattern and are determined by the Ewald-sphere.[28,30] The centre of the sphere is given by the coordinate $(\bar{k}_x, \bar{k}_y)$ which are the mean value of the k_{xi} and k_{yi} coordinates of all the k-vectors respectively. There after, the angles of each partial beam parallel to the interference plane (β_j) can be calculated as indicated in Fig. 3(c).[27]

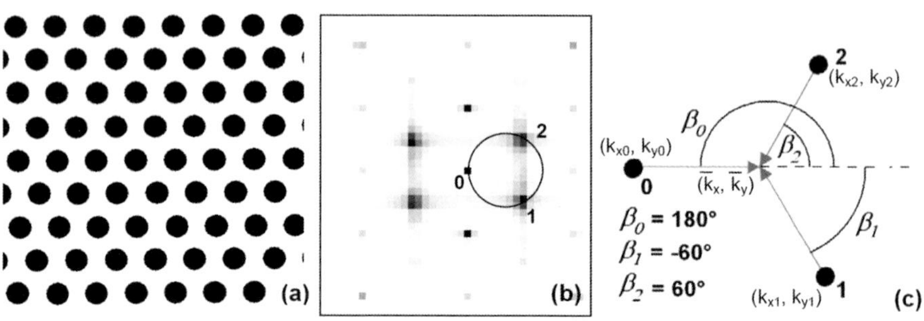

Fig. 3. (a) Desired interference pattern and (b) 2D-Fourier transformation of (a). The most relevant points of the FT are indicated with 0, 1, and 2. (c) Calculated β_j angles of each wave vector from (b).[27]

In our experimental setup the laser beams are not confined to 2D-space, but strike the target with a certain angle α_j (Fig. 2). The projection of the wave vector onto the diffraction plane must then be equal to the wave vector in 2D-space given in Fig. 3(c). This constraint defines the values of the angles α_i between the normal to the interference plane (Fig. 2(e)). In addition, the angles α_j of the beams are related to each other by

$$\alpha_j = \cos^{-1}\left(\cos\alpha_i\sqrt{(k_{xj}-\bar{k}_x)^2+(k_{yj}-\bar{k}_y)^2}\Big/\sqrt{(k_{xi}-\bar{k}_x)^2+(k_{yi}-\bar{k}_y)^2}\right). \tag{9}$$

The angles α_j given in Eq. (9) in the general case have to be calculated iteratively in order to obtain a specific period. As starting value, α_{ini} of the line-type pattern case has normally been used[10]

$$\alpha_{\text{ini}} = \sin^{-1}(\lambda/2P), \tag{10}$$

where P is the desired period. Depending if the period of the interference pattern calculated using Eq. (5) is larger or smaller than P, α_{ini} must be increased or reduced, respectively. However, the principal shape in x, y-coordinates of the intensity distribution of the desired interference pattern is defined by the angles β_j independently of the α_j.

Figure 2(b) shows the calculated interference pattern corresponding to the desired periodical pattern of Fig. 3(b) using the beam arrangement described in Fig. 3(e). Another example of a more complex periodical pattern is shown in Fig. 4. In Fig. 4(a), the desired interference pattern is shown together with the identified wave vectors (Fig. 4(b)), and the laser-beam configuration required to reproduce the pattern (Fig. 4(c)). The calculated interference pattern using this last arrangement is given in Fig. 4(d).

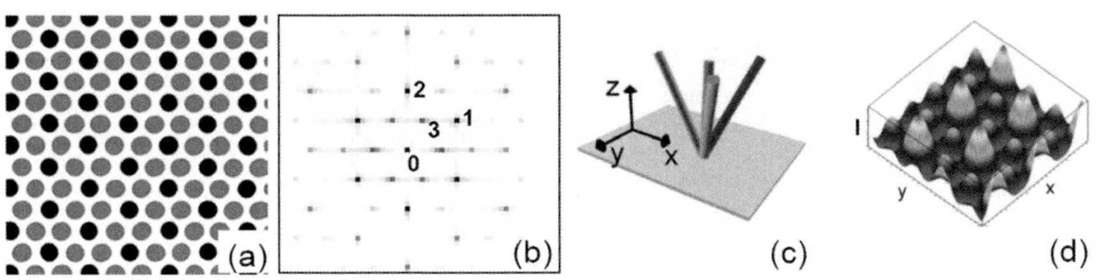

Fig. 4. (a) Desired periodical pattern; (b) 2D-Fourier transformation of (a); and (c) calculated laser-beams arrangement according to (b). (d) 3D-representation of the intensity distribution of the interference pattern required to obtain (a).[27]

4. Laser Interference Patterning System

In order to achieve high temperatures at the material's surface, a high-power pulsed laser must be utilized for the laser interference experiments. Additionally, absorption of laser light by materials is strongly dependent on wavelength. Consequently, for the laser experiments it is necessary to select an appropriate laser system which covers these requirements. A good option is the Nd:YAG laser system with an harmonic generator that permits to obtain laser beams with wavelengths ranging from the ultraviolet (UV) to the infrared spectrum (266, 355, 532, and 1064 nm). Additionally, to obtain interference patterns, a sufficient coherence length (L_c) is necessary. The coherence length is defined as the distance over which the output beam maintains a fixed phase relationship (Eq. (11)) and depends on the characteristic of the laser system[31]

$$L_C = \frac{c}{\Delta v}, \tag{11}$$

where c is the speed of light and Δv is the line-width.

The line-width of a laser is the width of its power spectral density. For example, for the Nd:YAG laser (Quanta-Ray PRO 290), the line-width is about 1 GHz which produces a coherence length of more than 1 m.

The schematic setup for the Laser Interference Experiments is shown in Fig. 5. The primary laser beam is split into two or more beams to interfere with each other on the sample surface. By the combination of a polarizer and a half-wave plate, it is possible to adjust the intensity impinging on the sample. The intensity can also be adjusted by a mechanical shutter to choose the number of pulses. By using a two-lens consisting telescope, one can change the diameter of the laser beam and therefore

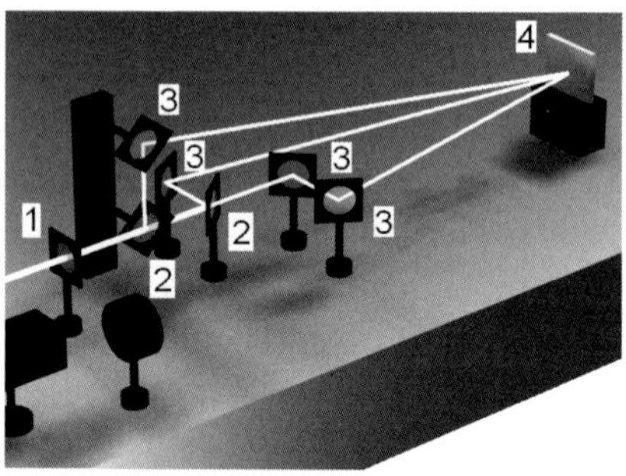

Fig. 5. Schematic setup for the laser interference system with the optical elements: (1) lens; (2) beam-splitters; (3) mirrors; (4) sample. Note that the primary beam is divided into three sub-beams which are overlapped at the sample surface.[21] Reproduced with permission — Copyright Carl Hanser Verlag GmbH & Co. KG.

control the energy density received on the sample surface. To reduce influences from the optical elements, high flatness of the splitters and mirrors are needed to construct the optical system. The angles in between the laser beams are changed in order to produce different periodical structures as was explained in Sec. 2.

5. Thermal Simulation

Thermal simulations are of great help to calculate the temperature distribution for different laser intensities (laser fluence) induced by the interference patterns, as well as the depth of the molten or even vaporized regions.[25,32] The thermal simulation is based on the heat diffusion equation[33]

$$\rho c_p \frac{\partial T}{\partial t} = q_a - q_m - q_v + \nabla(k \nabla T), \tag{12}$$

where $T = T(x, z, t)$ is the temperature at the position (x, z) at time t, q_a the heat added, q_m and q_v the heat required to melt and vaporize the material respectively, c_p the specific heat, k the thermal conductivity, and ρ the density.

For the analysis the following assumptions were made: (1) no radiation loss from surface, (2) no convection above the surface, (3) when the material melts there is no convection due to gravitational effects, (4) the energy of the laser is considered as ten periods of a sinusoidal function including a Gaussian distribution in time, (5) the heat conduction problem includes solid–liquid and liquid–vapor transitions, considering the latent heat required to melt and vaporize the metals, and (6) the interaction of the material with the laser is considered to be pure photothermal.

The energy of the laser is introduced into the metals following the intensity distribution of the interference pattern I (see Eq. (6)) given by Eqs. (13) and (14)[10]

$$q_a = \alpha \frac{I}{\sigma\sqrt{2\pi}} \exp\left(-\frac{(t-t_p)^2}{2\sigma_0^2}\right)(1-r)\exp(-\alpha z), \tag{13}$$

$$\sigma = \frac{\tau_p}{2\sqrt{2\ln 2}}, \tag{14}$$

where α is the absorption coefficient, t the time, t_p the pulse-time, τ_p the pulse-duration (FWHM), and r is the reflectivity of the surface.

In general, for metals with relatively low reflectivity values (<70%) at room temperature, the reflectivity can be considered to be constant with temperature.[20] This is the case of metals like copper and stainless steel (SS) irradiated at 355 nm of wavelength showing reflectivity values of 41.5% and 55.6%, respectively. However, aluminum presents a reflectivity of 93.5% at the same wavelength.[34] Experimental observations have shown that the reflectivity of aluminum can reduce to values of 65% during laser irradiation at the melting point.[35] Consequently, this must be considered for the thermal simulation using the following equation[36]:

$$r_{Al} = \begin{matrix} 1.048 - 3.989 \cdot 10^{-4} T & 298\text{ K} < T < T_{mAl} \\ 0.65 & T > T_{mAl} \end{matrix}, \tag{15}$$

where T_{mAl} is the melting point of aluminum.

We assume that the liquid–solid fraction, which represents the fraction of molten metal in the solid, is given by Eq. (16):

$$f_{SL} = 0.5 \text{ erfc} \left(\frac{T - T_m}{\Delta T_0} \right), \tag{16}$$

where T is the local temperature, T_m the melting temperature, ΔT_0 a temperature range over which the melting transition occurs, and erfc the complementary error function.[26] The value of f_{SL} varies from 0 to 1. For $f_{SL} = 1$, the metal is in solid state and for $f_{SL} = 0$ it is completely molten. The temperature range ΔT_0 must be defined in order to give the simulation a smooth function for the amount of molten material so that it is possible to calculate the local temperature at each element (for the simulation we assumed $\Delta T_0 = 50$ K).[37]

Then, the heat absorbed during the phase transition is

$$q_m = L_m \rho \frac{\partial \left[0.5 \text{ erfc} \left(\frac{T - T_m}{\Delta T_0} \right) \right]}{\partial t}. \tag{17}$$

The same analysis was performed for the liquid–vapor transition. More details about the thermal simulation have been already published in Refs. 17 and 25.

6. Practical Examples

6.1. *Topographic Design in Bulk Metallic Substrates*

Figure 6 shows different periodic line-type patterns fabricated on copper and SS substrates. The grating period was of 4.90 and 2.00 μm for SS and copper, respectively. The results indicate that interference patterns with relatively large periods (>3 μm) produce more homogeneous structures (Fig. 6(d)). If interference patterns with small periods are used, the obtained structures are less homogeneous and present discontinuities as observed in Fig. 6(a). However, periodic structures in SS substrates are more homogeneous compared to both copper and aluminum substrates for a specific period. This is attributed to the lower thermal diffusion length of SS compared to Cu and Al. The value of the thermal diffusion length provides an idea of the length in which the heat introduced by the laser radiation at the interference

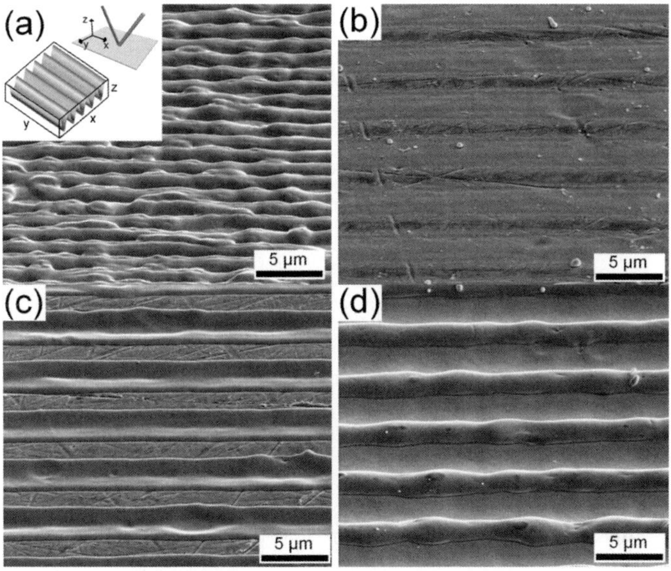

Fig. 6. SEM micrographs of structured (a) copper (Period = 2 μm, 2.15 J/cm^2) and (b–d) stainless steel substrates (Per. = 4.9 μm). (b) 0.9 J/cm^2; (c) 1.3 J/cm^2; (d) 1.8 J/cm.^{2} Tilt: 52°.[36] Reproduced with permission — Copyright Elsevier Limited.

maxima is evacuated. Thus, metals like Cu and Al will show a lower temperature gradient between interference maxima and minima resulting in less homogeneous patterns.

As observed in Fig. 6, the pattering results from the movement of molten metal from the hot regions (interference maxima) toward the cold regions (interference minima). This movement of molten metal is driven by the surface tension gradient.[38,39] If higher laser intensities (laser fluence) are used, the amount of molten metal and the time for which it remains in the liquid state increases.

Consequently, the liquid metal moving towards the interference minimum encounters fluid moving in the opposite direction (coming from the subsequent interference maximum) producing a sinusoidal-like profile as shown in Fig. 6(d). Using the lubrication approximation (quasi-steady state) and neglecting capillary forces,[39] a first estimate of the possibility of a mechanism based on the Marangoni flow can be made. This assumption is valid since instabilities generated by thermocapillary stresses produce the formation of protrusions along the ripples-line[39,40] and these are not observed in our substrates. The average velocity of lateral displacement from hotter to cooler regions can be written as[41]

$$\bar{v}(x,t) = \frac{h}{2\eta}\frac{d\gamma}{dT}\frac{\partial T}{\partial x}, \tag{18}$$

where h is the thickness of the molten surface layer, η is the dynamic viscosity, and γ is the surface tension. The time t^* required for fluid convection considering a line-type periodical pattern with period P is:

$$t^* = \frac{P}{2\bar{v}}. \tag{19}$$

Combining Eqs. (18) and (19), and considering that the thermal gradient $\partial T/\partial x$ is given by the temperature difference between maxima and minima over the half period ($2\Delta T/P$) we obtain

$$t^* = \frac{\eta P^2}{2h\Delta T}\left|\frac{d\gamma}{dT}\right|^{-1}. \tag{20}$$

The values of ΔT and h can be calculated using the thermal simulations as shown in Fig. 7(a), where the molten depth at the interference maxima positions was calculated for the three utilized substrates. In the case of SS, the surface tension of the liquid varies linearly with temperature ($\partial\gamma/\partial T = 4.9 \times 10^{-4}\,\mathrm{N\,m^{-1}K^{-1}}$).[41] The dynamic viscosity η of the molten metal was taken as an average within a temperature range ($\eta = 1.4 \times 10^{-2}\,\mathrm{Kg\,m^{-1}s^{-1}}$).[42] For a laser fluence of 1.25 J/cm,^{2} and a period P of 3.51 μm, the values of ΔT and h are 3010 K and 0.4 μm, respectively. Substituting these values in Eq. (20), a time of 146 ns is calculated. This estimate suggests that liquid convection driven by the surface tension gradient is possible since the melting range duration is on the same order (56 ns calculated using the thermal simulations for the conditions described above). At higher laser fluences (~2–3 J/cm^2), in the case where evaporation also takes place, the pressure of the boiled metal over the interference maxima positions also contributes to the removal of material. Similar results were calculated for copper and aluminum as depicted in Table 1.

Comparing the evolution of the *structure depth* (defined as the vertical size of the periodic structures, Fig. 7(b)) and the depth of the molten pool at the interference maxima (Fig. 7(a)) with the laser fluence, the following observations can be made. Firstly, both experimental results and thermal simulations indicate that SS samples are molten at lower laser fluences than aluminum and copper in that order (see insert in Fig. 7(a)). This result can be explained due to the high absorptance and the low thermal diffusivity of SS at the utilized wavelength (355 nm). Secondly, aluminum and copper start to melt practically at the same laser fluence, which confirms that

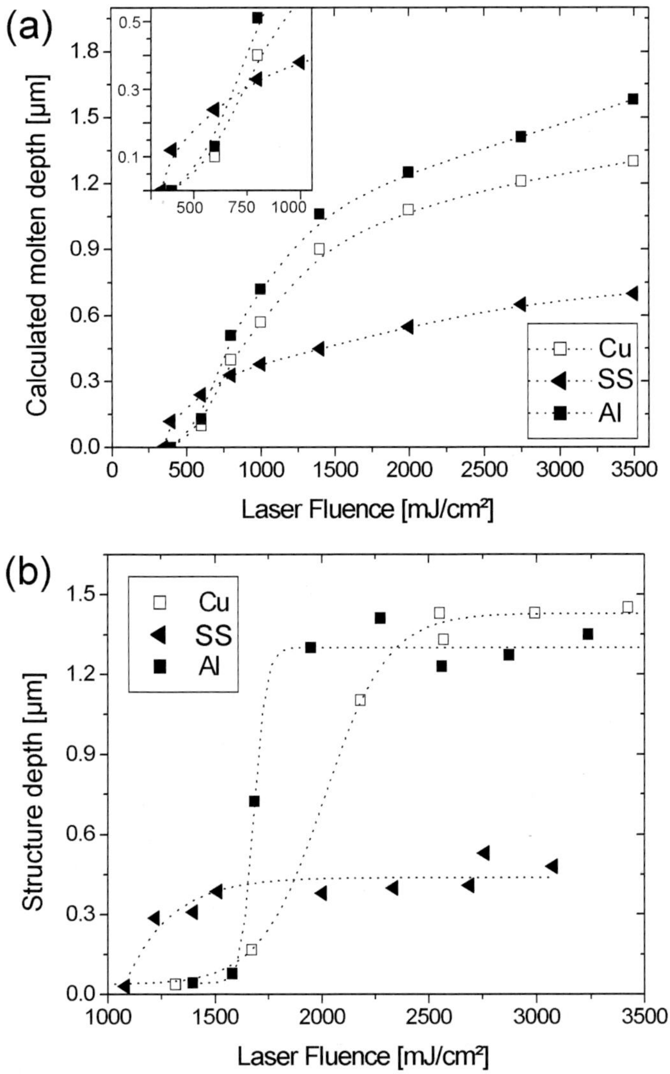

Fig. 7. (a) Calculated molten depth (thermal simulation) and (b) structure depth evolution (experimental results), for copper (Cu, period = 3.51 μm), aluminum (Al, period = 4.7 μm) and stainless steel (SS, period = 3.51 μm) substrates. The insert in (a) describes the laser fluence range from 400 to 1000 mJ/cm^2.[36] Reproduced with permission — Copyright Elsevier Limited.

Table 1. Parameters required for the calculation of the time (t^*) required for fluid convection.

	SS	Cu	Al
LF (J/cm^2)[a]	1.25	1.6	1.6
ΔT (K)[b]	3010	1950	1820
h (μm)[b]	0.4	0.92	0.97
$d\gamma/dT$ ($N \cdot m^{-1} K^{-1}$)[c]	−0.00049	−0.00014	−0.00035
η ($Kg\ m^{-1}\ s^{-1}$)[c]	0.014	0.002	0.0014
P (μm)[a]	3.51	4.38	4.7
t^* (ns)	146	76	25

[a] Laser experiment conditions.
[b] Thermal simulation parameters.
[c] Properties of the molten metals (SS: stainless steel).[36]

the reflectivity of aluminum strongly decreases during the laser irradiation process to values of about 65% (as described in Eq. (15)). Additionally, according to the thermal simulations, the laser fluences required to melt the metallic substrates at the interference maxima positions are lower than the laser intensities at which the periodical structures of the type shown in Fig. 6(d) are formed. Although it is possible to melt the metallic substrates at the interference maxima at these intensity values, both the time in which the material remains in liquid state and the thermal gradient are not large enough to permit the surface tension driven structuring mechanism (Fig. 6(b)).

6.2. *Microstructure Design in Thin Metallic Films*

6.2.1. *Grain-size distribution and texture*

As far as technical applications are concerned, microstructural randomness leads to highly inhomogeneous and nonoptimized device characteristics.[43] Consequently, circumventing the random grain alignment by a laser-interference induced localized recrystallization with specified threshold energies results in a control of nucleation sites and grain orientations and is therefore of basic importance.[44]

Application of the LIMET method to amorphous or nanocrystalline thin films can lead to changes in their structural properties. The films are molten during the exposure to a laser pulse with intensity above a certain threshold, and solidify again so that crystallization or recrystallization occurs at the local and periodic heated zones.

In the particular case of Super Lateral Growth regime (SLG), it is possible to induce different effects such as the formation of large-grained polycrystal microstructures, directionally solidified microstructures, or location controlled single-crystal regions in which the grain sizes are many times greater than the film thickness (up to several micrometers).[43,45]

Thin metallic films of Platinum deposited on Si substrate ([001] oriented) were irradiated with interference patterns in air at normal conditions of temperature and pressure using one single laser pulse. Using the two-beam configuration, a line-like pattern with a period of about 8 μm was produced (Fig. 8(a)). The dot-like pattern shown in Fig. 8(c) was obtained using three laser beams with a symmetric geometrical configuration (see Eq. (8)), whereas the cross-like patterns (Fig. 8(b)) were produced using two laser beams and a 90° sample rotation between the irradiation steps. In Figs. 8(d) and 8(f), three different regions can be identified. Region 1 corresponds to the unmodified surface at the interference minima positions with a *nanograined structure*. In region 2, a *fine-grained* microstructure is observed, while in region 3, *large*

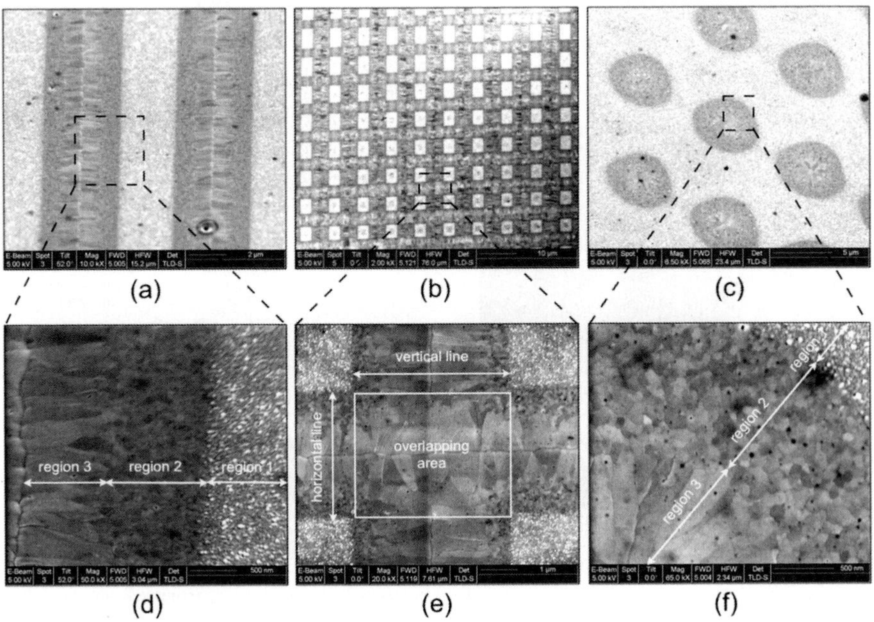

Fig. 8. Laser-induced crystallisation on Pt thin films. (a, d) line-like pattern (476 mJ/cm^2) indicating three different crystalline regions; (b, e) lattice-like pattern (709 mJ/cm^2); (c, f) dot-like structure (859 mJ/cm^2) with the grains radially distributed (region 3).[46] Reproduced with permission — Copyright Wiley-VCH Verlag GmbH & Co. KGaA.

grains oriented perpendicular to the interference fringes are produced (interference maxima regions). These long grains originate from the nonmolten regions of the film and meet each other at the centre of the laser-treated area (SLG).[43,47,48] For the dot-like structure obtained using three laser beams, the grains are radially distributed also following the thermal gradient (Fig. 8(f)). In addition, for the cross-like structures (Fig. 8(e)), the Pt layer is molten twice at the intersection between the interference lines, and thus a different grain-morphology can be obtained.

Changes in the film microstructure (e.g. texture and grain-size distribution) of the laser-treated Pt thin films were analyzed using Electron Backscatter Diffraction (EBSD). The determination of the orientation of the crystal ensemble and the grain-size distribution has been performed on the cross-like pattern showed in Fig. 9(c).

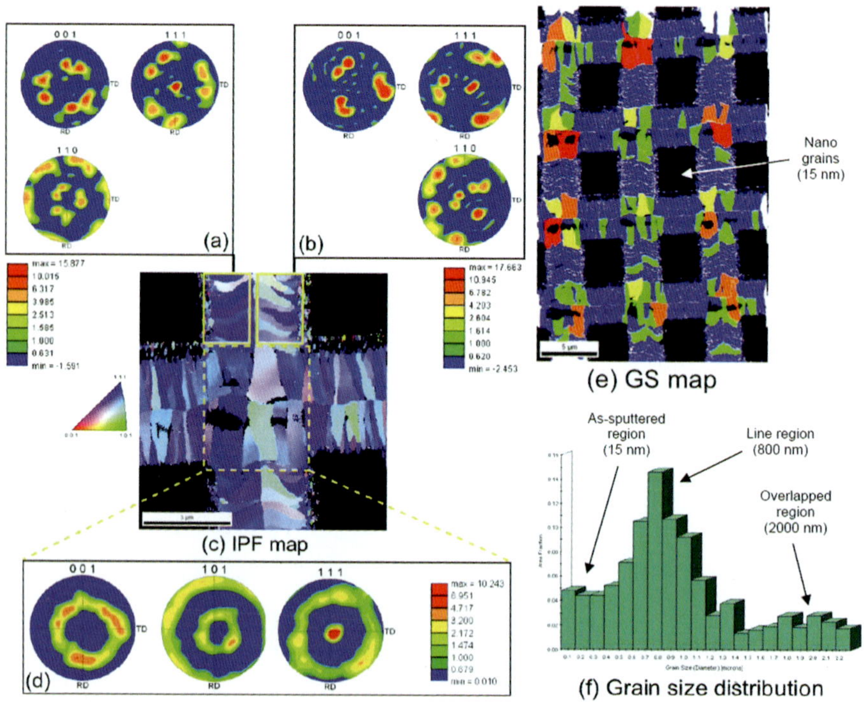

Fig. 9. Analysis of texture (a–d) and grain-size distribution (e–f) performed on cross-like structured Pt thin films. Two kinds of measurements are performed in order to find the orientation of grains in the overlapping regions and also in the vertical and horizontal regions. The grain-size distribution (f) evaluated from the GS map (e) reveals the grain-size difference between the overlapping area and the vertical and horizontal regions.[46] Reproduced with permission — Copyright Wiley-VCH Verlag GmbH & Co. KGaA.

At the interference maxima positions (excluding the overlapped zones due to the use of two laser pulses), the grains are oriented perpendicular to the interference lines (Fig. 9(c)). The grain orientation analysis performed at these positions shows that the most frequent grain-growth direction is the ⟨110⟩ direction and less frequently, the ⟨211⟩ direction. In addition, the grain size of these large grains is between 0.4 and 1 μm, with a maximum at 0.8 μm (Fig. 9(f)). On the other hand, in the regions where Pt layer was two times molten (see interrupted line in Fig. 9(c)), the grain size increases up to about 2 μm as shown in Fig. 9(f). The black regions in Fig. 9(e) describe the nanosized grains from the as-sputtered thin films, where grain-size determination is not possible due to the resolution limit of the EBSD detector.

Texture analysis on the cross-like pattern reveals a ⟨111⟩ direction, parallel to the normal of the sample (see pole figure in Figs. 9(a), 9(b), and 9(d)). Two domains were differentiated to perform the texture analysis. In Figs. 14(a) and 14(b), the two analyzed domains of both parts of one interference maximum are shown (see Fig. 9(c)). Analyzing these two domains separately leads to completely different texture as shown in Figs. 9(a) and 9(b). However, the texture analysis of the overlapped area of two interference maxima (see interrupted line in Fig. 9(c)), presents a fiber texture (see Fig. 9(d)).

The determination of mechanical properties of periodically re-crystallized Pt thin films has been performed using a nanoindentation test. The sample surface was subjected to a loading — unloading cycle with a Berkovich diamond tip. Using the calculation method developed by Oliver and Pharr,[49] local measurements of both hardness (H) and Young modulus (E) of the thin Pt films have been carried out.

The results indicate that hardness depends on the microstructure of the evaluated regions (Fig. 8(d)). At the interference maxima positions (with larger grains, R3 in Fig. 8(d)), the maximal hardness value is obtained (5.6 GPa). In the region of Fig. 8(d), the hardness lightly reduces to 5.51 GPa. However, at the interference minima positions (as-sputtered area) the hardness decreases up to 4.3 GPa, showing a difference of 23%. The differences between both systems can be explained by the inverse Hall-Petch behavior for the Pt layer.

6.2.2. *Long-range order intermetallic formation*

Due to the steep temperature gradient across the interference pattern, the process of laser interference structuring can also be used for local phase formation, specifically intermetallics. The intermetallic patterning permits the formation of surface composites and therefore, it can be a new kind of method for tailoring the mechanical properties of surfaces.[50–52]

On a substrate (silicon wafer c-Si(100)) a Ni–Al film was deposited with a film thickness of 900 nm at 400°C by magnetron sputtering technique. The atomic ratio

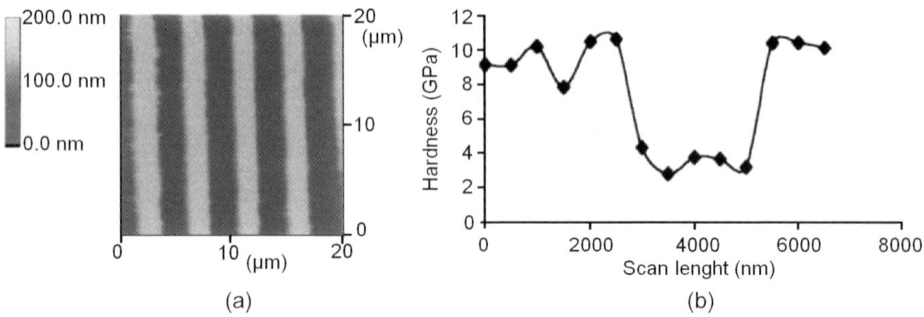

Fig. 10. AFM image (a) of the Ni–Al structured sample, and (b) the corresponding hardness distribution along the scan length in one period. The line is only indicated for the assistance of eyes.[54] Reproduced with permission — Copyright Elsevier Limited.

of the layer was kept 3:1 for Ni:Al, which is the stoichiometric ratio for Ni_3Al.[53] X-ray diffraction was applied to verify if an inter-metallic reaction has occurred between Ni and Al (for details see Ref. 54).

For characterization of the mechanical properties, a nanoindentational Atomic Force Microscope (AFM) was applied to check the local hardness of the structured surface directly within intervals of 200 nm. The hardness in the laser-treated area, where laser intensity exceeds a certain threshold level, is significantly higher than that of the area in between, where laser intensity is below this threshold level. The average hardness in the laser-treated area is close to 10 GPa, while for the "in between" area it is around 4 GPa, which is close to the untreated state (Fig. 10).

6.3. *Pattering of Polymeric Substrates*

Commercial polymer films can be easily microstructured using Laser Interference Patterning. In that way, the scope of the technique is increased since materials having well-known bulk and surface properties can be microstructured, allowing direct application, for example, in biomedical devices[55]: poly(etheretherketone) resists sterilization by radiation or heat treatment and it has been used to produce kidney dialysis machine components; poly(etherimide) is used in harmonic scalpels; polycarbonate (PC) is used in electrophysiology cathethers; and poly(imide) (PI) is used in off pump coronary artery bypass devices.[56] Moreover, the surface of already fabricated systems could be modified using this technique since it can be applied in air without altering the shape of the samples.

Figure 11 shows the surface topography of the irradiated polymethylmetacrylate (PMMA) and polystyrene (PS) substrates using 266 nm laser radiation. In the case of PMMA (Fig. 11(a)), the surface of the polymer is only ablated at high laser fluences

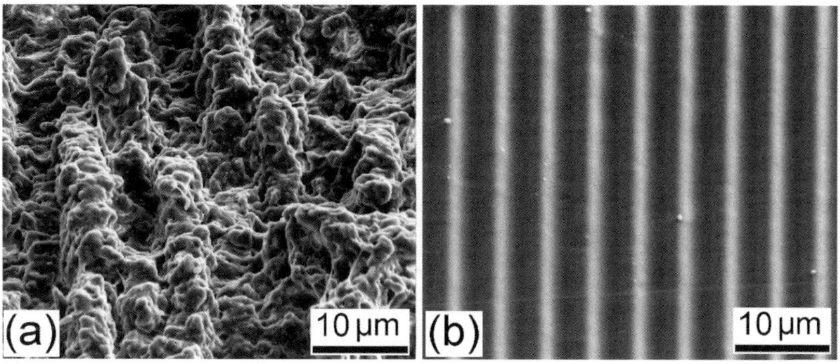

Fig. 11. Irradiated PMMA and PS samples using two laser-beams configuration. (a) PMMA, laser fluence = $2.88\,J/cm^2$; (b) PS, laser fluence = $1.80\,J/cm^2$.[60]

($2.88\,J/cm^2$) and its surface topography is completely different from PS (Fig. 11(a)). This is due to the weak absorption of PMMA at the utilized wavelength ($250\,cm^{-1}$ at 266 nm).[57] Since the chromophore in PMMA is an carbonyl group (C=O), and the transition from the unpaired electron pair (n) to the excited state of a π-bond (π^*) is forbidden, the absorption is low. PMMA ablates through subsurface superheating meaning that the material heats above its softening point below the surface producing acoustic waves due to the material's inherent elasticity.[58] These waves may lead to the development, aggregation, and subsequent bursting of microscopic bubbles resulting in the violent removal of surface material. This mechanism of ablation produces the overall destruction of the polymeric surface into a cratered structure with significant debris formation. Therefore, no periodical structures in this energy range were induced using single-pulse exposure (Fig. 2(a)). Moreover, at high intensities all of the steps needed for significant material removal through volume bubble formation can take place even using a single irradiation pulse.[58] On the other hand, for strongly absorbing polymers such as PS (Fig. 2(b)), the ablation rate is directly related to the laser fluence and pulse-duration.[59] The PS backbone has pendant phenyl groups. Those aromatic moieties have π conjugated orbitals where electronic transitions occur at lower energies (long wavelengths) than in methyacrylate monomer units. Consequently, relative low laser fluences are necessary to ablate the polymer surface. The periodical intensity modulation produced by the interference pattern results in a local and periodical removal of material obtaining a quite regular surface topography as indicated in Fig. 11(b). Similar results were observed for PC and PI substrates. PC and PI contain aryl carbonylic groups where the π-bond is conjugated with aromatic ring. The effect of conjugation is a decrease of transition energy (lower wavelength) and an increase of the absorptivity.[55]

In the case of PC, at 266 nm the laser fluence required to initiate the structuring process was 60 mJ/cm^2 while at 355 nm this value was ~1500 mJ/cm^2.[55] In comparison, for PI these laser fluences were quite similar at both wavelengths. The absorption coefficient of PC is about 100 times higher at 266 than at 355 nm while absorbance of PI does not differ very much. Therefore, these threshold laser fluences could be directly related with the absorption coefficient of the polymers.

In Fig. 12, the topographies of typical structures obtained by two- and three-beam interference patterning as a function of laser fluence in PI are shown. The period of the micropatterns was 6.2 μm. The width of the ablated lines varies from 3.3 to 6.2 μm for 312 and 1023 mJ/cm,^{2} respectively. For low laser fluences (300 mJ/cm^2), the profile of the micropatterns can be described by a sinusoidal-trapezoidal shape (Fig. 12(a)) while for high laser fluences (1000 mJ/cm^2), an U-shaped profile is observed (Fig. 12(b)).These geometries can be explained as follows. Laser ablation occurs when a certain laser intensity threshold value is exceeded. The intensity distribution of the interference pattern is described by a trigonometric function. At low fluences, only a very small area of the polymer can be ablated at the interference maxima positions (Fig. 12(a)) producing a sine-trapezoidal profile. As the laser fluence increases, a larger portion of the interference pattern surpasses the threshold energy needed to ablate the polymer resulting in wider ablated regions as shown in Fig. 12(b). For the three-beam configuration, the diameter of the ablated circular-regions can also be controlled by increasing the laser intensity.

A totally different behavior is observed for the PS–PMMA copolymers. Figure 13 shows the evolution of the surface topography for the copolymer PS-PMMA at two different laser fluence values (280 and 905 J/cm^2). Figure 13(a) indicates that a very good homogeneity of the periodical structure is obtained on a large area. The results also indicate that the copolymer PMMA-PS can be structured using the same laser fluence range of PS due to the addition of styrene monomer units with a high absorption

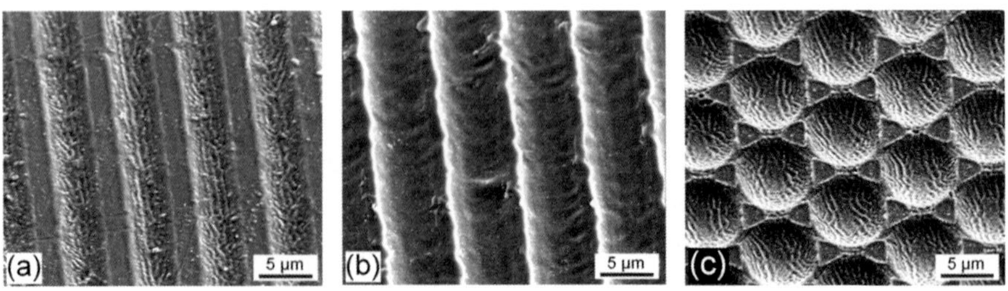

Fig. 12. Line- (a, b) and dot-like (c) in PI polymers. Laser fluences: (a) 312; (b) 1023; (c) 854 mJ/cm^2.[55] Reproduced with permission — Copyright Wiley-VCH Verlag GmbH & Co. KGaA.

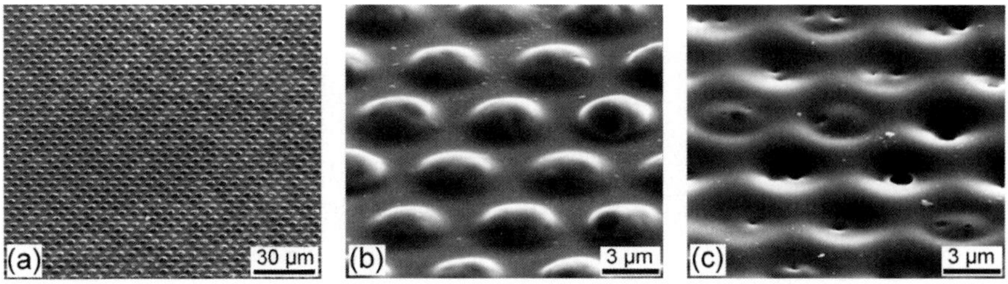

Fig. 13. Dot-like periodic structures in PMMA/PS copolymer (60:40%) substrates. (a, b) Low and (c) high energy topographic regimes. Laser fluences: (a, b) 280 and (c) 905 mJ/cm.[2]

coefficient. As observed in Fig. 13(b), when PMMA-PS is irradiated with relative low laser fluences ($\sim$280 mJ/cm^2), the regions at interference maxima positions inflate obtaining bubbles that cannot release from the material's surface due to the low local gas pressure.[61] The degradation of PMMA through photochemical reactions into methanol, carbon monoxide, and methyl formiate was confirmed by Krajnovich *et al.* Photochemical decomposition of the ester group ($-COOCH_3$), which absorbs most photon energy in this UV-light range, causes bond breakages and vaporous ablation products (CO and CO_2) that are responsible for the inflation of the polymer. Additionally, solid fragments (MMA-monomers) tend to be formed. However, according to Ref. 62, photothermal reactions also occur at a wavelength of 266 nm which produces the pyrolytic decomposition of the polymer. Thus, both photo-thermal and photo-chemical effects contribute to the ablation mechanism. Higher laser fluences will induce larger bubbles and thus larger craters as shown in Fig. 13(c). Nevertheless, a fraction of the produced bubbles remains under the polymer's surface. As observed in Fig. 13(c), at these high laser intensities values, the ablation mechanism is more similar to that of PS (or PI), which means that the addition of PS to the PMMA increases the contribution of the photothermal effect in the ablation mechanism of PMMA.

6.4. *In Vitro Cell Response of Micropatterned Polymer Surface*

Modification of polymer surfaces is used to improve biocompatibility or to provide medico-functionality of blood- and tissue-contacting medical devices.[63] Yu *et al.*[64,65] used Direct Laser Interference Patterning (DLIP) to prepare periodic micropatterns in polymers for study of contact guidance of mammalian cells.

It was observed that the cells cultured in an unpatterned surface show different forms and no orientation (Fig. 14(a)). The cells cultured in a line-like patterned substrate show elongation and orientation in the lines direction. For the mouse fibroblasts,

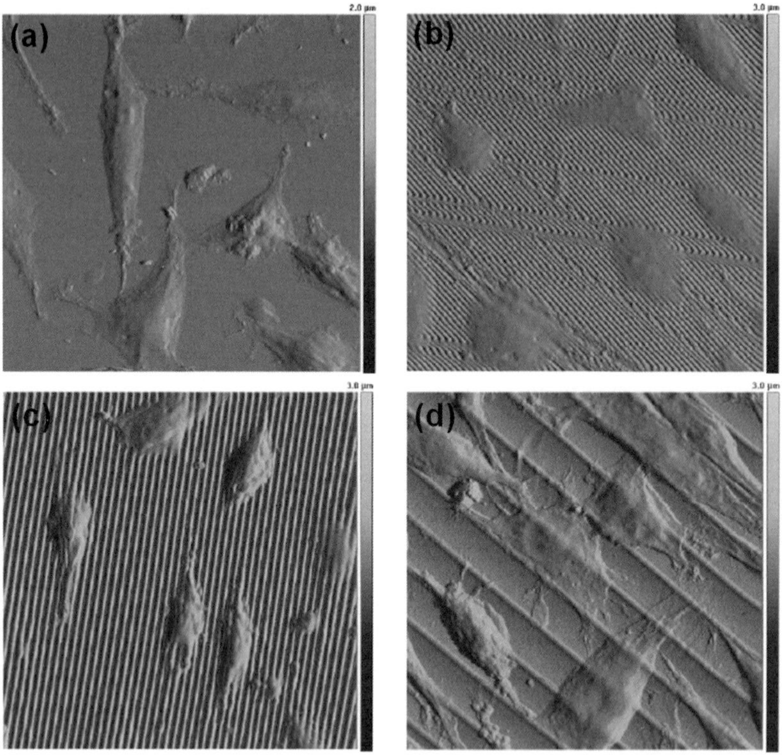

Fig. 14. Mouse fibroblasts cultured in Thermanox micropatterned by LIL. (a) Unpatterned polymer; Linear pattern with periods of (b) 1.3, (c) 2.1, and (d) 9.7 μm. The analyzed area was $100 \times 100\,\mu m^2$.[64] Reproduced with permission — Copyright IEEE.

when the period is small (1.26 μm, Fig. 14(b)) the cells are orientated but slightly elongated. If the period is larger (between 2 and 5 μm, Fig. 14(c)) the cells are highly orientated and better elongated. Cells grown in wider microgrooves (9.7 μm, Fig. 14(d)) are still orientated parallel to the microgrooves, but the effect was not as strong as for smaller periods.

To develop a model which can be generally used to explain this phenomenon, the following extended theory is suggested: cell orientation and alignment are initiated by surface chemistry, guided by the patterns available and suitable for adhesion, and occur as early as during the attachment stage. As soon as the cells are seeded to the substrates, they are sensing the environment and searching the surface positions in favor of attachments through chemical and physical signaling. On chemical micropatterned surfaces, they will find these favorable positions within the chemically modified strips, while on topographical micropatterns, the attachment is only possible on the top of ridges where the width is not large compared to the cell dimensions. Then, the cells will attach to the substrate, mainly on the area in favor of forming physicochemical linkages between cells and substrate. Protruding their plasma membrane to detect their environment, they will also find the next nearest position in pattern directions and extend along pattern features. This is the contact guidance initiated by surface chemistry.[64]

Numerous investigations reported that cell alignment was significantly influenced by the pattern depth. Clark *et al.*[66] observed that cell alignment increased with the depth of patterns whose width was in micrometer range and depth was 0.5 μm or larger. On patterns with small period, cells mainly lay on the ridges,[67,68] and tend to descend into the grooves and fill the bottom with increasing lateral pattern features.[67,69]

Acknowledgments

Based on research carried out over several years, the authors greatly acknowledge the contributions of: F. Liu, Z. Liu, D. Acevedo, R. Catrin, and C. Gachot. A. Lasagni acknowledges the following institutions for financial support: Alexander von Humboldt Foundation (2007–2008), the Deutsche Akademische Austauschdienst (2005), and the DFG Graduiertenkolleg GRK 232/3-2 (2003–2004). The research work of Prof. Mücklich was supported by the Alfried Krupp Prize for Young University Teachers awarded by the Krupp Foundation.

References

[1] Clapham PB, Hutley MC. (1973) Reduction of lens reflexion by the 'moth eye' principle. *Nature* **244**: 281–282.

[2] Wilson SJ, Hutley MC. (1982) The optical properties of "moth eye" antireflection surfaces. *Optica Acta* **29**: 993–1009.

[3] Weiner S, Traub W, Wagner HD. (1999) Lamellar bone structure-function relations. *J Struct Biol* **126**: 241–255.

[4] Chan CM, Ko TM, Hiraoka H. (1996) Polymer surface modification by plasmas and photons. *Surf Sci Report* **24**: 1–54.

[5] Xia Y. (2004) Soft lithography and the art of patterning. *Adv Mater* **16**: 1245–1246.

[6] Geissler M, Xia Y. (2004) Patterning: Principles and some new developments. *Adv Mater* **16**: 1249–1269.

[7] David C, Souvorov A. (1999) High efficiency Bragg-Fresnel Lenses with 100 nm outermost zone width. *Rev of Sci Instr* **70**: 4168–4173.

[8] Cumpston BH, Ananthavel SP, Barlow S, Dyer DL, Ehrlich JE, Erskine LL, Heikal AA, Kuebler SM, Lee I-YS, McCord-Maughon D, Qin J, Rockel H, Rumi M, Wu XL, Marder SR, Perry JW. (1999) Two-photon polymerization initiators for three-dimensional optical data storage and microfabrication. *Nature* **398**: 51–54.

[9] Jakubiak R, Bunning TJ, Vaia RA. (2004) Multi-component active photonic structures via holographic photopolymerization. *Nanotechnol E-Bull* **1.1**: 3–4.

[10] Mücklich F, Lasagni A, Daniel C. (2005) Laser Interference metallurgy — periodic surface patterning and formation of intermetallics. *Intermetallics* **13**: 437–442.

[11] Blatter A, Maillat M, Pimenov SM, Shafeev GA, Simakin AV. (1998) Lubricated sliding performance of laser-patterned sapphire. *Trib Lett* **4**: 237–241.

[12] Phillips HM, Callahan DL, Sauerbrey R, Szabó G, Bor Z. (1991) Sub-100 nm lines produced by direct laser ablation in polyimide. *Appl Phys Lett* **58**: 2761–2763.

[13] Heintze M, Santos PV, Nebel CE, Stutzmann M. (1994) Lateral structuring of silicon thin films by interference crystallization. *Appl Phys Lett* **64**: 3148.

[14] Kelly MK, Rogg J, Nebel CE, Stutzmann M, Kátai S. (1998) High-resolution thermal processing of semiconductors using pulsed-laser interference patterning. *Phys Stat Sol (a)* **166**: 651–657.

[15] Fukumura H, Kohji Y, Nagasawa K, Masuhara H. (1994) Laser implantation of pyrene molecules into poly(methyl methacrylate) Films. *J Am Chem Soc* **116**: 10304–10305.

[16] Shoji S, Kawata S. (2000) Photofabrication of three-dimensional photonic crystals by multibeam laser interference into a photopolymerizable resin. *Appl Phys Lett* **76**: 2668–2670.

[17] Lasagni A, Seyler M, Holzapfel C, Maier W, Mücklich F. (2005) Periodical gratings in mixed-oxide films by laser-interference irradiation. *Adv Mater* **17**: 2228–2232.

[18] Daniel C, Mücklich F, Liu Z. (2003) Periodical micro-nano-structuring of metallic surfaces by interfering laser beams. *App Surf Sci* **208–209**: 317–321.

[19] Daniel C, Mücklich F. (2004) Electrical behavior of periodically micro-structured Sn/$CuSn_4$ contacts under fretting conditions. *Wear* **257**: 266–270.

[20] Metev SM, Veiko VP. (1998) *Laser-Assisted Micro-Technology*, 2nd ed. Springer, Berlin, Germany.

[21] Mücklich F, Lasagni A, Daniel C. (2006) Laser interference metallurgy — using interference as a tool for micro/nano structuring. *Zeitschrift für Metallkunde* **97**: 1337–1344.

[22] Etspüler M. (2000) Working at industrial breakthrough of "moth-eye". *Fraunhofer Mag* **2**: 42.

[23] Hadobás K, Kirsch S, Carl A, Acet M, Wassermann EF. (2000) Reflection properties of nanostructure-arrayed silicon surfaces. *Nanotechnology* **11**: 161–164.

[24] Mele E, Pisignano D, Varda M, Farsari M, Filippidis G, Fotakis C, Athanassiou A. (2006) Smart photochromic gratings with switchable wettability realized by green-light interferometry. *Appl Phys Lett* **88**: 203124.

[25] Lasagni A, Holzapfel C, Mücklich F. (2005) Periodic pattern formation of intermetallic phases with long range order by laser interference metallurgy. *Adv Eng Mater* **7**: 487–492.

[26] Lasagni A. (2006) *Advanced Design of Periodical Structures by Laser Interference Metallurgy in the Micro/nano Scale on Macroscopic Areas*. Saarbrücker Reihe Materialwissenschaft und Werkstofftechnik, Band 7, Shaker-Verlag, Germany (PhD Thesis).

[27] Lasagni A, Holzapfel C, Mücklich F. (2006) Design and calculation of advanced microstructural patterns by laser interference metallurgy. *Proc Mat Res Soc Symp* **890**: Y04.02.1.

[28] Mützel M, Meschede D. (2005) Complex light-mask generation for atom guiding by time and frequency domain intensity superposition. *Las Phys* **15**: 95–98.

[29] Drodofsky U, Stuhler J, Schulze T, Drewsen M, Brezger B, Pfau T, Mlynek J. (1997) Hexagonal nanostructures generated by light masks for neutral atoms. *Appl Phys B* **65**: 755–759.

[30] Lipson S, Lipson H, Tannhauser D. (1997) *Optik*, 3rd ed. pp. 171–205. Springer, Germany.

[31] Poprawe R. (2005) *Lasertechnik fuer die Fertigung*. Springer-Verlag, Berlin, Heidelberg.

[32] Lasagni A, Mücklich F. (2005) Study of the multilayer metallic films topography modified by laser interference irradiation. *Appl Surf Sci* **240**: 214–221.

[33] Incropera FP, Dewitt DP. (2002) *Fundamentals of Heat and Mass Transfer*, 5th ed. John Wiley & Sons, New York, USA.

[34] Modest MF. (2001) *Handbook of Laser Materials Processing*. Magnolia Publishing Inc., Orlando, USA.

[35] Steen WM. (2003) *Laser Material Processing*, 3rd ed. Springer London Limited, USA.

[36] Lasagni A, D'Alessandria M, Giovanelli R, Mücklich F. (2007) Advanced design of periodical architectures in bulk metals by means of laser interference metallurgy. *Appl Surf Sci* **254**: 930–936.

[37] Lasagni A, Soldera F, Mücklich F, Kaiser T, Hrastnik K. (2004) Determination of the cathode erosion and temperature for the phases of high voltage discharges using FEM simulations. *Modell Simul Mater Sci Eng* **12**: 835–844.

[38] Anthony TR, Cline HE. (1977) Surface rippling induced by surface-tension gradients during laser surface melting and alloying. *J App Phys* **48**: 3888–3894.

[39] Yingfeng G, Pedraza AJ. (2004) Synthesis, Characterization and modeling of self-organized nanostructure arrays generated by laser irradiation. *Proc Mat Res Soc Symp* **818**: M11.47.

[40] Carles P, Troian SM, Cazabat AM, Heslot F. (1990) Hydrodynamic fingering instability of driven wetting films: hindrance by diffusion. *J Phys Condens Matter* **2**: SA477–SA482.

[41] Cazabat AM, Heslot F, Troian SM, Carles P. (1990) Fingering instability of thin spreading films driven by temperature gradients. *Nature* **346**: 824–826.

[42] Wijs GA, Kresse G, Vočadlo L, Dobson D, Alfè D, Gillan MJ, Price GD. (1998) The viscosity of liquid iron under Earth's core conditions. *Nature* **392**: 805–807.

[43] Im JS, Crowder MA, Sposili RS, Leonard JP, Kim HJ, Yoon JH, Gupta VV, Jin Song H, Cho HS. (1998) Controlled super-lateral growth of Si films for microstructural manipulation and optimization. *Phys Stat Sol (a)* **166**: 603–617.

[44] Arzt E. (1998) Size effects in materials due to microstructural and dimensional constraints: A comparative review. *Acta Mat* **46**: 5611.
[45] Nebel CE, Christiansen S, Strunk HP, Dahlheimer B, Karrer U, Stutzmann M. (1998) Laser-interference crystallization of amorphous silicon: Applications and properties. *Phys Stat Sol (a)* **166**: 667–674.
[46] Catrin R, Lasagni A, Gachot C, Schmid U, Mücklich F. (2008) Microstructural design of advanced architectures in titanium/platinum thin films by laser interference metallurgy. *Adv Eng Mater* **10**: 466–470.
[47] Aichmayr G, Toet D, Mulato M, Santos PV, Spangenberg A, Christiansen S, Albrecht M, Strunk HP. (1998) Lateral grain growth during the laser crystallization of a-Si. *Phys Stat Sol (a)* **166**: 659–666.
[48] Lee M, Moon S, Hatano M, Suzuki K, Grigoropoulos CP. (2000) Relationship between Fluence gradient and lateral grain growth in spatially controlled excimer laser crystallization of amorphous silicon films. *J Appl Phy* **88**: 4994–4999.
[49] Oliver WC, Pharr GM. (1992) An improved technique for determining hardness and elastic modulus using load and displacement sensing indentation experiments. *J Mat Res* **7**: 1564–1583.
[50] Daniel C, Lasagni A, Mücklich F. (2004) Stress and texture evolution of Ni/Al multi-film by laser interference irradiation. *Surf Coat Technol* **180–181**: 478–482.
[51] Daniel C, Mücklich F. (2004) Electrical behavior of periodically micro-structured $Sn/CuSn_4$ contacts under fretting conditions. *Wear* **257**: 266–270.
[52] Daniel C, Balk TJ, Wübben T, Mücklich F. (2005) Bio-mimetic scaling of mechanical behavior of thin films, coatings, and surfaces by laser interference metallurgy. *Adv Eng Mater* 7: 823–826.
[53] Meng XK, Vehoff H. (2000) Hard multilayered thin films of metal-intermetallic Ni/Ni_3Al. *J Mater Res* **15**: 2595–2597.
[54] Liu Z, Meng XK, Recktenwald T, Mücklich F. (2003) Patterned intermetallic reaction of Ni_3Al by laser interference structuring. *Mat Sci Eng A* **342**: 101–103.
[55] Lasagni A, Acevedo D, Barbero C, Mücklich F. (2006) One-step production of organized surface architectures on polymeric materials by direct laser interference patterning, *Adv Eng Mater* **9**: 99–103.
[56] Scientific Advisory Panel on Reprocessing of Medical Devices (SAP-RMD) (2005), Health Canada, February 10–11.
[57] Sutcliffe E, Srinivasan R. (1986) Dynamics of UV laser ablation of organic polymer surfaces. *J Appl Phys* **60**: 3315–3322.
[58] Efthimiopoulos T, Kiagias C, Heliotis G, Helidonis E. (2000) Evidence of volume bubble creation during laser ablation of PMMA organic polymer. *Can J Phys* **78**: 509–519.
[59] Krajnovich D. (1997) Incubation and photoablation of poly(methyl methacrylate) at 248 nm. New insight into the reaction mechanism using photofragment translational spectroscopy. *J Phys Chem A* **101**: 2033–2039.

[60] Acevedo DF, Lasagni A, Barbero CA, Mücklich F. (2008) Micro/nano fabrication of surface architectures on polymers and copolymers using direct laser interference patterning. *Proc Mat Res Soc Symp* **1054**: FF01–FF07.

[61] Srinivasan R, Braren B, Casey KG, Yeh M. (1990) The significance of a fluence threshold for ultraviolet laser ablation and etching of polymers. *J Appl Phys* **67**: 1604.

[62] Estler R, Nogar N. (1986) Mass spectroscopic identification of wavelength dependent UV laser photoablation fragments from polymethylmethacrylate. *Appl Phys Lett* **49**: 1175.

[63] Dadsetan M, Mirzadeh H, Sharifi N. (1999) Effect of CO_2 laser radiation on the surface properties of polyethylene terephthalate. *Radiat Phys Chem* **56**: 597–604.

[64] Li P, Bakowsky U, Yu F, Loehbach C, Muecklich F, Lehr CD. (2003) Laser ablation patterning by interference induces directional cell growth. *IEEE Trans Nanobioscience* **2**: 138–145.

[65] Yu F, Li P, Shen H, Mathur S, Lehr C, Bakowsky U, Mücklich F. (2005) Laser interference lithography as a new and efficient technique for micropatterning of biopolymer surface. *Biomaterials* **26**: 2307–2312.

[66] Clark P, Connolly P, Curtis ASG, Dow JAT, Wilkinson CDW. (1991) Cell guidance by ultrafine topography in vitro. *J Cell Sci* **99**: 73–77.

[67] Walboomers XF, Croes HJE, Ginse LA, Jansen JA. (1998) Growth behavior of fibroblasts on microgrooved polystyrene. *Biomaterials* **19**: 1861–1868.

[68] Teixeira I, Abrams GA, Bertics PJ, Murphy CJ, Nealey PF. (2003) Epithelial contact guidance on well-defined micro- and nanostructured substrates. *J Cell Sci* **116**: 1881–1892.

[69] Walboomers XF, Croes HJE, Gensel LA, Jansen JA. (1999) Early spreading events of fibroblasts on microgrooved substrates. *J Biomed Mater Res* **47**: 204–212.

Chapter 11

Electrodeposition — Fundamental Aspects and Methods

Stanko R. Brankovic

The metal electrodeposition is a complex phenomenon, which spans over different fields of chemistry and thermodynamics, mechanics, metallurgy, and material science. It takes place at the solid/liquid interface which makes the process somewhat unique and very attractive. Over the years, our knowledge in electrodeposition has significantly improved leading to better definition of the governing phenomena and development of new methods for electrodeposition. The general description of the most important aspects and methods of electrodeposition is presented in this chapter.

1. Introduction

The early applications of electrodeposition were mainly confined to situations where relatively thick polycrystalline metal deposits were needed. These included protective or sacrificial metal layers for corrosion protection, decorative applications, and for coatings with specific mechanical properties. However, in the last several decades, electrodeposition has been proven to be one of the enabling fabrication methods behind the train of hi-tech enterprises.[1,2] There are many examples where electrodeposition is used as convenient if not "*the only*" approach to deliver the desired structures, materials or catalytic surfaces. Today, technology development dictates ever smaller devices with length scale in sub-100 nm regime. Electrodeposition is often recognized as the standard processing operation in magnetic recoding heads fabrication,[3,4] in microprocessors and MEMS technologies.[5–8] In recent years, the scientific community has witnessed numerous examples where electrodeposition is used to grow multilayered metallic thin films and nanostructures,[9–11] nanoscale metallic architectures,[12–18] and a high quality single crystal overlayers.[19–23] In addition to these examples, electrodeposition has been used for epitaxial growth of thin film oxides and oxide nanodots,[24,25] and for synthesis of different semiconductor superlattices and compounds.[26,27] The most recent developments suggest that electrodeposition is becoming an attractive fabrication process for many emerging fields of technology. These new applications make the future of the electrodeposition research an interesting and exciting endeavor.

2. Electrodeposition Kinetics

The thermodynamic equilibrium potential of an electrode surface in contact with its ions in the solution is defined by the well-known Nernst equation,

$$E_{M^{n+}/M} = E^0 + \frac{RT}{2.303 \cdot nF} \log[a_{M^{n+}}]. \tag{1}$$

Here, E^0 represents the equilibrium potential of the electrode surface at standard conditions ($P^\circ = 101\,\text{kPa}$, $T^\circ = 298\,\text{K}$), R is the universal gas constant, F is Faraday's constant, and T is the absolute temperature. The term $a_{M^{n+}}$ represents the activity of the metal ions in solution, with n being their oxidation state. In dilute solutions, the ion activity is approximated by their concentration, i.e. $a_{M^{n+}} \approx C_{M^{n+}}$.[28] The above expression suggests that equilibrium potential of metal electrode shifts for $\approx 0.059\,\text{V/n}$ in negative direction if the metal ion concentration in the solution is decreased ten times. This means that the actual value of the equilibrium potential can be effectively changed by adjusting the concentration of the corresponding metal ions in the solution. If the applied potential (E) to the metal electrode is more negative than $E_{M^{n+}/M}$, ($\eta = E - E_{M^{n+}/M} < 0$) the electrode is at overpotential conditions, and η is called overpotential. Under this condition, metal electrodeposition occurs, ($M^{n+} + ne^- \rightarrow M$).

During metal electrodeposition, for a given value of overpotential, the corresponding cathodic current density j [$\text{A} \cdot \text{m}^{-2}$] is observed which can be correlated to the deposition flux [$\text{mol} \cdot \text{m}^{-2}\text{s}^{-1}$] using the expression, flux $= j/nF$. The relation between the deposition current density (j) and electrode surface overpotential $\eta_S[V]$ is described by *Butler Volmer* equation defined below[28]:

$$j(<0) = j_0 \cdot \left(\exp \frac{\alpha_a F}{RT} \eta_s - \exp \frac{-\alpha_c F}{RT} \eta_s \right). \tag{2}$$

Here, j_0, [$\text{A} \cdot \text{m}^{-2}$], α_a and α_c stand for the exchange current density, anodic, and cathodic transfer coefficients. For $\eta_S \ll 0$, Eq. (2) can be simplified, yielding the well-known *Tafel* relation between the current density and overpotential;

$$j(<0) = -j_0 \cdot \exp \frac{-\alpha_c F}{RT} \eta_s. \quad (3)$$

However, the transport limitations for the ions from the bulk solution, going through the diffusion layer, towards the electrode surface also play an important role in overall j vs n dependence. The maximum deposition current density (flux) that can be reached through the diffusion layer, j_L [$A \cdot m^{-2}$], is the function of the diffusion layer thickness $\delta[m]$, ion diffusivity $D_{M^{n+}}$ [$m^2 \cdot s^{-1}$], and concentration of depositing ions in the bulk solution $C_{M^{n+}}$ [$mol \cdot m^{-3}$];

$$j_L = \frac{nFD_{M^{n+}} \cdot C_{M^{n+}}}{\delta}. \quad (4)$$

As the depositing current density approaches j_L, the significant raise of electrode overpotential occurs, and this contribution to the overall electrode overpotential is called concentration overpotential η_C. The common formulation of the concentration overpotential as a function of the ratio between actual current density and limiting current density is described as[28]:

$$\eta_C = \frac{RT}{nF} \ln \left(1 - \frac{j}{j_L}\right). \quad (5)$$

For small ratios of j/j_L, (<0.4) the contribution of the η_C to the overall electrode overpotential is small and the η vs j dependence is considered to be under kinetic control ($\eta \approx \eta_S(j)$, activation polarization), Fig. 1. However, for j/j_L being in the range between 0.4 and 0.9, the contribution of the concentration overpotential to the overall electrode overpotential cannot be neglected and the η vs j dependence is considered to be under mixed control ($\eta \approx \eta_S(j) + \eta_C(j/j_L)$). For the values of j/j_L close to 1, the overpotential of the electrode surface is predominantly defined by the transport limitation effects, and the η vs j dependence is considered to be under transport control ($\eta \approx \eta_C$, concentration polarization, Eq. (5)). The graphical presentation of these three regimes of η vs j dependence for an arbitrary electrodeposition process is shown in Fig. 1, represented by the light gray line.

During the electrodeposition of metals that have equilibrium potential more negative than the equilibrium potential for hydrogen evolution reaction like Fe, Co, Ni, etc., the co-deposition of hydrogen (hydrogen evolution) occurs in parallel to the metal deposition reaction. This means that the observed current for a given potential of the electrode surface is the sum of the partial currents that correspond to the metal deposition, j_M, and hydrogen evolution j_H, ($j = j_M + j_H$). The relative contribution of the metal deposition current to the overall current is defined by the term called

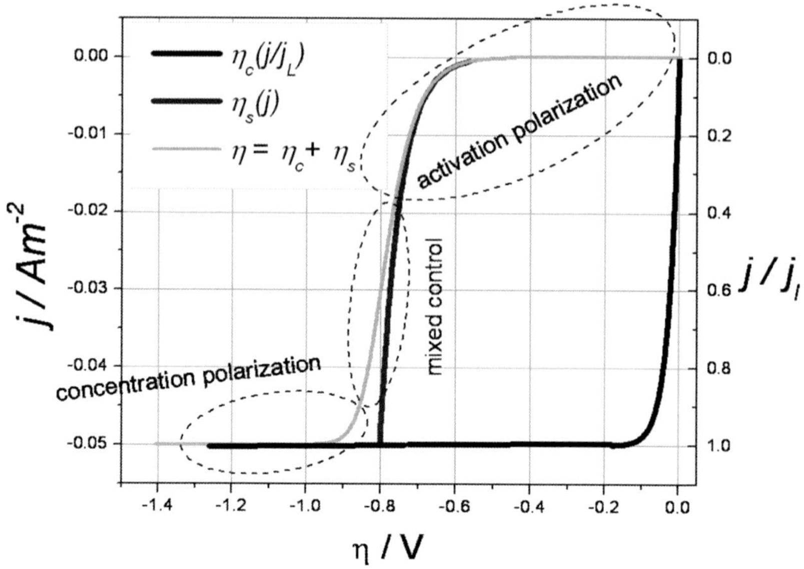

Fig. 1. Overpotential vs current density dependence (light gray) for electrodeposition reaction where $j_0 = 10^{-8}$ A m^{-2}, $\alpha_a = \alpha_c = 0.5, j_L = 0.05$ A m^{-2}. Black line is η_C vs j dependence (Eq. (5)), and the gray line is η_S vs j dependence (Eq. (2)).

current efficiency, defined as $\gamma = j_M/j$. For electrodeposition of ferromagnetic metals like Ni, Fe, or Co, $\gamma < 1$ is commonly observed and its value is directly dependent on the potential at which the electrodeposition is conducted, pH of the solution and, in some cases, on the type of additives used, their content, and characteristic of their adsorption processes.[29,30]

3. Overpotential Co-Deposition (OPCD) — Electrodeposition of Alloys

The generalized thermodynamic condition for electrodeposition of A_pB_{1-p} binary alloy is defined as[31]:

$$E_A + \eta_A = E_B + \eta_B. \tag{6}$$

Here, E_i ($i = A, B$) is the equilibrium potential of the metal-component i in the alloy A_pB_{1-p} defined by modified Nernst expression:

$$E_i = E_i^0 + \frac{RT}{2.303nF} \log \left[\frac{a_i^{n+}}{a_i^{\text{alloy}}} \right]. \tag{7}$$

The activity of the metal in the alloy is always less than 1 which yields different values of the equilibrium potential for the component in the alloy as compared to the corresponding elemental bulk metal electrode. According to Eq. (6), in order to form an AB alloy, the applied potential to the electrode surface has to be such that both components of the alloy are at overpotential conditions with respect to the equilibrium potentials of constituents in the alloy (Eq. (7)). In practice, the term OPCD is related to electrodeposition of alloys in the potential range where the applied potential is such that both components of the alloy are at overpotential conditions with respect to the equilibrium potentials of their elemental phase, defined by Eq. (1). In this case, the resulting composition of the electrodeposited alloys is controlled by the combination of several effects: (1) kinetics of the electrodeposition of each component itself, (2) transport limitations, (3) conditions at the electrochemical interface, and (4) mutual interaction of adsorbed intermediates.

The schematic indicating the potential region where OPCD occurs is shown in Fig. 2 using the example of CoFe alloy. The schematic considers electrodeposition process from the solution which is at standard conditions (P°, T°, $C_{Co^{2+}} = C_{Fe^{2+}} = 1$ mol). In order to obtain desired composition of CoFe (50:50) alloy, the concentrations of Co and Fe ions in the solution have to be appropriately adjusted together with the potential (overpotential) or current at which the alloy deposition occurs. Typical approach towards the solution and deposition potential (current) design involves experiments where the concentration of more noble metal, Co, is such that $C_{Co^{2+}} < C_{Fe^{2+}}$, so that Co deposition occurs under mixed control for a

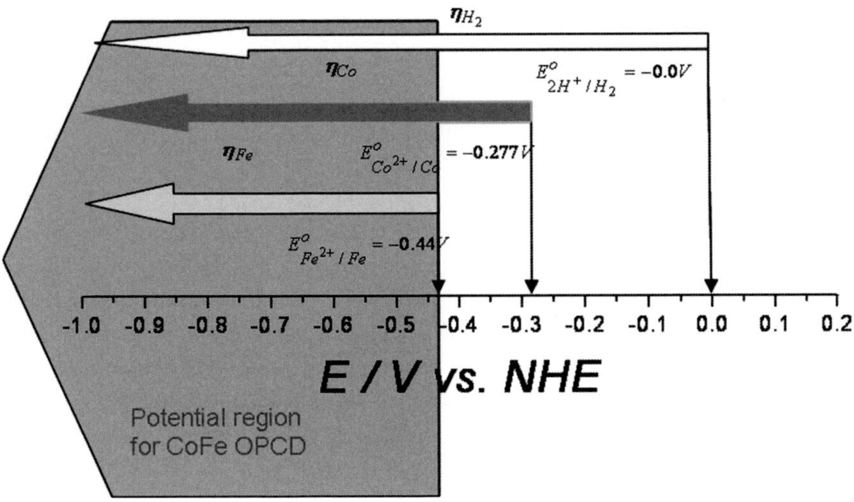

Fig. 2. Potential region for CoFe alloy OPCD.

wide range of potentials. The Fe^{2+} concentration is typically taken to be such that Fe deposition occurs under kinetic control for the chosen potential (current). The deposition potential (current) is then varied through different runs until the satisfactory composition of CoFe alloy is achieved.

4. Underpotential Deposition (UPD)

In many electrochemical systems that include a metal electrode in solution with different metal ions, an UPD phenomenon is observed. This is demonstrated by formation of one or two "*wetting*" monolayers (MLs) onto a foreign metal electrode (substrate) at potentials that are more positive than the equilibrium potential defined by Nernst equation (Eq. (1)). During 1960s and 1970s, the extensive studies of many UPD systems were performed using single crystal and polycrystalline electrodes. Good reviews from this period are offered by Trasatti[32] and Kolb.[33] The development of scanning probe methods (STM and AFM) and X-ray scattering techniques has renewed the interest for *in situ* UPD studies. In 1980s and 1990s, many UPD systems on single crystal substrates were re-examined in detail. The reviews summarizing the work from this period are offered by Budevski *et al.*[34] and more recently by Abruña *et al.*[35]

The characteristic voltammetry features associated with UPD process are demonstrated by existence of one or more deposition (stripping) peaks in the underpotential region observed during the potential sweep in cathodic (anodic) direction (Fig. 3). The complexity of the UPD peaks is often dependent on the perfection and preparation of the substrate surface,[34,35] existence of one or more UPD adlayer superstructures,[36,37] and the number of full UPD MLs formed.[38,39] The exemplary cyclic voltammetry for Pb UPD on Au(111) is shown in Fig. 3 where different stages and coverage of Pb UPD ML are indicated.

Different models were used to explain UPD as the potential dependent adsorption. *Langmuir*,[40,41] *Temkin*,[42] and *Frumkin*[43] isotherms have been proposed to model the UPD ML/MLs underpotential-coverage dependence (ΔE vs θ). This relation in most cases, depends on the attractive forces between the UPD metal and the substrate, as well as the repulsive interactions among the adatoms in the UPD ML/MLs. However, the effect of strain due to the lattice mismatch between the substrate and the UPD ML, the anion co-adsorption, as well as the double layer effects are found to be important for complete understanding of this relation.[33,35] If the UPD process represents a single energy state, and its electrosorption valence is equal to the oxidation state of metal ions in the solution, the underpotential-coverage dependence is described as[44]:

$$\Delta E = \Delta E^0_{\theta \to 0} - \frac{RT}{n \cdot F} \cdot \left[\ln \left(\frac{\theta}{1-\theta} \right) + f\theta + g\theta^{3/2} \right]. \tag{8}$$

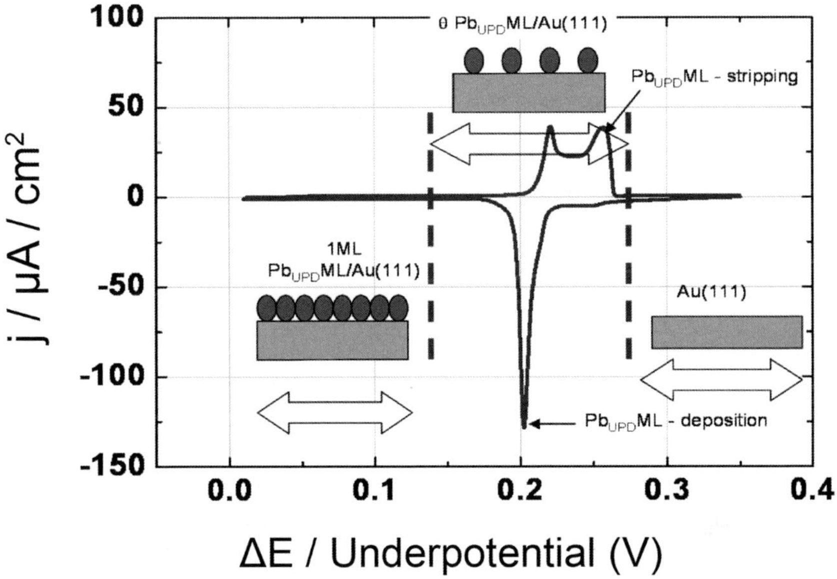

Fig. 3. Cyclic voltammograms for Pb UPD on Au(111). Solution: 10^{-3} M Pb^{2+} + 0.1 M $HClO_4$, sweep rate 10 mV · s^{-1}. Potential range of full Pb UPD ML, partial Pb UPD ML, and Pb free Au surface are indicated in the graph.

The isotherm has a *Langmuir* basis and implies additivity of the various UPD adatom–substrate and adatom–adatom interactions. The $\Delta E^0_{\theta\to 0}$ term represents the underpotential of the most positive stripping peak of the UPD adlayer where the UPD ML coverage approaches zero. The term "*f*" is the *Temkin* parameter describing the UPD layer–substrate interactions such as the electrode work function change with the UPD adlayer coverage. The term "*g*" is the *Frumkin* parameter representing the lateral adatom interactions within the UPD adlayer.

Different computational approaches were also considered in order to describe UPD as the potential dependent adsorption. For example, the Monte Carlo (MC) simulations of Pb UPD on Ag(111), preformed by Obretenov *et al.*,[45] showed that the experimental isotherms can be completely recovered from the MC simulations if the proper distribution of the adsorption energy states on substrate surface and the interaction energies among the adatoms in the UPD ML are taken into account. This work has also highlighted the importance of surface defects and strain in the UPD ML on specific voltammetry features observed experimentally.

In some UPD systems like Pb and Cd on Ag(111),[46–48] Pb on Cu(100)[49] and, Tl and Cd on Au(111),[50,51] the significant surface alloying and structural transformation

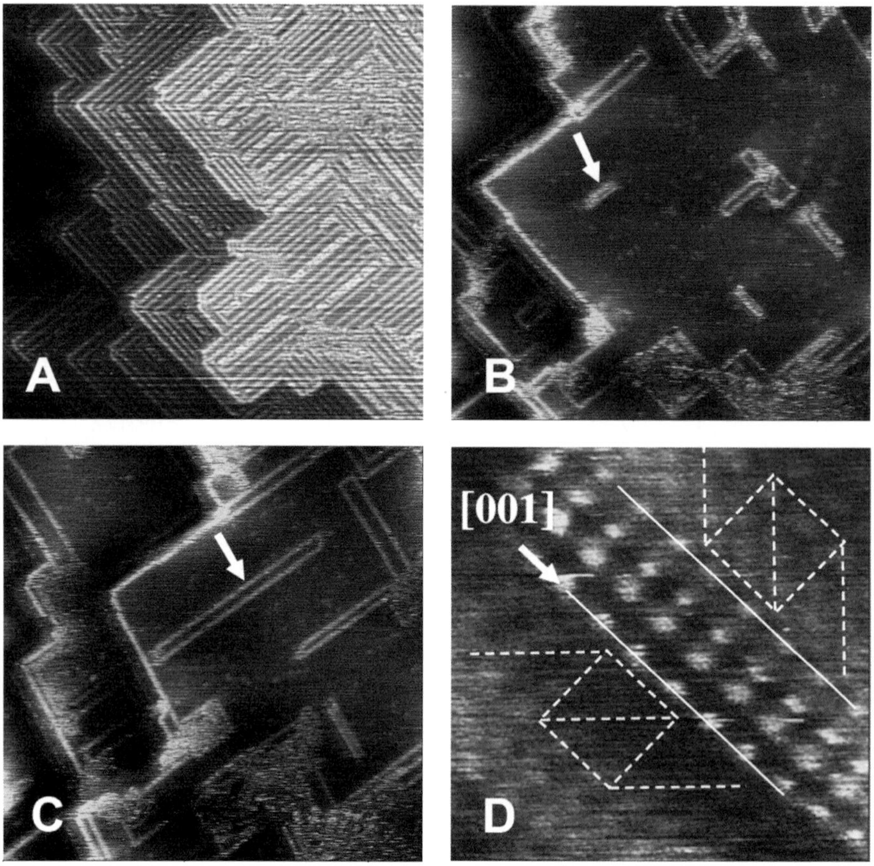

Fig. 4. *In situ* STM images of Pb UPD on Cu(100). The coverage of Pb is ~0.3 ML. Images (A)–(C) ordered Cu_4Pb_3 surface alloy phase and 2 × 2 adsorbed Cl adlayer are visible as bright and dark regions. The labyrinth structure in (A) has a period on 2.2 nm and forms as closed loops of the Cu_4Pb_3 phase elongates. Arrows mark the elongation of a loop in (B) and (C). Atomic resolution (D) shows that paired Cu_4Pb_3 stripes are of different variants rotated by 90°. Dashed lines mark the 0.72 nm unit cell and the direction of Pb rows. Reprinted with permission from Ref. 49. Copyright (2005) by the American Physical Society.

of the UPD ML were reported (Fig. 4). The alloying process was found dependent on UPD adlayer coverage, substrate step density, presence of substrate surface reconstruction, and on the type of co-adsorbed anions. The formation of self-organized surface patterns[49] and nanostructures[51] (Fig. 4) were also reported as a result of the UPD adlayer–substrate alloying.

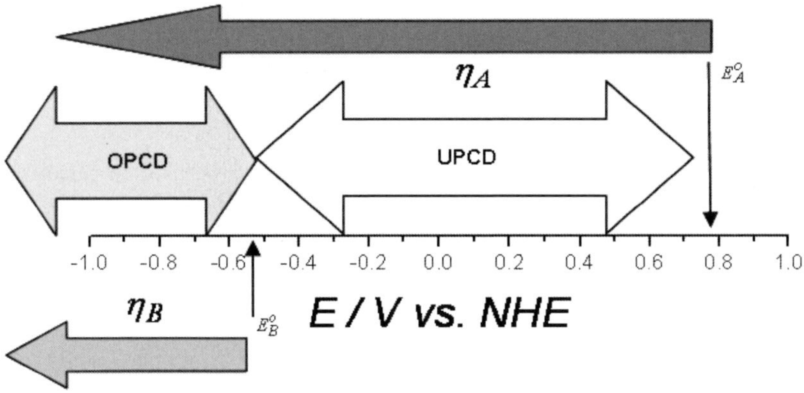

Fig. 5. Potential regime for underpotential co-deposition process of alloy AB (UPCD). The UPCD regime is indicated relative to OPCD potential range and equilibrium potentials of elemental A and B.

5. Underpotential Co-Deposition (UPCD)

The term UPCD is related to the deposition of alloys at potentials which are more positive than the equilibrium potential for the elemental phase of less noble component in the alloy, Eq. (1). The schematic of the potential range where UPCD occurs is shown in Fig. 5. The phenomenological reason for UPCD is that the equilibrium potential of less noble component B in AB alloy can be more positive than the corresponding equilibrium potential of its elemental phase E_B°. The origin of this condition is more transparent if Eqs. (1) and (7) are combined as;

$$E_B < E_B^{\text{alloy}},$$
$$\Rightarrow E^0 + \frac{RT}{2.303nF}\log\left[a_B^{n+}\right] < E^0 + \frac{RT}{2.303nF}\log\left[\frac{a_B^{n+}}{a_B^{\text{alloy}}}\right]. \tag{9}$$

For alloys that obey ideal solution model, the activity of the less noble component B in the alloy is equal to its molar/atomic fraction, i.e. $a_B^{\text{alloy}} = X_B$. The value of X_B in this case is always less than one and the conclusion expressed by Eq. (9) is straightforward. The typical system that does show this behavior is AuCu alloy.[52] However, the systems that obey a nonideal solution model are more common, and the typical examples of these alloys are CoPt, FePt, CoPd, and FePd. Because of their high magnetic anisotropy introduced by incorporation of Pt, or Pd in to Co hcp or Fe bcc lattice and ordering (L10 phase[53]) these alloys are studied intensively in the past several years as possible candidates for the bit pattern media application as well as MEMS and NEMS technologies.[54–59]

The departure of the alloy from ideality is modeled through the activity coefficient χ_B, where the activity of the less noble component in the alloy is expressed as

$$a_B^{\text{alloy}} = \chi_B \cdot X_B. \quad (10)$$

Depending on the type of the alloy, different models can be used to calculate the activity coefficient. For electrodeposited CoPt and FePt alloys which show strong negative departure from ideality,[53,54] an asymmetric regular solution model is used to define χ_B (B = Fe or Co):

$$RT \ln \chi_B^{\text{alloy}} = (W_{G1} + 2(W_{G2} - W_{G1})X_B) \cdot (1 - X_B)^2. \quad (11)$$

Here, W_{G1} and W_{G2} represent temperature and composition independent constants called Margules parameters. The substitution of Eq. (11) into Eq. (7) yields the equilibrium potential of the less noble component in the alloy. If this expression is subtracted from the equilibrium potential of the elemental phase defined by Eq. (1), the relation between the underpotentially co-deposited alloy composition and corresponding value of underpotential can be obtained. The example of this approach is shown in Fig. 6 where the composition of UPCD CoPt and FePt is measured as a function of deposition underpotential. The solid lines in the plot indicate the fit of the asymmetric regular solution model. It is important to note that Eq. (11) combined with Eqs. (7) and (1) suggest that composition of UPCD AB alloys (CoPt and FePt) is not dependent on A and B (Co and Fe) deposition kinetics (concentrations in the solutions),

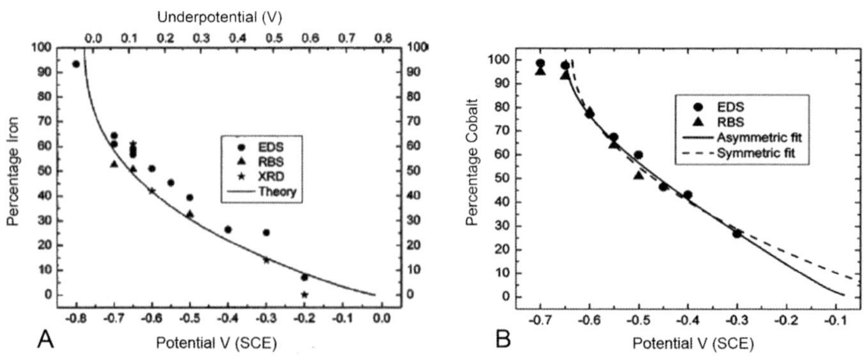

Fig. 6. (A) FePt UPCD alloy composition as a function of deposition underpotential. Reproduced by permission of ECS — The Electrochemical Society, see Ref. 54. (B) CoPt UPCD alloy composition as a function of deposition underpotential. Reproduced by permission of ECS — The Electrochemical Society, see Ref. 53. Solid lines represent the model discussed in the text fit to the experimental data.

but on the value of underpotential for co-deposition. The preceding statement is valid only if B ions (Co^{2+} and Fe^{2+}) do not encounter any transport limitations relative to A deposition (Pt) during UPCD. In practice, this is easily achieved by choosing the solution design where concentration of the more noble metal A ions is significantly lower than the concentration of the less noble metal B ions so that $j_A \ll j_B$. In this situation, the deposition rate of UPCD alloy is controlled by the deposition kinetics of the more noble metal A and the composition of the alloy is controlled by the underpotential used.[53,54]

6. Metal Deposition by Galvanic Displacement of UPD ML (MLS)

This method has been recently developed by Brankovic *et al.*[60] and is extensively used for deposition of noble metal MLs and catalyst materials.[61–66] The basic steps describing the concept behind this new deposition method are illustrated in Fig. 7.[60] The method is based on the idea of using the UPD ML of metal M on substrate $S(h, k, l)$ as a sacrificial material for deposition of more noble metal P. The deposition of P occurs as *an irreversible surface controlled red–ox reaction* where the amount of deposited metal P is controlled by reaction stoichiometry, structure, and coverage of the UPD ML/MLs,

$$(b \cdot \theta \cdot q) \cdot M^0_{\mathrm{UPD}}/S(h,k,l) + (b \cdot \theta \cdot q) \cdot \left(\frac{m}{p}\right) P^{p+}_{\mathrm{solv}} \rightarrow (b \cdot \theta \cdot q) \cdot M^{m+}_{\mathrm{solv}} + (b \cdot \theta \cdot q) \cdot \left(\frac{m}{p}\right) P^0_s/S(h,k,l). \tag{12}$$

Here, m and p are the oxidation states of UPD metal M and the more noble metal P. The factors b, θ, and q are introduced to accurately express the amount of deposited metal P in ML units with respect to atomic areal density of the substrate $S(h, k, l)$. They represent, respectively, the number of full UPD MLs, the UPD ML coverage, and the packing density of M atoms in complete UPD ML with respect to the substrate $S(h, k, l)$. The subscripts s and solv indicate the physical state of the metal (solv = solution phase and s = deposited). If sequence A–E (Fig. 7) is repeated an arbitrary number of times, a multilayer homo- or heteroepitaxial films can be obtained.[22,67] The thin film growth using this method can be completely automated with experimental apparatus for *Electrochemical Atomic Layer Epitaxy* developed by Stickney *et al.*[68]

The UPD represent a potential dependent adsorption process and the UPD ML coverage is effectively controlled down to a fraction of a ML.[44] According to the stoichiometry of Eq. (12), the same accuracy for deposition of metal P can be achieved as

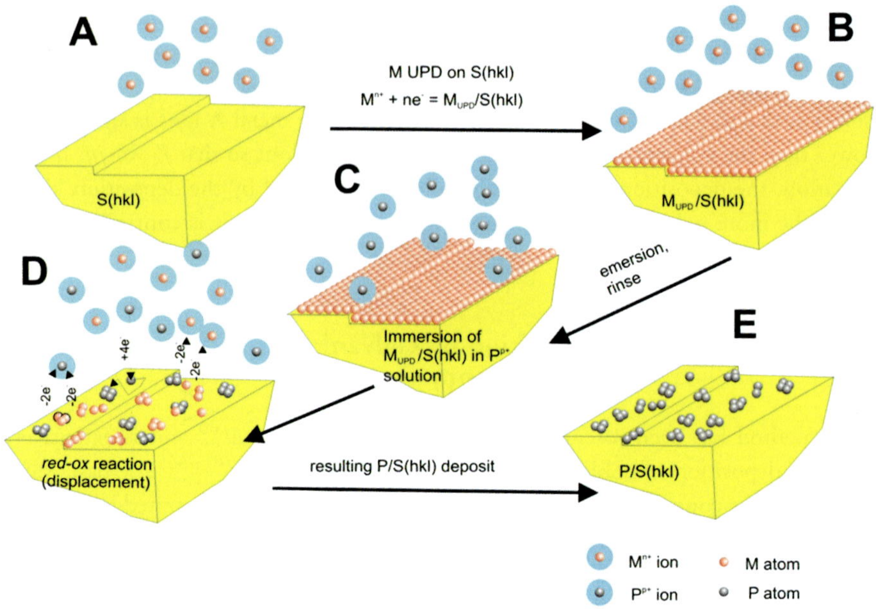

Fig. 7. The basic steps of the deposition method described in Ref. 60. A–B: formation of M UPD ML on $S(h,k,l)$, B: $M_{\text{UPD}}/S(h,k,l)$ surface emersion from solution at certain potential ensuring the desired M UPD ML coverage of $S(h,k,l)$ and/or number of UPD monolayers. C: transfer to P^{p+} containing solution at open circuit potential (OCP) conditions, D: *red–ox* or displacement reaction at OCP, E: final morphology of $P/S(h,k,l)$ deposit.

well. This important advantage gives a whole array of new opportunities for application of this method. Some examples are the design of catalyst ML,[61,62] the preparation of sub-monolayers (SML) of noble metals/alloys with different coverage and morphology, deposition of heteroepitaxial ultra thin films, and ultra thin film alloys.[63–69]

The electrochemical driving force for the displacement of the UPD ML is the positive difference between the equilibrium potential of P metal and the equilibrium potential of $M_{\text{UPD}}/S(h,k,l)$ at its coverage approaching a zero limit, $\theta_{\text{UPD}} \to 0$. This condition (Fig. 8) is defined as[60,69]:

$$\Delta E_{\text{red–ox}} = \Delta E^0_{\text{EMF}} - \Delta E^0_{\theta\to 0} - \frac{RT}{F}\ln\frac{[a_{M^{n+}}]^n}{[a_{P^{p+}}]^p} > 0. \tag{13}$$

Here, $\Delta E^0_{\text{EMS}}(\Delta E^0_{\text{EMF}} = E^0_{Pp^+/P} - E^0_{Mn^+/M})$ represents the electromotive force for the bulk M and P galvanic couple at standard conditions (Fig. 8). The $\Delta E^0_{\theta\to 0}$ represents

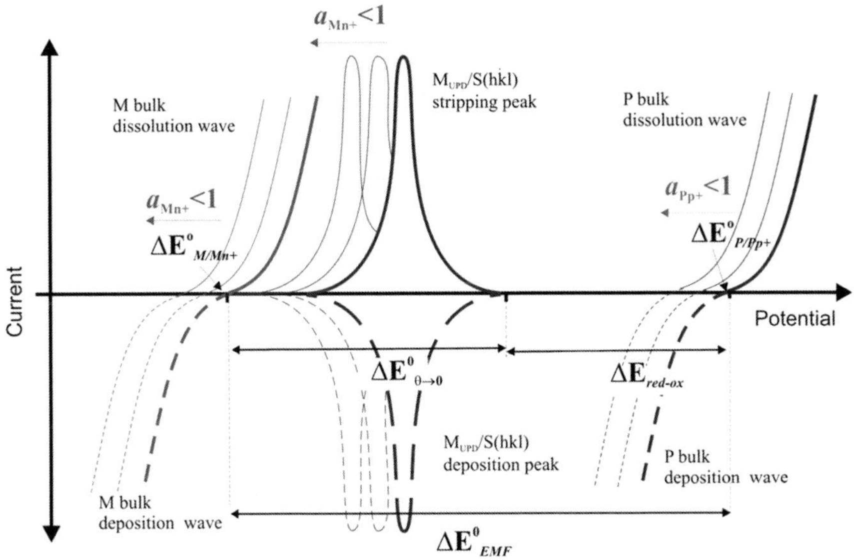

Fig. 8. The schematic of the thermodynamic entities and their relations determining the $\Delta E_{\text{red-ox}}$, Eq. (13) for displacement reaction. Gray lines are current–potential dependence for bulk M metal electrode and UPD ML of M on substrate S. The dark line is the current–potential dependence for bulk noble metal electrode. The terms from Eq. (13) are identified in the figure.

the equilibrium potential of $M_{\text{UPD}}/S(h,k,l)$ at $\theta_{\text{UPD}} \to 0$ limit at standard conditions ($a_{M^{n+}} = 1, a_{P^{p+}} = 1$, Fig. 8), while the logarithmic term provides the correction for the departure from standard conditions. Inspection of Eq. (13) offers the evidence that the driving force for the displacement reaction can be modified by adjusting the activities of M^{n+} and P^{p+} ions in the immersing (displacement) solution. If there are no M^{n+} ions in the immersing solution, the logarithmic term in Eq. (3) dominates the value of $\Delta E_{\text{red-ox}}$ (very large and positive), which means that extremely high overpotentials for nucleation could be achieved.

Depending on the combination of the $M_{\text{UPD}}/S(h,k,l)$ and P^{p+} ions involved in the *red–ox* reaction, a different morphology and amount of P deposit is formed. This is illustrated in Fig. 9 using the example of Pd and Pt deposit obtained by displacement of $\text{Cu}_{\text{UPD}}/\text{Au}(111)$. The deposit morphology varies from an atomic ML thin film to ML high nanoclusters with very narrow size distribution (Figs. 9(A)–9(B)).

A particularly interesting result in Fig. 9 is the spatially uniform coverage of Pt nanoclusters on Au(111) shown in image B. This suggest that nucleation of the depositing metal (Pt) is independent on thermodynamically favorable nucleation sites

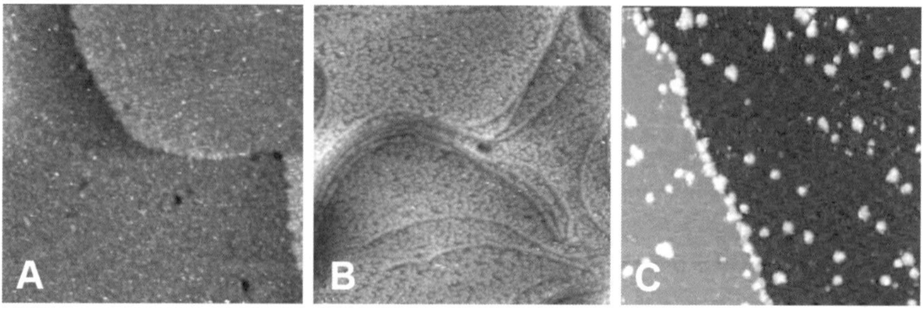

Fig. 9. The STM images of deposit morphology of (A) ~ 1 ML of Pd on Au(111) via galvanic displacement of $Cu_{UPD}/Au(111)$, displacement reaction: $Cu_{UPD}/Au(111)+Pd^{2+} = Cu^{2+} + Pd/Au(111)$, $b = 1, \theta = 1, q = 1$ $m/p = 1$, image size 200 nm × 200 nm, (B) ~ 0.5 ML of Pt on Au(111) via galvanic displacement of $Cu_{UPD}/Au(111)$, displacement reaction: $Cu_{UPD}/Au(111) + Pt^{4+} = Cu^{2+} + 0.5Pt/Au(111)$, $b = 1, \theta = 1, q = 1, m/p = 0.5$, image size 320 nm × 320 nm, (C) Pt submonolayer on Au (111), image size 150 nm × 150 nm. Size of Pt clusters is ~ 3 nm. The achieved nucleation density is ~ 10^{13} cm^{-1}. Images A and B reprinted from Ref. 60, with permission from Elsevier. Image C is reprinted from Ref. 70 and The Lancet, with permission from Elsevier.

on substrate surface such as steps for example. This is evident if one compares the morphology of Pt SML deposit on Au(111) obtained by electrodeposition at low overpotentials[70] (Fig. 9(C)) and the morphology of Pt sub-ML and Pd ML deposits shown in Figs. 9(A)–9(B). In the latter case, one can note that Pt and Pd nanoclusters are equally distributed across the surface regardless the presence of surface steps where preferential nucleation and growth is typically found during electrodeposition (Fig. 9(C)).

7. Spontaneous Noble Metal on Noble Metal (NMonNM) Deposition

The spontaneous noble metal deposition on noble metal surface occurs when freshly prepared and clean noble metal electrode is immersed in the solution containing different noble metal ions. This phenomenon has been reported in various systems like $Ru^{2+}/Pt(h,k,l)$,[71,72] $Pt^{4+}/Ru(h,k,l,m)$,[73,74] $Pd^{2+}/Ru(0001)$,[75] and $Pd^{2+}/Pt(h,k,l)$.[76,77] The morphology of the deposit varies from ML high nanoclusters to larger 3D structures. The spontaneous NMonNM deposition occurs as a result of an *irreversible surface controlled red–ox reaction* among depositing noble metal ions and noble metal substrate. The substrate surface in this reaction becomes oxidized by noble metal ions to form the certain type of hydroxide. The entire process is represented by following reaction[73]:

$$\frac{m}{n}P^{n+} + S^0(hkl) + m\mathrm{H_2O} \rightarrow \frac{m}{n}P^0_{dep} + [S^{m+}][\mathrm{OH}^-]_m + m\mathrm{H}^+, \qquad (14)$$

where P^{n+}, P^0, S^{m+}, and $S^0(h,k,l)$ are respectively, the noble metal ions, the noble metal deposit, the substrate in the surface hydroxide, and the noble metal substrate. The conductive nature of Ru-oxide/hydroxide and ability to form more than one ML without hindering the electron transport across the interface are the reasons for the spontaneous formation of multilayer Pt and Pd deposits on Ru (0001) surface.[73,75] The thermodynamic driving force for NMonNM deposition can be expressed in terms of the difference between the equilibrium potential of the noble metal electrode in contact with its ions in the solution (Eq. (1)), and the equilibrium potential for oxidation of the noble metal substrate[73,75]:

$$\Delta E_{\mathrm{red-ox}} = \Delta E_{P^{n+}/P^0} - \Delta E_{[S^{m+}]_x[\mathrm{OH}]_y/S^0} > 0. \qquad (15)$$

According to Eq. (15), the change in concentration of the ions of depositing noble metal, or pH of the solution alters the effective driving force for spontaneous NMonNM deposition. This fact is used to achieve different noble metal nucleation densities, regime of nucleation, and deposit morphology (Fig. 10). The NMonNM deposition offers advantage over the existing deposition methods due to the fact that 2D morphology of depositing noble metal is easily achieved without significant control of the deposition process itself (simple immersion). An example of this advantage is

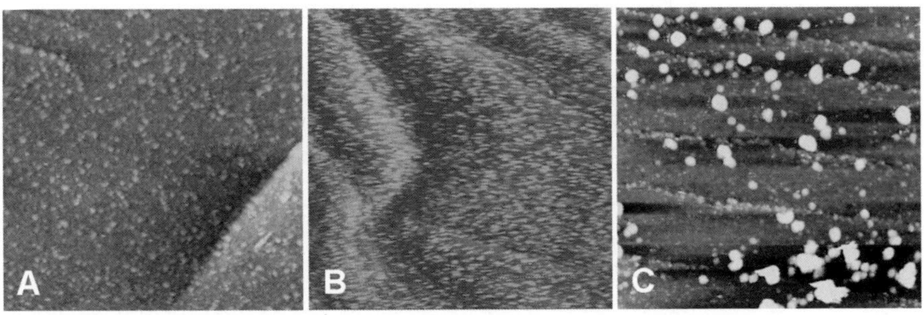

Fig. 10. Morphology of the Pt deposit on Ru(1010) (A) and Ru(0001) (B and C). The spontaneous NMonNM deposition is used in image (A) and (B),[78] and the potentiostatic deposition is used in image (C), $\eta = 30$ mV. The depositing solution was 10^{-2} M $\mathrm{PtCl_6^{2-}}$ + 10^{-2} M $\mathrm{HClO_4}$. The nucleation density in (A) and (B) is approximately ~500 times higher than in (C). (A) image size; 280 × 280 nm, 2D ~ 3 nm Pt clusters, (B) image size; 130 × 130 nm, 2D, ~ 5 nm Pt clusters, and (C) image size; 115 × 115 nm, 3D, ~ 10 − 15 nm Pt clusters. Reprinted from Ref. 78 and The Lancet, with permission from Elsevier.

illustrated in Fig. 10 by comparison of ~1 ML Pt/Ru(0001) deposit obtained by potentiostatic deposition, and ~0.7 ML Pt/Ru(0001) and 0.4 ML Pt/Ru(1010) obtained by spontaneous NMonNM deposition, Figs. 10(A)–10(C).

8. Pulse Current Deposition

The electrodeposition process can either occur under potential (potentiostatic deposition) or current (galvanostatic deposition) control. In practice, the common approach is to control the deposition flux (current) in order to obtain the deposit with desired thickness and properties. The electrodeposition under galvanostatic control can be executed as either the constant current process or using some more complex current-time function profile. In the latter case, the electrodeposition process is commonly called the pulse current deposition and it has been widely used in industrial and academic applications. The typical pulse current function has the simple on/off profile shown in Fig. 11.[4]

The successful application of this pulse current function is dependent on the proper determination of the pulse time t_{on}, rest time t_{off}, and magnitude of the pulse current density j_{pulse}. The optimum duration of the pulse current stage should be such that no limitations for metal ion transport (M^{m+}) towards the electrode–electrolyte interface is encountered. Finding these conditions requires the mathematical treatment of the

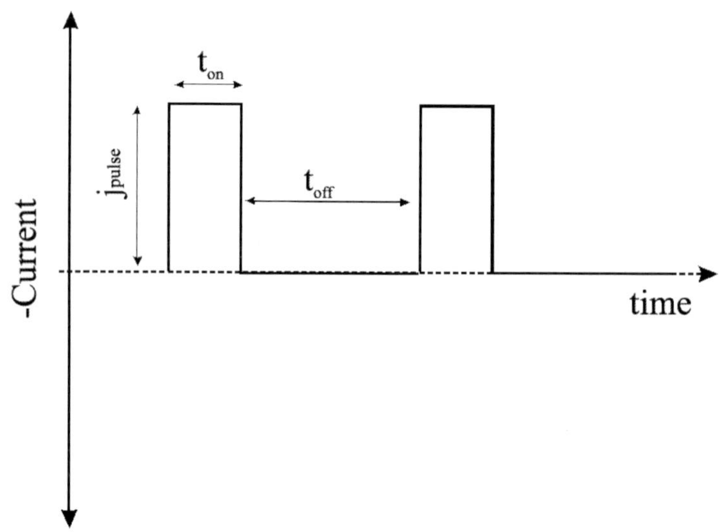

Fig. 11. The pulse deposition function profile. The parameters t_{on}, t_{off}, and j_{pulse} are indicated in the figure.

M^{m+} diffusion through the Nernst diffusion layer which means solving Fick's second law of diffusion with appropriate initial and boundary conditions:

$$\frac{\partial C_{M^{m+}}}{\partial t} = D_{M^{m+}} \cdot \frac{\partial^2 C_{M^{m+}}}{\partial x^2}. \tag{16}$$

The exact mathematical solution of Eq. (16) is somewhat cumbersome, presented in the form of a Fourier series expansion.[79] For practical application, it is more convenient to consider the approximate solutions that are dependent on the value of the term $a \cdot t$ where a is defined as:

$$a = \frac{\pi^2 D_{M^{m+}}}{4\delta^2}. \tag{17}$$

The evaluation of this term is dependent on an appropriate estimate of $D_{M^{m+}}$ and δ and for many practical applications, these parameters can be estimated or measured experimentally.[80] There are two limiting situations which define the shape of the approximate solution for Eq. (16). The first one, if $a \cdot t < 0.2$, the approximate solution for the concentration-time profile of the metal ions at the electrode–solution interface is given as[79]:

$$C^{\infty}_{M^{m+}} - C^{i}_{M^{m+}} = \frac{\gamma \cdot J_{\text{pulse}} \cdot \delta}{mF \cdot D_{M^{m+}}} \cdot \left\{1 - \frac{8}{\pi^2} \cdot \exp\left(-\frac{\pi^2 \cdot D_{M^{m+}}}{4\delta^2} \cdot t\right)\right\}, \tag{18}$$

and the second case, $a \cdot t > 0.2$, the approximate solution for the concentration-time profile of metal ions at the electrode–solution interface is given by Sand's equation[81];

$$C^{\infty}_{M^{m+}} - C^{i}_{M^{m+}} = \frac{2\gamma \cdot J_{\text{pulse}}}{mF} \cdot \left(\frac{t}{\pi \cdot D_{M^{m+}}}\right)^{0.5}. \tag{19}$$

In both equations, $C^{\infty}_{M^{m+}}$ and $C^{i}_{M^{m+}}$ stand for the concentration of metal ions in the bulk and concentration of the metal ions at the electrode–solution interface. The term m indicates the oxidation state of the metal ion, i.e. the number of electrons transferred across the interface per metal ion during deposition process. Equations (18) and (19) can be solved explicitly for the time t, for which the concentration of the metal ions at the interface are completely depleted ($C^{i}_{M^{m+}} = 0$) for a given pulse current magnitude. This estimate should represent an upper limit for the duration of the pulse stage commonly called transition time (τ). In order to avoid any transport limitations during pulse current stage, the rule is that t_{on} should be always less than τ, ($t_{\text{on}} < \tau$).

The determination of the duration of the rest stage, t_{off}, Fig. 11, is concerned with solving Eq. (16) for different initial and boundary conditions.[80] The t_{off} is defined as the time necessary for the concentration of the metal ions at the electrode–solution interface to be recovered to some arbitrary portion of its initial (bulk) value (before the pulse stage). For example, if $C^i_{M^{m+}}$ is recovered to 99% of $C^\infty_{M^{m+}}$ and assuming that the same approximations apply as in the case of Eq. (18), the obtained estimate of $t_{\rm off}$ is $\approx 3\tau$.[82,83] For practical pulse current function profile estimates, the commonly used term is called duty cycle, ω, which represents the ratio between $t_{\rm on}$ and $t_{\rm off}$ defined as:

$$\omega = \frac{t_{\rm on}}{t_{\rm on} + t_{\rm off}} < \frac{\tau}{\tau + t_{\rm off}}. \tag{20}$$

The $t_{\rm off}$ values are typically taken to be longer than 3τ limit and the common values of ω found in literature are usually less than 0.25.[83]

The magnitude of the pulse current is very often dependent on the application of the electrodeposition process. If deposit with smaller grains and better surface finish is desired, the values of the pulse current density that are used could be very high.[83] In general, the rule is that the potential of the electrode surface during the pulse current stage should not exceed the value of -1.2 V vs SHE. The reason is that below this potential, the electrochemical reduction of water occurs ($2H_2O + 4e^- = 2OH^- + H_2\uparrow$) and most of the current increase below this potential is related to the described reaction rather than to the metal deposition. This leads to undesirable effects such as low current efficiency, hydrogen embrittlement of the deposit, high deposit stress, local increase of pH, and hydroxide precipitation at the electrode surface.[4] If additives are used in the electrodeposition process, the pulse current magnitude should be such that the potential of the electrode surface during the pulse stage is where the maximum additive adsorption occurs in order to get full benefit of their action.[84] This is particularly important in pulse current deposition processes used for fabrication of nanostructures where the morphology of the deposit as well as the additives used are of essential importance.[4] A good example of this effect is shown in Fig. 12 where CoFe nanostructures produced with the same pulse current, but slightly different pulse current magnitudes. The potential of the electrode surface during the pulse current stage in Figs. 12(A) and (B) corresponds to different coverage of saccharin additive. This results in distinctively different CoFe nanostructure morphology.

9. Additive Effect

In order to obtain electrodeposits with better surface finish and adequate material properties, different additives are used in the plating solutions.[85] The common action of additives is expressed through leveling and brightening of the deposit (Fig. 13), improvement of the crystal structure,[86] smaller grain size,[87] and reduction of the residual stresses.[88] A good example of an additive is saccharine which is one of the

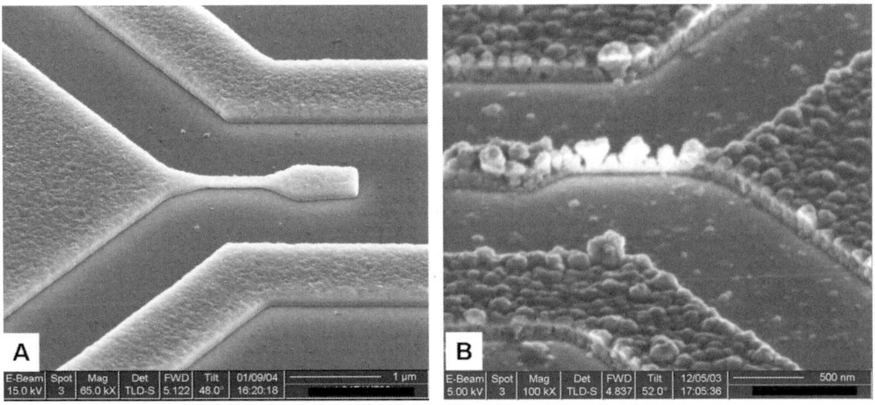

Fig. 12. The $Co_{37}Fe_{63}$ pole tip nanostructures with ~ 50 nm critical dimensions produced with two different pulse currents; (A) $j_{\text{pulse}} = 18\,\text{mA cm}^{-2}$, E ~ -1.24 V, (B) $j_p = 25\,\text{mA cm}^{-2}$, E ~ -1.35 V. Potentials are quoted vs SCE, the saccharin coverage in A is 95% and in B $\sim$ 30% during the pulse current stage. Duty cycle 20%. Reprinted from Ref. 4 with permission. Copyright (2006) IEEE.

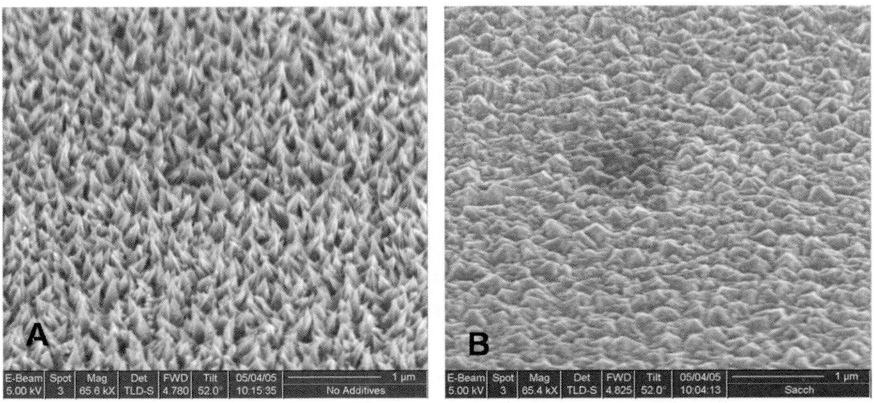

Fig. 13. SEM of the CoFe deposit surface produced during galvanostatic deposition, $j = 4\,\text{mA cm}^{-2}$. (A) No saccharin in solution, RMS of the 5 μm $\times$ 5 μm area is 27 nm, (B) with 10^{-3} mol saccharine in solution, RMS of the 5 μm $\times$ 5 μm area is 7 nm.

most common additives used in electrodeposition of alloys. During electrodeposition, saccharin adsorbs on the metal surface forming some type of condensed phase (film).[29] The density and coverage of the adsorbed saccharin phase are dependent on the potential of electrode surface and the concentration of saccharin in the plating solution.[89]

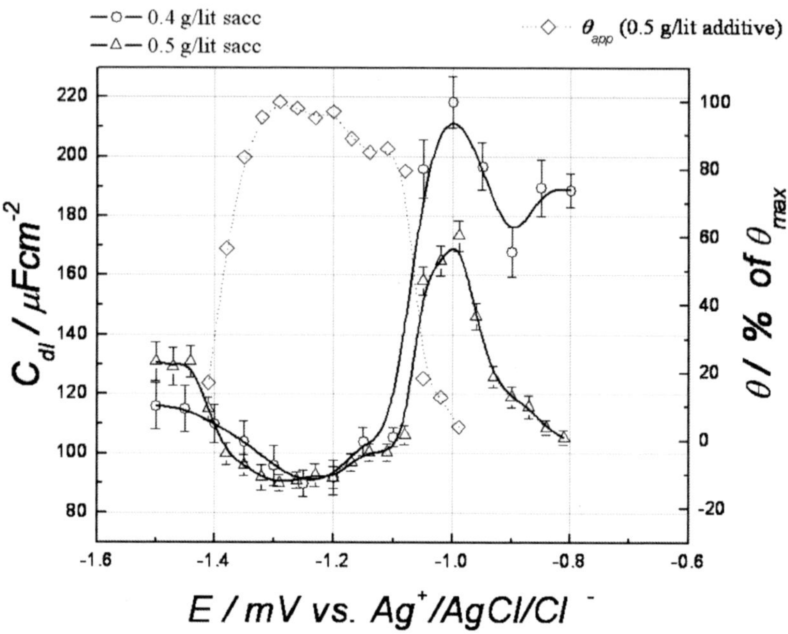

Fig. 14. The double layer capacitance vs potential dependence extracted from the impedance measurements for CoNiFe surface. The corresponding coverage of saccharine is calculated from double layer capacitance measurements are shown as dashed line. The additive coverage is estimated using following expression[90]: $\theta = \frac{C_{dl} - C_{dl-\min}}{C_{dl-\max} - C_{dl-\min}}$. Reproduced by permission of ECS — The Electrochemical Society, see Ref. 29.

Useful information about saccharine adsorption and coverage can be deduced from the impedance spectroscopy.[90] An example of these measurements is shown in Fig. 14.[29]

Using additives to enhance the properties of electrodeposits is possible because of their incorporation into the deposit. The two main mechanisms by which additives incorporate into deposit are adsorption–electroreduction and physical incorporation.[91,92] The first mechanism represents the chemical route responsible for incorporation of molecular fragments and formation of inter-metallic compounds like metal sulfides while the second one represents the incorporation of entire molecules of additives. Recently, it was shown by Brankovic *et al.* that the additive incorporation (saccharin) into deposit as a function of their concentration in the plating solution is well described within the scope of following analytical model[92]:

$$R = K_1 \cdot \frac{b \cdot C_a}{1 + b \cdot C_a} + K_2 \cdot \frac{b \cdot C_a}{(1 + b \cdot C_a)^2}. \tag{21}$$

In the above expression, the R represents the additive incorporation rate in flux units $[\mathrm{mol \cdot m^{-2} s^{-1}}]$ and C_a stands for concentration of the additive in the plating solution $[\mathrm{mol \cdot m^{-3}}]$. The first term on the right side represents the additive incorporation via electroreduction mechanism and the second term represents the additive incorporation by physical entrapment of the entire molecules. The K_1 and K_2 are the electroreduction and incorporation rate constants expressed in $[\mathrm{mol \cdot m^{-2} s^{-1}}]$ units and b is the additive adsorption constant $[\mathrm{m^3 \cdot mol^{-1}}]$. The above model can predict the concentration of the different interstitials/inclusions in electrodeposited films originating from incorporated additives as well as to correlate the additive incorporation effect with the deposit properties.[92]

An example of this analysis is shown in Fig. 15. Image (A) shows the model fit to the experimentally determined S content in CoFe deposit, and image (B) shows correlation between the decrease of CoFe alloy coercivity and incorporation rate of sulfur/saccharin in the deposit. The constants K_1, K_2, and b (Eq. (21)) are obtained from the fit of the model to data in image (A) and then used to calculate the incorporation rate of sulfur for particular concentration of saccharine in the bath from which CoFe films are deposited. Different coercivities of CoFe films are correlated to sulfur incorporation rate yielding a linear regression.

One of the most important benefits that additives bring to the electrodeposition process is the effective reduction of the residual stresses in the deposit. The stress in electrodeposited films is dependent on many parameters. Very often, the sign of the

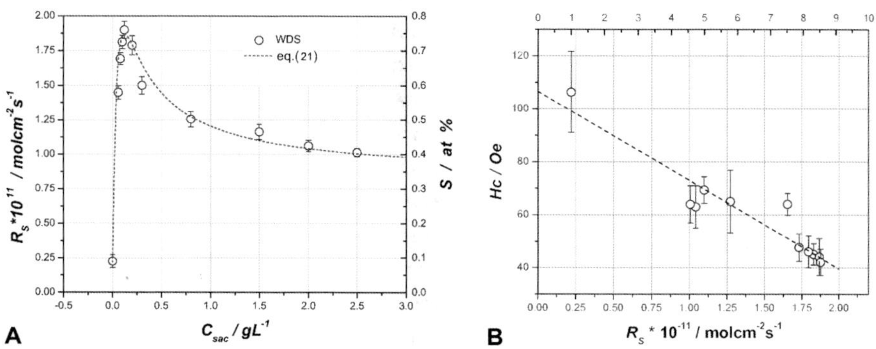

Fig. 15. (A) Model fit (Eq. (21) to experimental data on sulfur incorporation rate and at % in electrodeposited CoFe alloys. Reproduced by permission of ECS — The Electrochemical Society, see Ref. 92. (B) Correlation between the decrease in coercivity of CoFe alloys and incorporation rate of sulfur coming saccharine. Reproduced by permission of ECS — The Electrochemical Society, see Ref. 92.

stress (compressive or tensile) is dependent on the final thickness of the electrodeposited film, crystal structure, substrate, and film texture.[93] The stress represents a great technological problem for functional thin films because it causes de-lamination, substrate bending, creep, electromigration, and other undesirable effects. The sulfur burring additives were found to be very effective as stress relieving agents in the electrodeposited films.[3] The true mechanism of their action is still unknown however; some common conclusions are derived based on existing knowledge. From previous metallurgical studies, the sulfur related inclusions or inter-metallic compounds are known to create distortion of the metal lattice, and to contribute to the local disorder.[94] By precipitating on the grain boundaries during the deposition process as low surface energy phase,[95] additives and additive related inclusions act as effective mediators and stress relievers during the grain coalescence process, yielding a deposit with lower stress levels.

The beneficial additive effect on stress evolution during electrodeposition is shown in Fig. 16. In (A), from *in situ* measurements, one can see that for Permalloy films of 100 nm thickness, the significant stress reduction in deposit is achieved if saccharin (stress reduction 50%) or l-Methionine (stress reduction 85%) is used as additives in plating solution. The effect of saccharine concentration on stress in electrodeposited CoNiFe films is shown in Fig. 16(B).[96] These data suggest that stress reduction is decreasing towards asymptotic value for increasing saccharin content in the solution.

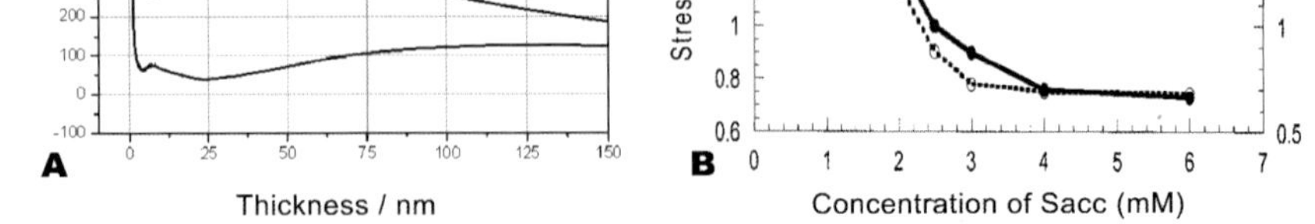

Fig. 16. (A) *In situ* stress measurements for Permalloy films as a function their thickness. Saccharine (middle curve) and l-methionine (low curve) concentration in solution is 0.5×10^{-4} mol. (B) Stress in CoNiFe films as function of saccharin concentration in the plating solution. Reproduced by permission of ECS — The Electrochemical Society, see Ref. 96.

10. Specific Aspects of Electrodeposition into Nanotemplate Electrodes

The particular advantage of electrodeposition onto a nanotemplate electrode ($l <$ 100 nm) is that the minimized electrode dimensions promote a perfect mononuclear layer-by-layer growth. The origin of this effect comes from the nucleation aspect of thin film growth treated theoretically by Venables *et al.*,[97] Markov *et al.*,[98] and Tersoff *et al.*[99] According to these considerations, the nucleation density for room temperature thin film growth is represented as[100]:

$$n \approx \left(\frac{F}{a^2 4D}\right)^{1/3}. \tag{22}$$

Here, F, a, and D stand for deposition flux, in-plane nearest neighbor's distance, and surface diffusion coefficient of depositing atoms. The nucleation density is defined as 1/area, and square root of its inverse value represents the approximate in-plane distance between two adjacent nuclei ($d \sim (a^2 \cdot 4D/F)^{1/6}$). Using the known data in the literature for surface diffusion coefficients, (for example Au; $D_{\text{Au}} \sim 10^{-15}\,\text{cm}^{-2}\text{s}^{-1}$,[101] $a_{\text{Au}} \sim 3 \times 10^{-8}$ cm) the appropriate deposition flux for which $d > l$ is achieved for low deposition current. The same is valid for other metals like Pt, Ag, Ru, Pd, Ni, Cu, etc. which means that formation and growth of single crystal deposit in nanotemplate electrode is expected during a regular electrodeposition process. The experimental evidence of these considerations for deposition of Ni nanocontacts is shown in Fig. 17. The cuboctahedral shape of Ni deposit on top of the Al_2O_3 layer is obtained by bottom-up pulse current deposition ($j_p \sim 10\,\text{mA cm}^{-2}$) into an etched nanotemplate

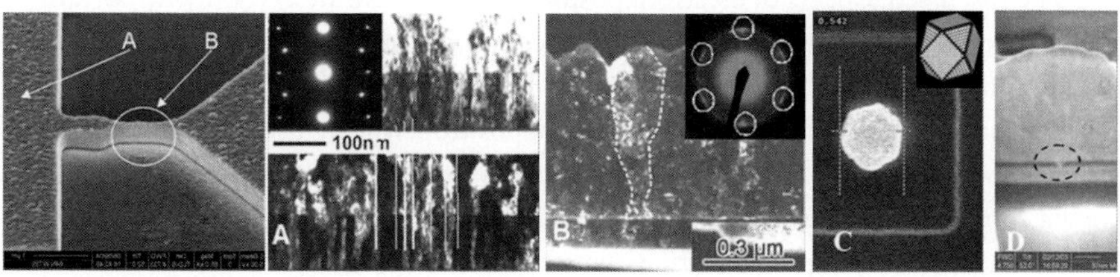

Fig. 17. Image on the left; FIB–SEM side view of CoFe deposit obtained by through mask electrodeposition. Image (A); FIB–TEM of the CoFe deposit indicated as A region on left, Image (B); FIB–TEM of the CoFe deposit indicated as the B region on the left. The lines in (A) and (B) are drawn along the grain boundaries to guide the eye. Image (C); Ni cluster electrodeposited on top of the Ni nanocontact, inset shows the model of cuboctahedra. Image (D); FIB cross section of Ni 30 × 30 nm nanocontact corresponding to the structure in (C). Images (A) and (B) are from Ref. 4. Reprinted from Ref. 4 with permission. Copyright (2006) IEEE.

with diameter $d \sim 30$ nm (Fig. 17(C)). The representative FIB cross section is shown in Fig. 17(D). The symmetry and facets of Ni deposit is characteristic for the shape of equilibrium single crystal fcc metal defined by its low-index planes.[102] This indicates that the Ni nanodeposit (indicated in (D)) from which this overgrowth evolves is single crystal as well.

The additional aspect of electrodeposition at nanoscale is discussed using the example of CoFe alloy deposition into trench geometry with critical dimension of ~ 50 nm.[4] In Fig. 17, on the very left, the CoFe deposit in thin film (shown as A) and nanostructure configuration (shown as B) are indicated. Both deposits have bcc crystal structure with predominant $\langle 110 \rangle$ fiber texture. The crystal grain dimensions in (A) are approximately ~ 10–15 nm, and all (110) diffraction peaks are visible. The TEM image of the representative CoFe nanostructure is shown in Fig. 17(B). The most appealing difference between A and B is that the grain shape and lateral grain size are significantly different. The grain size in B gradually increases with the thickness of the structure, which contributes to the overall hybrid type of the grains (Fig. 17(B)). There are several differences in the diffraction patterns of A and B samples. The 90° diffraction peak is missing in B (weak spot), which is, in FIB–TEM setup configuration, the (110) plane parallel to the incident electron beam. This indicates that grain orientation in nanotrench electrode geometry with (110) plane facing the resist walls is dominating. The other grain orientations gradually die out as the thickness of deposit is increasing.[4] If the dimensions of the photoresist pattern are reduced to the level of the lateral size of the CoFe grains, the properties of CoFe alloy in nanoconfinement will be dominated only by one type of grain orientation. This could result in quite different properties of nanodeposit as compared to the one in thin film configuration.

The discussed aspects of electrodeposition at nanoscale are certainly important to be considered if the future nanodevice fabrication and material design would involve deposition onto nanotemplate electrodes. The current trends in fabrication of magnetic recording heads indicate that electrodeposition will be used for fabrication of future devices with critical dimensions below 50 nm.[103] The emerging bit pattern media concept as a roadway to increase stability of magnetic data and magnetic recoding density certainly represents the situation where the concept of electrodeposition onto nanotemplate electrodes will reach its full potential.

11. Electrodeposition vs Surface Hydrophobicity

The electrodeposition of thin films, in general, occurs at the potentials far from equilibrium resulting in kinetic roughening of the surface. Information regarding the evolution of the surface morphology during electrodeposition is obtained by following the surface width (roughness) over the time and space ordinate. In this analysis, the surface is treated as self-affined fractal geometry where the surface width (w), describing the average surface corrugation in statistical sense, is measured for different length scale (l) and deposition time (t). The surface width (Eq. (23)) is defined as root mean square of fluctuations of surface height, $h(r,t)$, with respect to some reference plane where the term within the "$\langle\rangle$" brackets indicates an average over the system of size l and $\bar{h}(r,t)$ indicates an average of $h(r,t)$ measured over the N different points:

$$w(l,t) = \sqrt{\langle (h(r,t) - \bar{h}(r,t))^2 \rangle}. \tag{23}$$

Although the kinetic roughening during the electrodeposition is a complex phenomenon involving many different processes, most of the electrodeposition systems are well described by normal scaling law,[104] (Eq. (24)). However, it has been shown recently that for some systems,[105] the anomalous scaling law has to be used in order to properly describe the kinetic roughening of particular films (Eq. (25)). The common formulations of these scaling laws are presented below:

$$w(l,t) = l^H f(t/l^{H/\beta}), \tag{24}$$

$$w(l,t) = l^H t^{\beta_{\text{loc}}} f(t/l^{H/\beta}). \tag{25}$$

Here, the scaling function f is such that $w(l,t) = l^H$ (normal scaling) or $l^H t^{\beta_{\text{loc}}}$ (anomalous scaling) for $(t/l^{H/\beta}) \gg 1$ and $w(l,t) = l^\beta$ (normal scaling) or $l^{\beta+\beta_{\text{loc}}}$ (anomalous scaling) for $(t/l^{H/\beta}) \ll 1$. The constants H, β, and β_{loc} are Hurst, growth, and local growth exponent, respectively. The values of these exponents are very much dependent on the conditions of the electrodeposition process such as j/j_L ratio, depositing potential, diffusion layer thickness, presence of additives, and others.[105,106] In general, the higher values of scaling exponents for particular electrodeposition process mean that the roughness of electrodeposited films develops faster over the space and time ordinates.

Considering that the surface wetting properties are dependent on the average surface corrugations or roughness, (see Chap. 3) it is quite straightforward to conclude that electrodeposition could be used as an effective approach to modify the roughness of the initial surfaces and produce rougher surfaces with modified wetting behavior. This concept has been explored by many authors for different analytical and technological applications (self-cleaning surfaces).[107–109] The rough Ag films with super hydrophobic behavior (Lotus effect), for example, were reported by Zhao *et al.*[109] They were produced by electrodeposition through the adsorbed organic phase on ITO. The organic phase has hindered diffusion of Ag ions (effective decrease in $j_L \Rightarrow j/j_L \rightarrow 1$) which promoted Ag dendrite growth resulting in significant roughening of Ag–ITO surface. Such surface has shown a major increase of the hydrophobicity with respect to initial ITO surface having a wetting angle of 154°.

The similar experimental approach has been recently described by Schirtcliffe *et al.*,[108] where the electrodeposition of Cu at extremely high current densities was explored to create rough surfaces with super hydrophobic behavior (wetting angle >150°). Here, the condition leading to rough Cu surfaces is $j/j_L \rightarrow 1$ which is the same as in previous case. However, this is achieved by using high current densities close to the diffusion limiting one (j_L) for the given composition of electrolyte.

Perhaps, the most interesting approach to create surface morphology with specific wetting behavior is electrodeposition through templates. These surfaces will then have a well-defined amplitude and wave length of corrugations which are defined by the template shape, pattern and thickness of the electrodeposit. Controlling these two parameters is relatively easy and allows us to tune wetting properties to desired level, and to engineer surface hydrophobicity through surface structuring. Different methods could be applied to generate patterns for this purpose, including standard lithography[107] or some more exotic approaches.[110] A good example of nanostructured surface with gold deposit is shown in Fig. 18.[110] The templates for gold electrodeposition in this case are generated by self-assembly polystyrene latex spheres (500 nm diameter) on cysteamine coated Au. Gold deposition was carried out from cyanide free solution to obtain the deposits with different thicknesses. The polystyrene latex

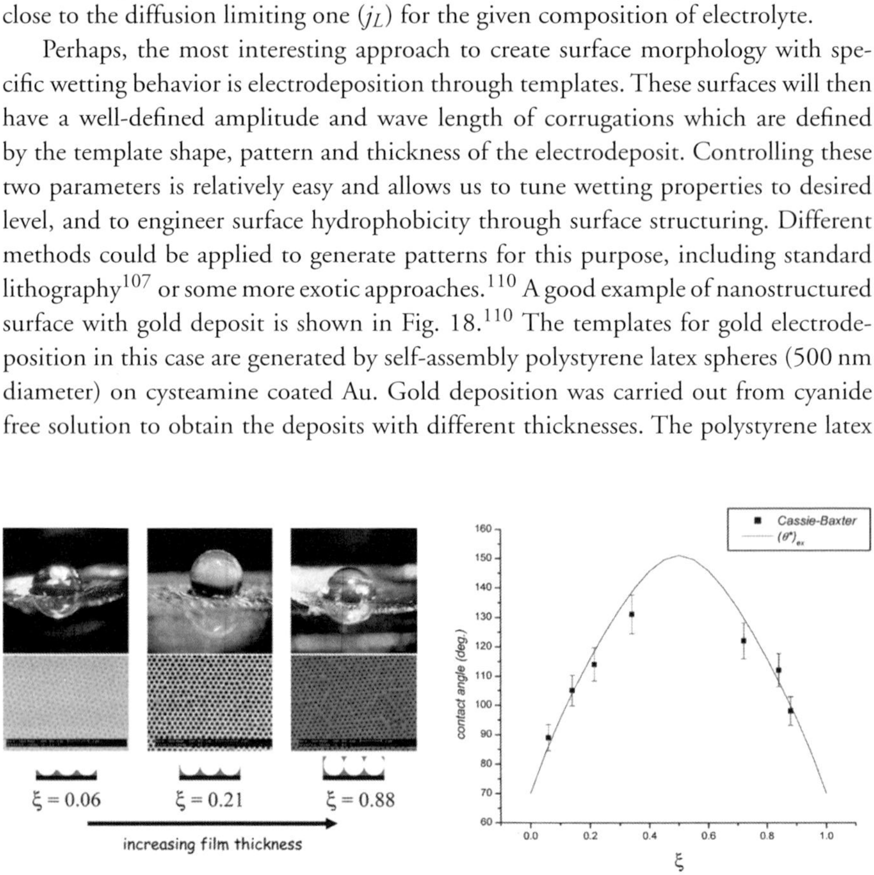

Fig. 18. Images on the left (upper portion); wetting of 5 μL water droplet of the macro-porous Au films with different thickness, (bottom portion) Top-down SEM of the Au deposit morphology corresponding to the wetting of water droplets shown in upper portion of the images. Image of the right; wetting angle as a function of the dimensionless thickness parameter ξ. Reprinted and adapted with permission from Ref. 110. Copyright 2005 American Chemical Society.

spheres are dissolved upon deposition in THF and resulting Au deposit morphology is shown in the lower portion of each image with corresponding sketch of the deposit cross section. The snap-shots on the left side in Fig. 18 show the wetting action of the 5 μL water droplet on surface with Au morphology corresponding to different deposit thicknesses. The figure on the right shows the wetting angle dependence on the dimensionless thickness parameter ξ (ξ = film thickness/diameter of the sphere). The flat gold surface has wetting angle of 70°. Interestingly, for nanostructured gold surface, the wetting angle increases as the ξ goes from 0 to 0.5. The $\xi = 0.5$ represents the thickness for which the wetting angle reaches maximum (~150°, corresponds to the most open Au cavities with largest roughness amplitude). The further increase of the Au deposit thickness results in gradual closure of the nanocavities and decrease of the roughness which results in the subsequent decrease in wetting angle. Evidently, the control of the Au deposit thickness allows the precise alteration of the wetting behavior of nanostructured gold surface and engineering of the Au surface hydrophobicity for a given template shape and size.

Acknowledgment

The author would like to acknowledge the support of the University of Houston and Texas Center for Superconductivity during preparation of this article.

References

[1] Edelstein A, Cammarata R. (1996) *Nanomaterials: Synthesis, Properties and Applications.* IOP Publishing, Bristol, UK.

[2] Cao G. (2004) *Nanostructures and Nanomaterials: Synthesis, Properties and Applications.* Imperial College Press, London, UK.

[3] Cooper E *et al.* (2005) Recent developments in high moment electroplated materials for recording heads. *IBM J Res Dev* **49**: 103–125.

[4] Brankovic S *et al.* (2006) Electrodeposition of 2.4 T Co37Fe63 alloys at nanoscale for magnetic recoding application. *IEEE Trans Magn* **42**: 132–139.

[5] Data M *et al.* (1995) Electrochemical fabrication of mechanically robust PbSn C4 interconnections. *J Electrochem Soc* **142**: 3779–3785.

[6] Andricacos P. (1999) Copper on-chip interconnections: A breakthrough in electrodeposition to make better chips. *Electrochem Soc Interface* **8**: 32–37.

[7] Deligianni H. (2006) Electrodeposition and microelectronics. *Electrochem Soc Interface* **15**: 33.

[8] Ehrfeld W. (2003) Electrochemistry and microsystems. *Electrochim Acta* **48**: 2857–2868.

[9] Nicewarner-Peña S *et al.* (2001) Submicrometer metallic barcodes. *Science* **294**: 137–141.

[10] Schwarzacher W. (1999) Metal nanostructures, a new class of electronic devices. *Electrochem Soc Interface* **8**: 20–24.

[11] Pauling H, Juttner K. (1992) Top-on-top monolayer formation of foreign metals on gold single crystal surfaces. *Electrochim Acta* **37**: 2237–2244.

[12] Whitney T *et al.* (1993) Fabrication and magnetic properties of arrays of metallic nanowires. *Science* **261**: 1316–1319.

[13] Sung M *et al.* (2001) Electrodeposition of magnetic nanoparticle arrays with ultra-uniform length in ordered alumite. *Appl Phys Lett* **78**: 2964–2966.

[14] Bartlett P. (2004) Electrodeposition of nanostructured films using self-organizing templates. *Electrochem Soc Interface* **13**: 28–33.

[15] Whitaker J, Nelson J, Schwartz D. (2005) Electrochemical printing: Software reconfigurable electrochemical microfabrication. *J Micromech Microeng* **15**: 1498–1503.

[16] Kolb D, Ullmann R, Will T. (1997) Nanofabrication of small copper clusters on gold(111) electrodes by a scanning tunneling microscope. *Science* **275**: 1097–1099.

[17] Zach M, Ng K, Penner R. (2000) Molybdenum nanowires by electrodeposition. *Science* **290**: 2120–2123.

[18] Li C *et al.* (1999) Fabrication of stable metallic nanowires with quantized conductance. *Nanotechnology* **10**: 221–223.

[19] Yong F *et al.* (1999) Large magnetoresistance of electrodeposited single-crystal bismuth thin films. *Science* **284**: 1335–1337.

[20] Sieradzki K, Brankovic S, Dimitrov N. (1999) Electrochemical defect-mediated thin-film growth. *Science* **284**: 138–141.

[21] Brankovic S, Dimitrov N, Sieradzki K. (1999) Surfactant mediated electrochemical deposition of Ag on Au (111). *Electrochem Solid State Lett* **2**: 443–445.

[22] Vasilic R, Dimitrov N. (2005) Epitaxial growth by monolayer-restricted galvanic displacement. *Electrochem Solid State Lett* **8**: C173–C176.

[23] Hwang S, Oh H, Kwak J. (2001) Electrodeposition of epitaxial Cu (111) thin films on Au (111) using defect-mediated growth. *J Am Chem Soc* **123**: 7176–7177.

[24] Switzer J *et al.* (2003) Enantiospecific electrodeposition of a chiral catalyst. *Nature* **425**: 490–493.

[25] Allred D *et al.* (2005) Electrochemical nanofabrication using crystalline protein masks. *Nano Lett* **5**: 609–613.

[26] Stickney J. (2002) Electrochemical atomic layer epitaxy (EC-ALE): Nanoscale control in the electrodeposition of compound semiconductors. In RC Alkire & DM Kolb (eds), *Advances in Electrochemical Science and Engineering,* Vol. 7, pp. 1–105. Wiley-VCH, Weinheim.

[27] Varazo K *et al.* (2002) Morphology in electrochemical atomic epitaxy. In MP Soriaga, JL Stickney, LA Bottomley & YG Kim (eds), *Thin Films: Preparation, Characterization and Applications,* pp. 83–94. Kluwer Academic, New York.

[28] Prentice G. (1991) *Electrochemical Engineering Principles.* Prentice Hall International Series, New Jersey.

[29] Brankovic S *et al.* (2005) Influence of additive adsorption on properties of pulse deposited CoFeNi alloys. *J Electrochem Soc* **152**: C196–C202.
[30] Popov N, Yin K, White R. (1993) Galvanostatic pulse and pulse reverse plating of nickel–iron alloys from electrolytes containing organic compounds on a rotating disc electrode. *J Electrochem Soc* **140**: 1321–1330.
[31] Brenner A. (1963) *Electrodeposition of Alloys I.* Academic Press, New York.
[32] Trasatti S. (1971) Work function, electronegativity, and electrochemical behaviour of metals: II. Potentials of zero charge and electrochemical work functions. *J Electroanal Chem* **33**: 351–378.
[33] Kolb D. (1978) Physical and electrochemical properties of metal monolayers on metallic substrates. In Gerischer H & Tobias W (eds), *Advances in Electrochemistry and Electrochemical Engineering*, pp. 125–271. Wiley & Sons, New York.
[34] Budevski E, Staikov G, Lorenz W (eds). (1996) *Electrochemical Phase Formation and Growth: An Introduction to the Initial Stages of Metal Deposition.* Wiley VCH, Berlin.
[35] Herrero E, Buller L, Abruña H. (2001) Underpotential deposition at single crystal surfaces of Ag, Au, Pt and other materials. *Chem Rev* **101**: 1897–1930.
[36] Will M, Dietterle M, Kolb D. (1995) The initial stages of electrolytic copper deposition: An atomistic view. In A Gewirth & H Siegenthaler (eds), *Nanoscale Probes of the Solid/Liquid Interface*, Vol. 288, pp. 137–162. Springer, Berlin.
[37] Garcia S, Salinas D, Staikov G. (2005) Underpotential deposition of Cd on Ag (111): An in situ STM study. *Surf Sci* **576**: 9–18.
[38] Corcoran S, Chakarova G, Sieradzki K. (1994) An in-situ STM investigation of the underpotential deposition of Ag on Au (111) electrodes. *J Electranal Chem* **377**: 85–90.
[39] Esplandiu M, Schneeweiss M, Kolb D. (1999) An in-situ scanning tunneling microscopy study of Ag electrodeposition on Au (111). *Phys Chem Chem Phys* **1**: 4847–4854.
[40] Engelsmann K, Lorenz W, Schmidt E. (1980) Underpotential deposition of lead on polycrystalline and single-crystal gold surfaces: Part I. Thermodynamics. *J Electroanal Chem* **114**: 1–10.
[41] Schmidt E, Gygax H, Bohlen P. (1966) Polarographie an Festelektroden IV. Einfluss des Leitelektrolyten auf die submonoatomare Bleibedeckung von Silberelektroden. *Helv Chim Acta* **49**: 733–740.
[42] Adzic R, Yeager E, Cahan B. (1974) Optical and electrochemical studies of underpotential deposition of lead on gold evaporated and single-crystal electrodes. *J Electrochem Soc* **121**: 474–484.
[43] Bewick A, Tomas B. (1977) Optical and electrochemical studies of the under-potential deposition of metals: Part III. Lead deposition on silver single crystals. *J Electroanal Chem* **84**: 127–140.
[44] Swathirajan S, Burckenstein S. (1983) Thermodynamics and kinetics of underpotential deposition of metal monolayers on polycrystalline substrates. *Electrochim Acta* **28**: 865–877.

[45] Obretenov W, Dimitrov N, Popov A. (1996) Monte Carlo simulation of underpotential deposition in the Ag (111)/Pb^{2+} system. *J Cryst Growth* **167**: 253–259.

[46] Popov A *et al.* (1995) Modelling transformation processes in underpotential lead adsorbate on Ag (111). *Electrochim Acta* **40**: 1495–1499.

[47] Popov A *et al.* (1997) Structural transformation process in underpotential monolayer adsorbates: Two types of adsorption sites on the electrode. *J Cryst Growth* **171**: 250–258.

[48] Jovic V, Jovic B. (2002) Underpotential deposition of cadmium onto (111) face of silver from chloride containing solution. *Electrochim Acta* **47**: 1777–1785.

[49] Kukta R *et al.* (2005) Self-assembly of paired nanoribbons. *Phys Rev Lett* **95**: 186103, 4 p.

[50] Sorenson T *et al.* (2001) Formation of and phase transitions in electrodeposited tellurium atomic layers on Au (111). *Surf Sci* **470**: 197–214.

[51] Lay M, Stickney J. (2003) Electrodeposition of Au–Cd alloy nanostructures on Au (111). *J Am Chem Soc* **125**: 1352–1355.

[52] Lu D, Tanaka K. (1997) Cu + Au alloy particles formed in the underpotential deposition region of copper in acid solutions. *J Electroanal Chem* **430**: 69–76.

[53] Mallet JJ *et al.* (2005) Compositional control in electrodeposited Co*x*Pt1À*x* films. *Electrochem Solid State Lett* **8**: C15–C18.

[54] Mallet JJ *et al.* (2004) Compositional control in electrodeposition of FePt films. *Electrochem Solid State Lett* 7: C121–124.

[55] Cavallotti P, Bestetti M, Franz S. (2003) Microelectrodeposition of Co–Pt alloys for micromagnetic applications. *Electrochim Acta* **48**: 3013–3020.

[56] Huang Y *et al.* (2002) CoPt and FePt nanowires by electrodeposition. *J Appl Phys* **91**: 6869–6871.

[57] Zana I, Zangari G. (2003) Electrodepositon of Co–Pt films with high perpendicular anisotropy. *Electrochem Solid State Lett* **6**: C153–C156.

[58] Zana I, Zangari G, Shamsuzzoha M. (2004) Influence of Cu (100) substrate on growth, morphology, structural and magnetic properties of high coercivity electrodeposited Co-Pt-P thin films. *J Electrochem Soc* **151**: C637–C644.

[59] Wang F *et al.* (2004) Nanostructured $L1_0$ Co–Pt thin films by an electrodeposition process. *Electrochem Commun* **6**: 1149–1152.

[60] Brankovic S, Wang J, Adzic R. (2001) Metal monolayer deposition by replacement of metal adlayers on electrode surfaces. *Surf Sci* **474**: L173–L179.

[61] Brussel V *et al.* (2003) Oxygen reduction at platinum modified gold electrodes *Electrochim Acta* **48**: 3909–3919.

[62] Zhang J *et al.* (2005) Controlling the catalytic activity of platinum-monolayer electrocatalysts for oxygen reduction with different substrates. *Angew Chem Int Ed* **44**: 2132–2135.

[63] Zhang J *et al.* (2005) Mixed-metal Pt monolayer electrocatalysts for enhanced oxygen reduction kinetics. *J Am Chem Soc* **127**: 12480–12481.

[64] Zhang J *et al.* (2007) Stabilization of platinum oxygen-reduction electrocatalysts using gold clusters. *Science* **315**: 220–222.

[65] Park S *et al.* (2002) Transition metal-coated nanoparticle films: Vibrational characterization with surface-enhanced Raman scattering. *J Am Chem Soc* **124**: 2428–2429.

[66] Kowal A *et al.* (2009) Ternary Pt/Rh/SnO2 electrocatalysts for oxidizing ethanol to CO_2. *Nat Mat* **8**: 325–330. Advanced Online Publication http://www.nature.com/nmat/index.html

[67] Viyannalage L, Vasilic R, Dimitrov N. (2007) Epitaxial growth of Cu on Au (111) and Ag (111) by surface limited Redox replacement — An electrochemical and STM study. *J Phys Chem* **C 111**: 4036–4041.

[68] Huang B *et al.* (1995) Preliminary studies of the use of an automated flow-cell electrodeposition system for the formation of CdTe thin films by electrochemical atomic layer eEpitaxy. *J Electrochem Soc* **142**: 3007–3016.

[69] Gokcen D, Bae S, Brankovic S. (2007) Nucleation and growth of low-dimensional noble metal structures using galvanic displacement of UPD monolayers. Abstract 1381, 212th ECS Meeting, Washington DC, October 7–12.

[70] Waibel H *et al.* (2002) Initial stages of Pt deposition on Au (111) and Au (100). *Electrochim Acta* **47**: 1461–1467.

[71] Chrzanowski W, Wieckowski A. (1997) Ultra-thin films of Ruthenium on low index platinum single crystal surfaces: An electrochemical study. *Langmuir* **13**: 5974–5978.

[72] Chrzanowski W, Kim H, Wieckowski A. (1998) Enhancement in methanol oxidation by spontaneously deposited Ruthenium on low index platinum electrodes. *Catal Lett* **50**: 69–75.

[73] Brankovic S, McBreen J, Adzic R. (2001) Spontaneous deposition of Pt on Ru (0001) surface. *J Electroanal Chem* **503**: 99–104.

[74] Brankovic S *et al.* (2002) Electrosorption and catalytic properties of bare and Pt modified single crystal and nanostructured Ru surfaces. *J Electroanal Chem* **524–525**: 231–241.

[75] Brankovic S, McBreen J, Adzic R. (2001) Spontaneous deposition of Pd on Ru (0001). *Surf Sci* **479**: L363–L368.

[76] Attard G, Bannister A. (1991) The electrochemical behaviour of irreversibly adsorbed palladium on Pt (111) in acid media. *J Electroanal Chem* **300**: 467–485.

[77] Llorka M *et al.* (1993) Electrochemical structure-sensitive behaviour of irreversibly adsorbed palladium on Pt (100), Pt (111) and Pt (110) in an acidic medium. *J Electroanal Chem* **351**: 299–319.

[78] Brankovic S *et al.* (2002) Carbon monoxide oxidation on bare and Pt-modified Ru(1010) and Ru(0001) single crystal electrodes. *J Electroanal Chem* **532**: 57–66.

[79] Rosebrugh T, Millar W. (1910) Mathematical theory of the changes of concentration at the electrode brought about by diffusion and by chemical reaction. *J Phys Chem* **14**: 816–885.

[80] Bard A, Faulkner L. (2001) *Electrochemical Methods, Fundamentals and Applications*, 2nd ed. Wiley, NewYork.

[81] Crank J. (1999) *The Mathematics of Diffusion*, 2nd ed. Oxford University Press.

[82] Brankovic S *et al.* (2003) Pulse plating of soft high magnetic moment materials — properties and application. In S Krongelb *et al.* (eds), *Magnetic Material, Processes and*

Devices VII and Electrodeposition of Alloys, 202nd ECS Meeting Proceeding, pp. 269–281. Pennington, NJ.

[83] Puippe J, Leaman F. (1986) *Theory and Practice of Pulse Plating*. AESF, Orlando, FL.

[84] Brankovic S *et al.* (2006) Influence of additive adsorption on properties of pulse deposited high magnetic moment CoNiFe and CoFe alloy. In Krongelb *et al.* (eds) *Magnetic Materials Processes and Devices VIII and Electrodeposition of Alloys*, PV 2004-23, pp. 347–356, Pennington NJ.

[85] Schlesinger M, Paunovic M (eds). (2000) *Modern Electroplating*, 4th ed. John Willey & Sons, NY.

[86] Osaka T *et al.* (1999) Effects of saccharin and thiourea on sulfur inclusion and coercivity of electroplated soft magnetic CoNiFe film. *J Electrochem Soc* **146:** 3295–3299.

[87] Osaka T *et al.* (1998) A soft magnetic CoNiFe film with high saturation magnetic flux density and low coercivity. *Nature* **392:** 796–798.

[88] Hsu J and Lin K, (2005) The effect of saccharin addition on the mechanical properties and fracture behavior of electroless Ni-Cu-P deposit on Al. *Thin Solid Films* **471:** 186–193.

[89] Buess-Herman C. (1994) Self-assembled monolayers at electrode metal surfaces. *Prog Surf Sci* **46:** 335–375.

[90] Damaskin B, Petri O, Batrakov V. (1971) *Adsorption of Organic Compounds on Electrodes*, p. 259. Plenum Press, New York.

[91] Brankovic S, Haislmaier R, Vasiljevic N. (2007) Molecular incorporation of saccharin into 2.4 T CoFe alloys. *Electrochem Soc Trans* **3**: 71–80.

[92] George J *et al.* (2008) Sulfur and saccharin incorporation into electrodeposited CoFe alloys. *J Electrochem Soc* **155**: D589–D594.

[93] Freund L, Suresh S. (2003) *Thin Film Materials: Stress, Defect Formation, and Surface Evolution*. Cambridge University Press, NewYork.

[94] Dieter G. (1986) *Mechanical Metallurgy*, 3rd ed., p. 145. McGraw Hill, New York.

[95] Rice J. (1976) Hydrogen and interfacial Cohesion. In A Thompson & IM Bernstein (eds), *Effect of Hydrogen on Behaviour of Materials*, pp. 375–368. The Metallurgical Society of AIME, Warrendale, PA.

[96] Tabakovic I, Inturi V, Riemer S. (2002) Composition, structure, stress and coercivity of electrodeposited soft magnetic CoNiFe films: Thickness and substrate dependence. *J Electrochem Soc* **149**: C18–C22.

[97] Venables J, Spiller G, Hanbücken M. (1984) Nucleation and growth of thin films. *Rep Prog Phys* **47**: 399–459.

[98] Stoyanov S, Markov I. (1982) On the 2D–3D transition in epitaxial thin film growth. *Surf Sci* **116**: 313–337.

[99] Tersoff J, Denier van der Gon A, Tromp R. (1994) Critical island size for layer-by-layer growth. *Phys Rev Lett* **72**: 266–269.

[100] Tang L. (1993) Island formation in submonolayer epitaxy. *J Phys I* **3**: 935–950.

[101] Dursun A, Pugh D, Corcoran S. (2003) Dealloying of Ag–Au alloys in halide-containing electrolytes. *J Electrochem Soc* **150**: B355–B360.

[102] Wulff G. (1901) Zur Frage der Geschwindigkeit des Wachstums unter Auflosung der Kristallfiachen. *Z Kristallogr* **34**: 449–530.

[103] Childress J, Fontana R. (2005) Magnetic recording read head sensor technology. *C R Physique* **6**: 997–1012.

[104] Family F, Vicsek T. (1985) Scaling of the active zone in eden process on percolation networks and the a ballistic deposition model. *J Phys A* **18:** L75–L81.

[105] Huo S, Schwarzacher W. (2001) Anomalous scaling of the surface width during Cu electrodeposition. *Phys Rev Lett* **86**: 256–259.

[106] Schwarzacher W. (2004) Kinetic roughening of electrodeposited films. *J Phys Condens Matter* **16**: R860–R878.

[107] McHale G, Shirtcliffe N, Newton M. (2004) Super-hydrophobic and supper-wetting surfaces: Analytical potential. *The Analyst* **129**: 284–287.

[108] Shirtcliffe N, McHale G, Newton M, Perry C. (2005) Wetting and wetting transitions on copper-based super-hydrophobic surfaces. *Langmuir* **21**: 937–943.

[109] Zhao N *et al.* (2005) Combining layer-by layer assembly with electrodeposition of silver aggregates for fabricating superhydrophobic surfaces. *Langmuir* **21**: 4713–4716.

[110] Abdelsalam M *et al.* (2005) Wetting of regularly structured gold surfaces. *Langmuir* **21**: 1753–1757.

Chapter 12

Surface Modification by Plasma-Based Processes

Evangelina De Las Heras, Gabriel Ybarra, Iñigo Braceras and Pablo Corengia

In this chapter, the use of plasma processes technology in the biomedical field is presented. The introduction focuses on plasmas and the associated techniques employed to modify surfaces of a wide variety of materials.

The versatility of plasma-based treatments allows the design of special function-oriented surfaces for medical devices, which are presented next. An overview of the plasma-based thin films and surface treatments is included for a number of applications ranging from the improvement of mechanical properties and the modification of the surface chemistry of materials to regulate the interaction with living tissues, diagnostics, and drug delivery.

1. Introduction

Chemically reactive plasmas are widely used to modify the surface properties of materials. The technologies of plasma assisted processes, which are commercially used, allow a convenient modification of the resulting surfaces through a strict control of process parameters and variables.

Irving Langmuir was the first one to use the term plasma to refer to the gas obtained in an electric discharge in a tube. Overall, plasma can be defined as an ionized medium composed by electrons, ions, and neutral particles in random motion which, on average, is electrically neutral and has a collective behavior.[1] Collective behavior is related to the long range electromagnetic interactions that determine the system statistical properties. This behavior holds a substantial difference with respect to ideal gases, where particles interact with each other only through elastic collisions. Each of the plasma species has its own velocity distribution function, from which a temperature for the local or total thermodynamic balance case can be assigned. Thus, a temperature can be assigned for electrons, ions, and neutral species.[2]

Plasma is produced by applying energy to a gas; for instance, when a high voltage is applied between two electrodes, the gas molecules are broken down into electrons and positive ions. The acceleration of electrons and their subsequent collision with other particles induces ionization, excitation, and fragmentation processes of molecules. These processes lead to a complex mixture of active species, which, under the effect of electric and/or magnetic fields, take part in chemical reactions and interact with surfaces.

The species found in plasmas have a chemical reactivity considerably higher than the same species in the absence of electric discharge. They can promote the formation of new species, both in the gaseous phase as well as over the surfaces of a specimen immersed in the discharge. Reaction and recombination mechanisms are quite different from the ones found in conventional chemical reactions.[3] Likewise, a series of events which take place on the surface of the parts immersed in the discharge can influence the final condition of the surface. The processes occurring include emission of particles (neutral particles, secondary electrons, secondary ions, reflected incident particles, desorption of gases), emission of radiation (photons, X rays), ion implantation, altered surface layers and diffusion, dissociation processes (sublimation, chemical dissociation), and chemical sputtering.[4] These mechanisms are the basis of the reactive processes of plasma material treatments.

2. Classification of Plasmas

Artificially generated plasma may have different origins depending on the source of energy used. Plasmas can be generated by different means: mechanical (adiabatic compression), thermal (heating in electric ovens), chemical (through exothermal reactions), radiant (electromagnetic and particle radiation, e.g. electron beams), nuclear (controlled by nuclear reactions), electric (discharges), and through the combination of some of the above.[3] With the exception of some special cases, electric discharges constitute the most widely used method for the plasma processing of materials, as they can be easily implemented and allow a precise control of working conditions within a wide range of parameters of interest.

Plasma properties, irrespective of their origin, are modified by variables such as electronic density, pressure, and temperature; these, among other parameters, make plasma fall into different categories. The most common are *hot* or *thermal plasmas* (near equilibrium plasmas), and *cold plasmas* (nonequilibrium plasmas). Hot plasmas include electric arcs, plasma jets of rocket engines, and thermonuclear reaction generated plasmas. On the other hand, cold plasmas include low-pressure direct current (DC) and radio frequency (RF) discharges as well as discharges from fluorescent (neon) tubes.

Hot plasmas are characterized by the high temperature of both electrons, ions and neutral particles and are close to the maximal degree of ionization (100%). On the other hand, cold plasmas are formed by low temperature heavy particles and high temperature electrons and are associated with low degrees of ionization (between 10^{-4} and 10%).[3]

Plasma sources can be classified by their excitation mode.[5] Three groups are then highlighted: DC (or pulsed in the range of kHz), RF (MHz), and microwave (MW) discharges (GHz). Table 1 shows a rough classification of plasmas.

2.1. *DC Discharges*

These are commonly called *glow discharges* or cold cathode discharges. The term glow discharge originated in the discharge of a capacitor in a circuit in which there is a gap between two electrodes immersed in a gaseous medium. If the applied voltage is high enough, a dielectric breakdown of the gas occurs, producing an ionized state. In this way, the circuit is closed and the capacitor discharges.

Plasma sheaths are thin regions present in every gaseous discharge perpendicular to the surface of a solid in contact with a plasma and have a net electrical charge, usually positive, surrounding the body. Plasma sheaths are important from the viewpoint of materials treatment since they control the transport process of the chemically activated species generated in the plasma towards the surface of the substrate, as well as the bombardment with positive ions of that surface and the chemical reactivity of surface induced by it.

Table 1. Classification of plasmas according to their origin, power source, temperature, and pressure.

Plasma types			
By origin	By sources (electromagnetic origin)	By temperature	By pressure
• Mechanical • Thermal • Chemical • Radiant • Nuclear • Electromagnetic	• DC • pulsed DC (~kHz) • RF (~MHz) • MW (~GHz)	• Hot, thermal, or near equilibrium plasmas • Cold or nonthermal plasmas	• Low-pressure discharges: *Glow discharge plasmas Capacitively/inductively coupled plasmas* • Atmospheric: *Arc discharge Corona discharge*

2.2. *RF Discharges*

In the case of alternating fields, the typical frequencies range from 100 Hz in the low frequency region, to 13.56 MHz in the RF range, and 2.45 GHz in the microwave range.

Some of the RF discharges are similar to DC discharges. This is the case of systems of RF capacitive coupling where there are two electrodes with sheaths on both of them, which separate them from the quasi neutral central ionized gas. In certain conditions, one of the electrodes behaves as a cathode with continuous negative reference potential with respect to the plasma that allows a quasi permanent flow of positive energetic ions towards the cathode/substrate. In other cases, however, RF discharges are totally different to DC discharges. For example, there is no net flow of electrons between the electrodes in one discharge cycle and the secondary electrons emitted by the electrodes do not have a dominant role in the discharge.

3. Modification of Functional Surface Properties by Plasma-Based Processes

Plasma surface processing is an effective and economical surface treatment technique for many materials and of growing interest in the biomedical field. One of the most important advantages of plasma surface modification is that the surface properties and biocompatibility can be modified selectively while the bulk characteristics of the materials remain unchanged.

The biomedical field requires biomaterials which, besides satisfying the physical and mechanical requirements for their application, must be biocompatible with the biological environment in which they are used. Since biocompatibility involves reactions at the interface of the biomaterial and the biological environment, surface modification techniques can be of great help to solve this problem, avoiding costly changes of materials.[6] By using plasma surface technology, it is possible to change the chemical composition and physical properties such as wettability, adhesion, dyeability, refractive index, hardness, chemical inertness, lubricity, and the subsequent biological interactions of surfaces.

Plasma processes may have different actions on a surface, involving the addition, removal, or changes of the material. Accordingly, surface modification by plasma-based processes can be classified as follows:

(1) *Surface functionalization* and *grafting* involve the introduction of new chemical functionalities at the most superficial level of the substrate, at a molecular scale.
(2) *Coating deposition* implies the formation of a new phase on top of the substrate. The surface properties of the material are then defined by the coating (e.g. wear

and corrosion resistance). The coating might in some cases undergo a subsequent surface functionalization process.

(3) *Formation of a surface gradient* involves the transformation of the chemical composition of the region nearby the surface through the addition (diffusion) of one or more elements, which are present in the plasma, such as C, N, O, S, etc. Thermochemical treatments are widely used with this aim in different industrial fields. The main advantage of this process is the improvement of wear, friction, corrosion, and fatigue properties of materials such as steels, titanium, and its alloys.

(4) *Coating + surface gradient formation* is a duplex treatment consisting in a combination of the two previous treatments. In many cases, applying a thermochemical treatment before a top-coating deposition might be advantageous. The hardened substrate provides mechanical support and thus prevents the coating at the top from breaking if the substrate below is deformed.

(5) *Etching* and *ablation*, in opposition to the previous processes, involve the removal of material from the substrate surface due to the aggressive action of the plasma rather than the addition of new material.

All these processes generate a new surface, as schematically shown in Fig. 1, with a different morphology and surface chemistry, and consequently new physicochemical properties are to be expected. The competition between recombination–deposition of plasma species and the destructive action of plasma, i.e. sputtering, determine the prevalence of surface functionalization, film formation, or ablation processes. It is the mastering of the plasma action on the substrate which allows tailoring a surface towards a specific end.

Figure 2 presents the most common plasma-based surface modification techniques for biomedical applications, described in more detail later: plasma assisted chemical vapor deposition or PACVD (RF, MW), physical vapor deposition or PVD (sputtering, cathodic arc), plasma polymerization and grafting, plasma-based thermochemical treatments (e.g. plasma nitriding), ion implantation, plasma immersion ion implantation or PIII, and plasma spraying. Each technique has unique advantages and applications, and the choice of the more adequate technique often depends on the

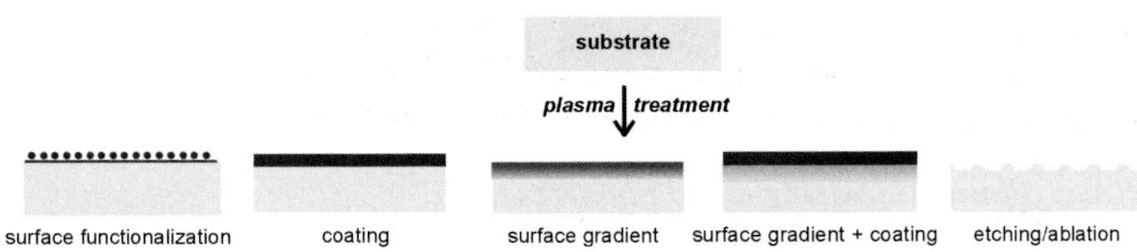

Fig. 1. Schematic representation of surface modification by plasma processes.

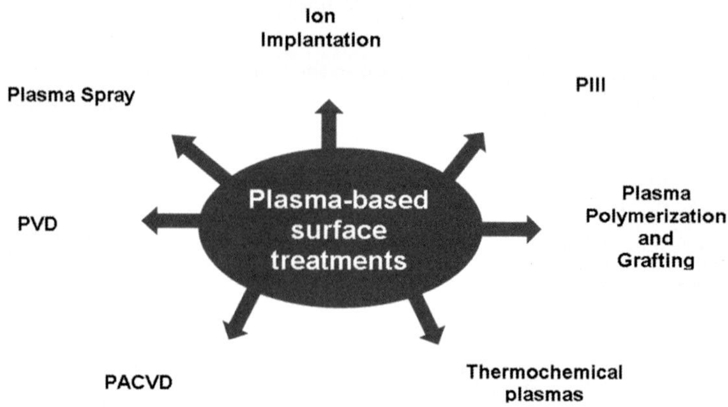

Fig. 2. Plasma-based processes for biomedical applications (PACVD: Plasma Assisted Chemical Vapor Deposition; PVD: Physical Vapor Deposition; PIII: Plasma Immersion Ion Implantation).

Table 2. Applications and examples of plasma treatment in biomaterials engineering.

Properties	Medical devices: examples
Sterilization	Surgical tools and implantable devices
Hydrophobic and hydrophilic surfaces	Catheters, implants
Biomolecule immobilization	Biochips, implants
Antireflective surfaces	Surgical tools
Surfaces with improved mechanical properties	Joint prostheses
Antimicrobial surfaces	Implants, surgical tools, catheters
Interaction	Load bearing implants, osseointegration
Diagnostics	Biochips such as DNA chips
Therapies and drug release	Stents, scaffolds for tissue regeneration

reliability, reproducibility, and productivity.[7] Overall, plasma surface modification has become a hot research topic in biomaterials. Table 2 shows some typical applications and examples of plasma-based treatments in the biomedical engineering field.

3.1. *Plasma Grafting and Polymerization*

Plasma functionalization and plasma-polymerized thin films offer attractive possibilities for developing new biomaterials and improving the performance of a wide range of medical devices. Many plasma grafting and polymerization methods have been published.[4,8,9]

The most frequently used plasma functionalization and polymerization systems can be classified into three basic groups[10]:

(1) Internal parallel plate metal electrodes: commonly, the discharge can be excited by an AC or RF source.
(2) External ring electrodes and/or external coil: RF power may be coupled into the plasma inductively, but in most cases the coupling is of capacitive nature.[11]
(3) Electrode-less microwave or high frequency reactors: usually composed from a silica tube that passes through the resonant cavity coupled to a microwave power supply (typically 2.45 GHz).

Most of those systems are carried on in a bell jar and tubular-type reactors. In addition, others systems have been used (e.g. Ion Gun Inverse Magnetron[12]), but most of them have been used in laboratory or pilot scales.

The most important parameters for describing and controlling the conditions of the process are the monomer flow rate (in case of plasma polymerization), system pressure, discharge power (or current density and voltage), distance between the electrodes and the sample, and nature of gas or gases mixtures.

This section is devoted to plasma grafting and polymerization processes, which are usually carried out at sub-atmospheric pressure and low temperature plasma. However, it is worth mentioning that since the beginning of the 1990s, many research groups and industries are working on other new plasma treatments, such as pulsed plasmas.[13]

3.1.1. *Grafting*

In the case of plasma functionalization, functional groups of the substrate, typically a polymer, are converted to other chemical groups by substitutive radical reactions, a process also called grafting, without the deposition of a proper coating.

Surface functionalization by plasma is a complex process. The plasma chemistry governs the generation of activated, chemically reactive species in the gas phase. These species may induce surface modifications in what is called plasma surface interaction. Rearrangements of a modified surface and heterogeneous reactions of activated sites, remnants of the plasma process, give rise to the so-called post-plasma surface process that may drastically change the surface composition.

Many functional groups can be grafted. Table 3 depicts some organic functional groups. These plasma formed functional groups critically determine the chemical and physical properties of the surface. If only a thin surface layer is being functionalized, the specific nature of the bulk material can be preserved.

Table 3. Name and structure of organic functional groups containing nitrogen and oxygen.

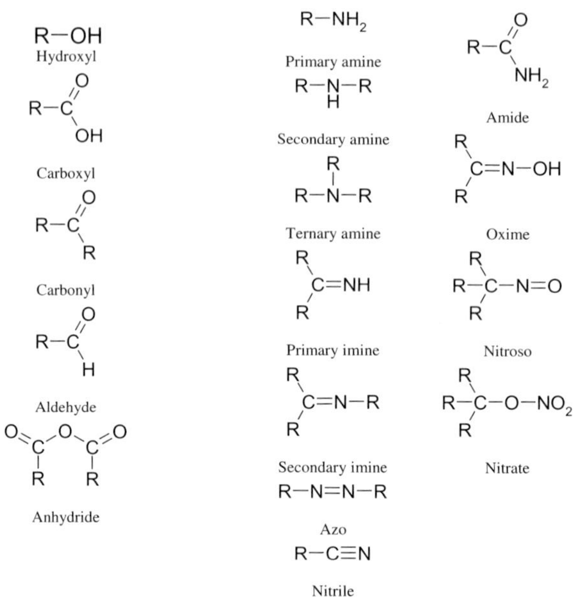

In order to introduce functional groups like amino or carboxylic acid groups on polymer surfaces, which can be used for subsequent covalent bonding, grafting is preferred over plasma polymerization. In this way, intrinsic adhesion problems between substrate and plasma-deposited coatings can be circumvented. Grafting plasma treatments using gases like NH_3, CO_2, O_2, H_2, H_2O, Ar, and air are commonly used for the functionalization of polymers such as polytetrafluoroethylene (PTFE), polyvinyl fluoride (PVF), polyethylene (PE), polypropylene (PP), polyacrylonitrile, etc.

Besides grafting or functionalization, both etching and crosslinking may occur simultaneously on a polymer surface exposed to plasma. The gas used primarily determines which process is dominant. When oxidizing gases like O_2, carbon monoxide, carbon dioxide, or water are used, etching and functionalization of the surface dominate.

3.1.2. *Plasma polymerization*

The term "plasma polymerization" is widely used to denote a specific type of plasma chemistry process resulting in the deposition of polymer films by a passage of an organic gas or vapor through the plasma. From the early 1960s, plasma polymerization has become a very useful method for surface modification and thin film deposition.[14] Lately, the use of this processing technique in the biomaterials field has attracted the interest of many researchers.

Plasma polymerization is usually a simpler and more versatile process than the traditional methods of polymer film production, since fewer fabrication steps are necessary.[15] In addition, the plasma process can be used to produce polymeric films (called *plasma polymers*) that do not polymerize under normal chemical polymerization conditions.[16]

Plasma polymerization involves reactions between plasma species, between plasma and surface species, and between surface species.[17] Usually a free radical mechanism is considered and two cases are distinguished:

(a) Plasma-state polymerization — occurs only in plasmas produced by an electric discharge in the presence of organic molecules. Energetic particles produced in the collisions with organic compounds undergo chain reactions and free radical termination reactions. As a result, any organic compound in the gaseous state may be plasma polymerized.
(b) Plasma-induced polymerization is essentially free radical-induced polymerization of molecules, including carbon–carbon bonds.

The plasma polymerization process takes place through several reaction steps: initiation, propagation, termination, and re-initiation. While in conventional polymerization the termination step finishes the process, in plasma polymerization neutral products formed in the termination steps can undergo re-initiation and propagation reactions.

Plasma polymerization is a strongly system-dependent process, which is not determined only by the monomer used but by plasma parameters. The structure and composition, physical and chemical properties of a plasma polymer and its deposition rate depend on many parameters for a given monomer or gas mixture: type of reactor, frequency of discharge (RF, MW), excitation voltage, power delivered, flow rate of monomer, working gas pressure, substrate temperature, substrate size and its position, etc. Detailed discussion of plasma polymerization processes can be found in several reviews and books.[4,18,19] Only the basic phenomena of plasma polymerization and plasma polymers for biomedical applications are described in this section.

Plasma polymer is a new type of material which has a different structure than the conventional polymer that has regularly repeating units. In the case of the plasma polymer, the chains are short and randomly branched, and terminated with a high degree of crosslinking. In many cases, a great number of free radicals are trapped within the network. The rather disordered structure of the plasma polymer depends on the

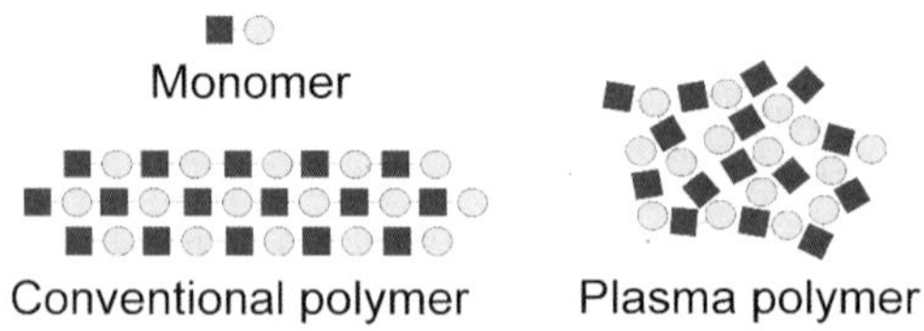

Fig. 3. Schematic representation of a conventional polymer and a plasma polymer.

intensity and energy of the species bombarding the growing film. Even if a "traditional" monomer is used, the resulting polymer film from the plasma polymerization process is very different from the corresponding "conventional" polymer film (Fig. 3).

Depending on the plasma processes conditions, it is possible to obtain hybrid (organic–inorganic) films. In addition, gradient films from the outer surface to the substrate can be obtained changing the process conditions. For instance, the O_2/hexamethyldisiloxane (HMDS) ratio has a great influence on the resulting plasma polymers obtained in the polymerization of HMDS, as determined by infrared spectroscopy. The infrared spectra of the HMDS monomer and two siloxane plasma

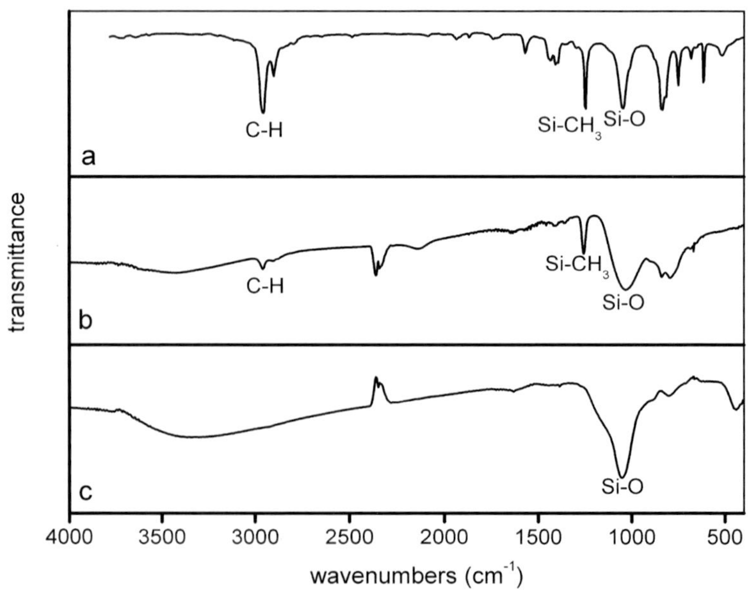

Fig. 4. Infrared spectra: (a) HMDS monomer; (b) sample synthesized with low oxygen/HMDS ratio plasma; and (c) sample synthesized with high oxygen/HMDS ratio plasma.

polymer films are shown in Fig. 4. One of the polymers was obtained in low oxygen/HMDS ratio plasma and the other one in high oxygen/HMDS ratio plasma. The bands corresponding to C–H and Si–CH_3 can be observed in the spectrum of the film obtained in the low oxygen plasma. However, in the spectrum of the polymer obtained with a higher oxygen/HMDS ratio, the band corresponding to methyl groups bonded to Si decreases and the Si–O bond increases, clearly indicating that the resulting film has a more inorganic character. Actually, by using oxygen plasma gas and manipulating the process conditions (power, flow rate), an inorganic product may be deposited.

3.2. *Physical Vapor Deposition (PVD) and Chemical Vapor Deposition (CVD)*

Coatings allow the surface properties of materials (e.g. wear resistance, lower friction, chemical barrier, corrosion resistance, etc.) to be improved while the intrinsic bulk properties (especially mechanical) remain unchanged. Among the most widely used techniques, plasma enhanced chemical vapor deposition (PECVD) and plasma enhanced physical vapor deposition (PEPVD) are widely applied. The difference between PVD and CVD is basically the mechanism used to generate and transport the material as vapor, and the way in which it is later deposited on the substrate.

PVD encompasses a variety of methods to deposit thin films by the condensation of a vaporized form of the material, involving purely physical processes. Michael Faraday was already using PVD to deposit coatings as far back as 1854.[20] Unlike the conventional PVD technique, where physical media are used to obtain the species directly from a target, evaporate and deposit them on the substrate, plasma assisted techniques, e.g. PEPVD, allow generating reactive species at relatively low temperatures, and increase the deposition rate due to the presence of ions that can be electrically accelerated towards the substrate. In the case of PEPVD, the efficiency or rate of the process is measured through a variable called sputtering yield,[21] which depends, among other parameters, on the energy, mass, angle of incidence of the ions, atomic number, constituent elements, and the crystalline structure of the target.

On the other hand, the PECVD technique consists in the formation of a solid film on some surface derived from the reaction of some species (molecules, atoms, ions, excited molecules and atoms, free radicals, etc.) in the gas state.[4]

Both techniques have spread quickly owing to the versatility they offer regarding to the synthesis and deposition of materials with a wide range of stoichiometric conditions that can be adjusted in a precise way, the versatility to deposit a great number of elements and compounds, and the possibility of obtaining multilayer structures and concentration gradients, among others.

3.3. *Ion Beam Processes*

The ion beam applications presented later in this chapter are basically related to (conventional) ion implantation and PIII (alternatively known as plasma source ion implantation or PSII).

Conventional ion implantation is a materials modification process, where firstly an ion beam is extracted from the plasma, and then specific ions might be selected, accelerated (typically from a few keV to MeV), focused, and scanned before hitting the target being surface treated. The ion beam is controlled by applying electromagnetic fields at consecutive stages under high vacuum (pressure lower than 10^{-5} mbar).

PIII uses plasma surrounding the target, where a high negative voltage and high current pulses are applied to accelerate ions into the target surface through a plasma sheath from all directions. Unlike conventional ion implantation, the process is not "line-of-sight."

Ion beam-based treatments typically involve addition (the implanted ions), removal (sputtering) and in some cases change (formation of new compounds) of the target material at its outermost surface. As a consequence, ion implantation treatments induce a change in the structure and nanotopograpy of the treated surface,[22] in the physical properties, chemical composition, and overall in its biological properties.[23,24] Ion implantation is used in semiconductor device fabrication, metal finishing, and materials science research as well as the biomedical applications presented later.

An alternative technique combining PVD and ion implantation is known as Ion Beam Assisted Deposition (IBAD), where the film is bombarded by an ion beam during its deposition, thus typically improving the density, structure, and adhesion of the coating.

The processes described in the previous pages present different advantages for biomedical applications. A comparison of different techniques is presented in Table 4.

4. Biomimetics

The excellence achieved by the surfaces of some living organisms towards a specific end, such as the adhesiveness of gecko feet, involves more than the simple modification of the physicochemical properties at a single scale. In those cases, the extraordinary material performance for a desired end function is obtained by Nature through the combination of an optimum surface physical chemistry and a hierarchical morphological structure, covering different size scales (in the case of the gecko feet, the optimization of contact surface area is achieved by different morphological patterns, with scales ranging from millimeters to nanometers). Artificial bioinspired surfaces often take only one aspect of the complex natural surface, which can nevertheless lead to a significant improvement of the material performance. Surface modification by

Table 4. Comparison of different plasma-based surface modification processes.

Characteristic	Polymerization	Grafting	PVD/CVD	Ion beam processes
Conformal (feasibility of treating complex geometries)	Medium	High	Medium	Medium/low
Coating without defects (pinhole-free, dense...)	High	—	High	—
Adhesion coating-substrate	High	—	High	—
Nontoxic (environmentally friendly)	Yes	Yes	Yes/?	Yes
Monomer/target cost	Low	Low	High	Low
Scaled up to industrial production (biomedical sector)	Medium	Medium	Medium	Low
Reproducibility	High	High	High	High
Low lixiviation (highly crosslinked nature)	Yes	Yes	?	?
Tailored hybrid films, e.g. organic–inorganic	High	—	Low	—
Variety of achievable surface properties	High	High	High	Medium
Permeation barriers	Yes	Yes	Yes	—
Dry process	Yes	Yes	Yes	Yes
Easiness of sterilization	Medium	Medium	High	High
Low temperature process	Yes	Yes	No	Yes
Feasibility of functionally graded surfaces	High	—	High	Low
Compatible with masking techniques, i.e. treatment of selected areas	Yes	Yes	Yes	Yes
Thermal stability	Yes	Yes	Yes	Yes
Process at atmospheric pressure	No	?	No	No
Process cost overall	Medium	Low	Low	High
Equipment cost	Medium	Medium	High	High
Some biomedical applications	Catheters, scaffolds, …	Biosensors, …	Surgical tools, lead bering prostheses	Stents, dental implants, radioactive seeds, …

?: often depends on singular process characteristics, i.e. treatment parameters.

plasma-based processes is found to be most useful for the chemical modification of surfaces, thin film deposition and, in some cases, for the morphological modification at the micro and/or nanoscale.

It is worth mentioning that biological surfaces can always be a source of inspiration, but the most convenient answer for a particular problem might be found by researchers with a completely different approach. In this sense, the example of the self-cleaning surfaces is illustrative; while Lotus leaves have self-cleaning properties due to their highly hydrophobic surface, TiO_2 coated glasses have self-cleaning properties due to their highly hydrophilic surface,[25] with the additional advantage that TiO_2 participate in the photo-oxidation of organic molecules, improving the cleaning by an independent mechanism.[26] Another example where a specific function can be attained by either very hydrophilic or very hydrophobic surfaces is the biocompatibility between a polymer and blood fluids; while proteins should not adsorb to hydrophilic surfaces and thus remain unaltered, hydrophobic surfaces become coated by firmly bound albumin and consequently the polymer appears as an endogenous material to the biological system.[27] Again, both surfaces are very different, but comply with a desired function by different means.

5. Applications

In the following sections, an overview of the plasma-based treatments of materials is presented stressing the applications in the biomedical field. The issues considered start with the improvement of some mechanical and physical properties and continue with the modification of the surface chemistry of materials. Finally, plasma treatments are considered for applications closely related to the biological and medical field, such as sterilization, antimicrobial surfaces, diagnostics, therapy, drug delivery, and interaction with living tissues and organisms.

5.1. *Surfaces with Improved Mechanical Properties*

Biomaterials, mainly those used in orthopedics and traumatology, are submitted to large mechanical loads. The materials and designs which must undergo the most extreme conditions are the ones used for joint prostheses, such as hip, knee, and shoulder implants. Therefore, wear resistance, fatigue, and corrosion resistance is of paramount importance as well as biocompatibility requirements which must be fulfilled.[28] Also, surgical tools, like drill pits and endmills, must comply similar requirements, although for shorter periods.

Austenitic stainless steel, titanium and its alloys, zirconium and Cr–Co alloys are found to be the metallic materials of choice for joint prostheses. The first two materials show a deficient wear behavior and, consequently, a wide variety of surface processes and coatings have been applied to circumvent this drawback without affecting their corrosion resistance.

Superhard coatings are of special interest in this regard. They can roughly be classified in four groups[29]:

(1) Intrinsically superhard materials, such as diamond, diamond-like carbon (DLC), and cubic boron nitride (c-BN). DLC actually refers to a wide range of materials whose properties depend on the nature of C bonds, hydrogen content, inclusion of dopants, etc.[30] For instance, nanocrystalline diamond produced by microwave plasma chemical vapor deposition has been deposited on Ti6A14V, a commonly used biomedical alloy, with the objective of being applied in articulating surfaces of prosthetic joints. This application requires low roughness and high hardness in order to diminish wear particle debris generation, good electrochemical corrosion resistance, and exceptional film adhesion.[29] A variety of DLC coatings have also been developed for the same purpose (Fig. 5).[31,32]
(2) Materials hardened by ionic bombardment, where the hardness is due to a complex synergic effect generated by the ionic bombardment during the plasma deposition process. In this case, the hardness increase through ion bombardment is due to formation of compounds, a decrease in the grain size, densification on the grain edges, formation of specific defects, and the increase of compression internal tensions.[28]
(3) Nanocomposite coatings, which require thermo-dynamically driven phase segregation. The superhardness of nanocomposites lies in the nanostructured condition of the coatings, whose structural stability could result from a spinodal

(a) (b)

Fig. 5. (a) DLC coated knee prostheses (top, left) and femoral heads for hip prostheses (bottom, right) (b) SEM image of a DLC coating, × 5000. Reproduced with permission from INASMET-Tecnalia.[35]

decomposition and promoted by the energetic ion bombardment. Furthermore, the compression tension has to be added, as well as the reduction in grain size, the densification on the grain edges, and the re-sputtering of weakly bonded atoms.[33]

(4) Multilayer structures, where the interfaces and the different mechanical properties of each layer impose restrictions to the dislocation movement. The hardness increases owing to the nanometric scale of the different layers that form the coating.[34]

Alternatively, plasma assisted treatments for surface hardening include thermo-chemical treatments such as glow discharge, active screen, ion beam, post-oxidation, etc. As usual with coatings, the resulting thermochemically treated surface depends on process parameters such as utilized gases, gas pressure, applied potentials, etc. On the other hand, the substrate material and their constituents determine the phases present on the surface, the diffusion areas, the thickness of those phases, and consequently their mechanical and tribological properties, performance in corrosive environments, etc. Examples of thermochemical treatments include plasma nitriding of martensitic and austenitic stainless steel. High surface hardness can thus be obtained, as well as a remarkably improved wear resistance.[36–39]

Currently, duplex treatments, i.e. thermochemical treatments plus deposits of mono- and multi-PEPVD layers, are an alternative to solve the limitations of depositing high hardness coatings on substrates of lower hardness. The phases obtained by thermochemical treatment can form a relatively thick layer that act as a mechanical support to the multilayer on top and prevents it from failing due to the deformation of the substrate. Several works have shown that duplex or multilayer coatings can offer attractive properties, with the advantage of being applicable on different substrate materials, e.g. TiN coatings on austenitic stainless steel,[40] nitrided/TiN coatings on AISI W1 80C,[41] etc.

Typically, not only a high surface hardness is desired, but also a lower coefficient of friction, which among other factors affects the temperature in the friction zone. Examples include surgical tools, where high temperatures in the bone tissue must be avoided to minimize bone necrosis.[42,43] The lowering of the friction of biomaterials is also of interest in many other medical devices,[44] such as catheters, needles, stents, joint prostheses.

5.2. *Antireflective Surfaces*

Antireflective coatings are those applied on different devices to reduce light reflection. These coatings are used in a wide range of areas, from sophisticated telescopes and solar panels to surgical tools. There are different ways to obtain surfaces with antireflective properties by using plasma assisted techniques, among them, etching and mono- and multi-layer deposition.

Etching permits the modification of surfaces by bombardment with high energy ions emitted by the plasma source. In this way, moth-eye-like surface morphologies can be obtained by plasma treatment. For instance, antireflective properties are obtained with argon bombardment on PMMA[45] by reaching a surface structure smaller than the length of the incident light wave and provoking a gradient in the refractive index (Fig. 6). Antireflective coatings of TiAlN and TiO_2 with a columnar morphology have also been obtained by plasma treatment.[46] The column diameters of the magnetron sputtered TiAlN near the surface were between 100 and 300 nm (Fig. 7). Unlike TiAlN coatings, TiO_2 coatings showed different structures. However, TiO_2 structures similar to those found in TiAIN coatings, which presented antireflective properties on steel substrates and PMMA, were obtained with low deposition rates and elevated O_2 flow.

Another methodology to obtain antireflective coatings is the superposition of films with different refraction index according to the so called bottom antireflective coating technique. In these cases, the coating design is made in such a way that the refraction index gradually decreases from the layer closest to the substrate towards the outer layer.[47]

Some coatings obtained through plasma assisted techniques such as ZrCN, TiA1N, TiCN, and other coatings, sometimes developed originally for other purposes such as DLC, incidentally lower light reflection and act as antireflective surfaces. The most common coatings include SiN coatings, frequently used in solar panels.[48] DLC coatings, which present high hardness, chemical stability, high electrical resistivity, low IR

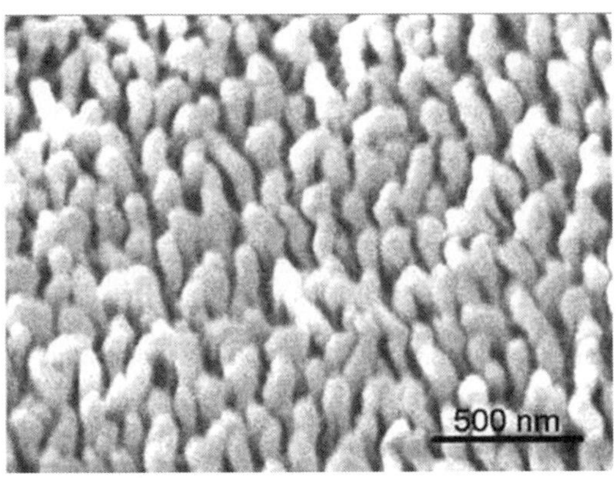

Fig. 6. SEM image of a PMMA sample after plasma treatment (tilt angle: 25°). Copyright by Elsevier. Reproduced with permission.

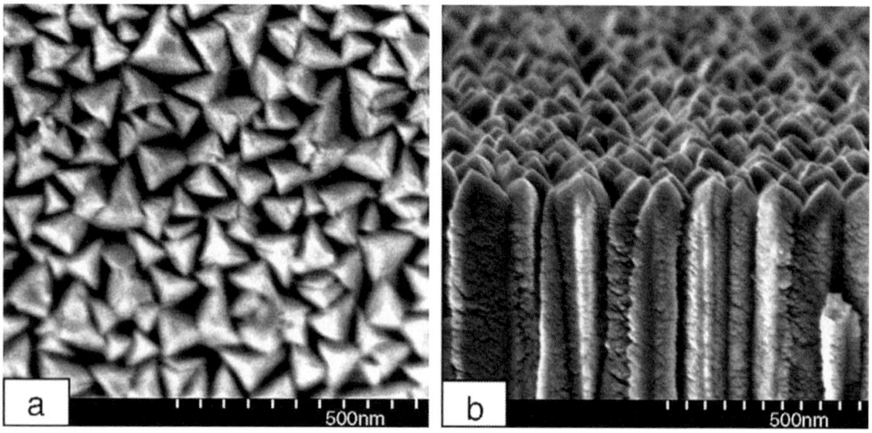

Fig. 7. SEM micrograph of deposited TiAlN coating with typical columnar structure and pyramidal tips: (a) top view of coating and (b) coating edge. Copyright by Elsevier. Reproduced with permission.

absorption,[49] high abrasion resistance, and high fracture toughness are also currently used as antireflective coatings.

5.3. *Hydrophobic and Hydrophilic Surfaces*

The hydrophobic/hydrophilic behavior of a surface is determined by its chemical composition, which can be modified with the use of plasma processes through the surface functionalization with organic molecules bearing hydrophilic or hydrophobic groups. As compared with conventional chemistry, plasma technologies have several advantages in this regard, allowing the modification of the very top layers of the material with dry, energy efficient processes.

There is a great interest in these treatments because the hydrophobic/hydrophilic behavior conditions the compatibility with other materials. This is especially important for the development of composites, in which dissimilar materials are employed with the aim of obtaining a product that combines the desired properties of the components. For instance, the incorporation of starch to synthetic polymers renders a biodegradable, water resistant material;[50] however, the components must be made compatible by modifying the hydrophobicity or hydrophilicity of one or both of them. Hydrophobic starch has been obtained with a $SiCl_4$ plasma treatment, followed by a reaction with either ethylene diamine or dichlorodimethysilane;[3] thus, a hydrophilic material was converted into a hydrophobic one. There are also many examples of the opposite conversion, as plasma treatments, e.g. oxygen plasma, are known to provoke the formation of hydrophilic functional groups in the surface of synthetic polymers such as PE, PP, and PTFE, allowing the production of composites in processes where the compatibility is achieved through plasma treatments, e.g. in cellulosic fiber reinforced PE.[3]

5.4. *Biomolecule Immobilization*

Although the immobilization of biomolecules on a surface can be achieved by different means, including encapsulation and the action of a variety of physicochemical forces, this section is focused on the immobilization via covalent bonds, which is very attractive due to their high chemical stability.

In order to covalently bond a biomolecule to a surface, the latter must be provided with organic groups which can act as anchoring sites for biomolecules. Different processes of surface modification can be employed to achieve this goal, ranging from surface functionalization and the deposition of thin films to the formation of multiphase systems, all of which involve the incorporation of free, reactive organic groups.

In these cases, plasma surface modification is often used in combination with conventional chemistry, usually as the first step in a more complex sequence of a surface design. To illustrate this point, the surface functionalization of PTFE is presented in Fig. 8. Firstly, the polymer is treated with oxygen plasma, which promotes the formation of peroxide groups and radicals, followed by a wet attack of acrylic acid.[51,52] The resulting surface contains carboxylic groups, which can satisfy a desired end function, e.g. a hydrophilic surface and higher wettability, or can be used for further functionalization, such as the bonding of molecules bearing amino groups by means of wet chemistry through a molecular linker. Finally, these groups can act as anchoring sites for biological macromolecules, such proteins, enzymes, and growth factors, via well-known chemical routes. Similarly, the deposition of films bearing reactive organic groups has also been used.

A different approach for biomolecule immobilization involves the formation of multiphase systems on top of a substrate. Multiphase systems are usually developed to carry on more complex tasks, such as detection and sensing, and will be considered in the following section.

5.5. *Biosensors*

A powerful tool in diagnostics consists of biosensors, devices that use a biological reaction for detecting target analytes by coupling a biological recognition element with a physical transducer that translates biorecognition events into a quantifiable electronic signal. Mediator-enzyme biosensors for the detection of blood sugar are a

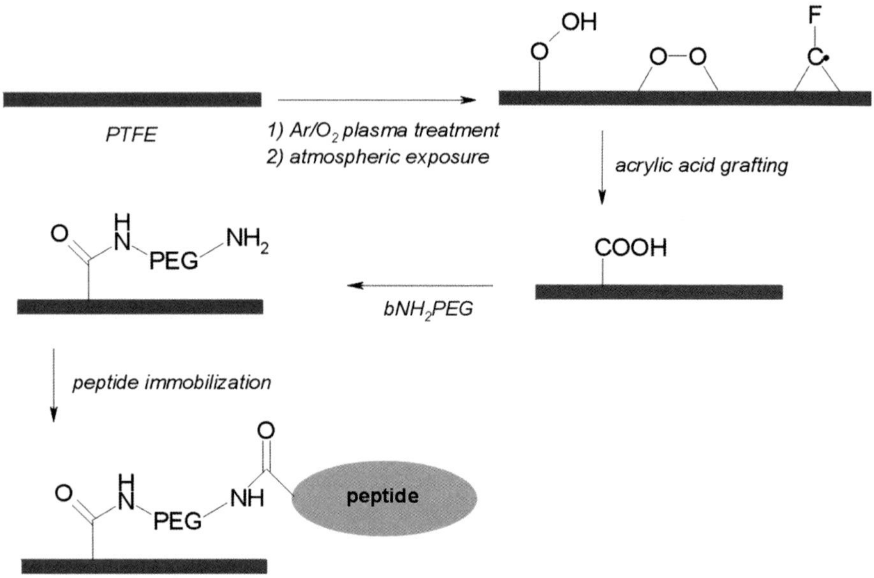

Fig. 8. Scheme of surface modification of PTFE by low-pressure plasma and biomolecule immobilization by wet chemistry. bNH_2PEG stands for 0,0′-bis(2-aminopropyl)-polyethylenglycol. Adapted from Ref. 52.

good example of such devices. Plasma-polymerized thin films have shown potential for use as the interface between the two components of biosensors, as Muguruma[53] has reported in detail. Actually, depending on the mode of signal transduction, several possible strategies have been explored.

In the case of electrochemical biosensors, they include immobilization of enzymes in porous membranes by plasma-polymerized films, e.g. plasma-polymerized allyl alcohol or hexamethyldisilazane as barriers on the membrane, providing a significant operational stability;[54] plasma treatment of the membrane to attach covalently recognition molecules, e.g. cellulose acetate membranes treated with amylamine plasma to determine the glucose content.[55] Another strategy for electron transfer mediated biosensors involves depositing a plasma-polymerized film onto a microsensor, which is subsequently plasma treated to introduce functional groups, where the enzyme is attached, e.g. a vinylferrocene plasma-polymerized film deposited directly onto the surface of an electrode and a subsequent plasma process with acetonitrile, introducing amino functional groups to make it possible to attach large amounts of enzymes.[56]

Plasma-polymerized films have also been applied to quartz crystal microbalance biosensors, which are based on the measurement of changes in resonant frequency associated with biospecific binding events. Examples include carboxyl terminated plasma-polymerized coatings where antibodies are immobilized either directly or via proteins.

Many biosensors are based on surface plasmon resonance. In this case, the excitation by light of surface plasmons, surface electromagnetic waves that propagate parallel along a metal/dielectric interface, results on oscillations in the wave that are very sensitive to the adsorption of molecules to the metal surface. Plasma-polymerized nanocoatings might be applied to immobilize such biomolecules.

Another important application field comprises biochips, such as DNA or protein chips. A DNA chip (Fig. 9) is a collection of microscopic DNA spots arrayed on a solid surface by covalent attachment to a chemical matrix, which utilize the selective nature of DNA–DNA or DNA–RNA hybridization and fluorophore-based detection for expression profiling, i.e. monitoring expression levels of thousands of genes simultaneously. Plasma polymerization techniques have been applied to achieve larger densities of the DNA probes attached, e.g. HMDS[57] or allylamine coatings.[58]

5.6. *Sterilization*

Sterilization consists in the destruction or elimination of microbial life, including bacteria, endospores, viruses, and fungi. It is applied to food, medicines, vaccines, textiles, medical devices, surgical tools, etc. Several regulations and standards deal with the analysis, control, and validation of the sterilization processes of medical devices. They typically specify that the probability to find organisms, i.e. the bioburden, should be smaller than a certain colony forming unit value.

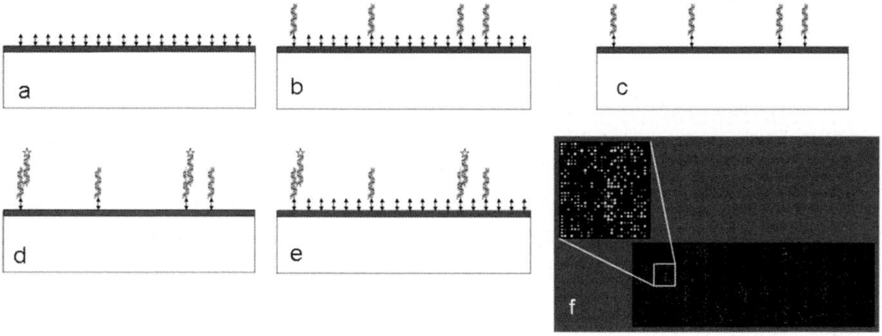

Fig. 9. Basic typical process steps of a DNA-array: (a) functionalization; (b) DNA probe immobilization; (c) blocking of reactive groups; (d) hybridization with second DNA sample; (e) washing; (f) reading by laser scanning.

There are different sterilization methods, the most widely used being chemical (ethylene oxide, aldehydes, hydrogen peroxide) or physical (heat, radiation, filtration). However, sterilization can also be achieved with atmospheric pressure plasma by means of physical, chemical, and physical–chemical processes.[59] The basic mechanisms are the following: destruction by UltraViolet (UV) irradiation of the genetic material of the microorganism; erosion of the microorganism, atom by atom, through intrinsic photodesorption; and erosion of the microorganism, atom by atom, through etching. However, it is important to highlight that in all cases the intensity of the irradiation with UV photons, sputtering, and etching will depend on process parameters such as pressure, the utilized gas (Ar,[60] O_2,[61] N_2, air, H_2, halogens, N_2O, H_2O, H_2O_2, CO_2, O_2, O_2/H_2O_2,[62] aldehydes, organic gases, etc.), and temperature.

In general, plasma sterilization proceeds as follows: the items to be sterilized are placed in the chamber, which is then closed and vacuumed. The sterilization gas, e.g. water and hydrogen peroxide, is injected in the chamber in such a way that it penetrates the package and surrounds the items. After some pressure reduction, low temperature plasma is generated by radio frequency application. In some cases, the gas injection and the plasma generation take place two or more times during the sterilization cycle.[63] The articles sterilized by plasma assisted techniques do not require ventilation and there is no emission of toxic residues. The preparation of the medical device for the sterilization process is similar to conventional practices: the devices are washed, dried, assembled, and wrapped in a packaging material which is permeable to sterilization (e.g. PP, or a perforated aluminum container). However, in some cases, the packaging of medical tools may reduce the effect of reactive species in the atmospheric plasma.[64]

Plasma sterilization effectiveness has been compared to the most conventional methods,[63,65–67] as well as its influence on some materials properties, such as wettability and surface energy,[68] resistance to fatigue,[69] and the mineralization of titanium implants.[70]

5.7. *Antimicrobial Surfaces*

Systemic antibiotic treatment is a very common medical procedure all over the world. Nevertheless, it presents certain limitations and drawbacks, such as systemic toxicity, poor penetration in certain tissues, and poor control of local drug levels. Additionally, in the case of implantable medical devices, if bacteria (typically *S. aureus*, *S. epidermidis*, *Pseudomonas aeruginosa*, *E. coli*, etc.[71]) adhere and proliferate, colonizing the implant surface and forming a biofilm, the patient may develop an infection despite systemic antibiotic treatment, which may lead to the rejection or removal of the implant (Fig. 10). Actually, an implant represents a challenge to the immune system,

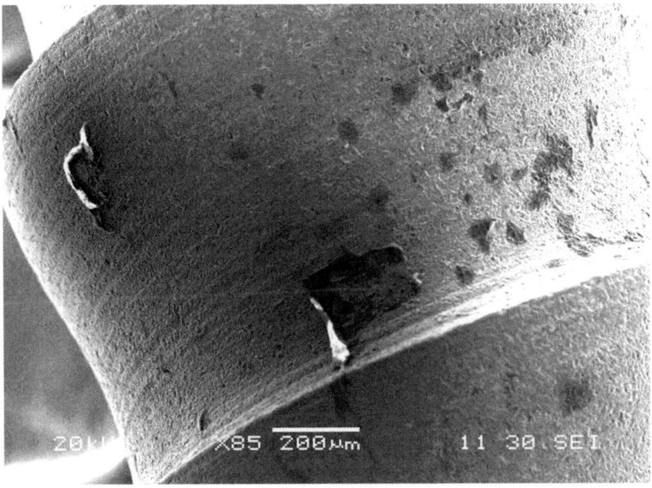

Fig. 10. ESEM image of a rejected dental implant due to infection (×85).

impairing an adequate local response, and reducing the number of bacteria required to cause infection. Therefore, the use of intrinsically antimicrobial surfaces is relevant on medical devices, where its role will usually be prophylactic rather than therapeutic.

Current solutions and developments in the field of antimicrobial surfaces include, among others, silver (or other antibacterial element) containing coatings and surfaces, photocatalytic TiO_2 surfaces, bacterial adhesion inhibiting surfaces, and antibiotic loaded coatings.[72]

The antibactericidal properties of silver and other metals have been well known for long time, but it has not seen widespread use due to concerns about its efficacy and toxicity. Nevertheless, the emergence of nanotechnology is deemed to represent a promising new possibility, allowing for maximizing the active surface area while keeping the total amount of the antibacterial element at a minimum. Ag and other antibacterial elements can be present either as a coating, e.g. deposited by PVD,[73] as nanoconstituents of a coating, e.g. by PVD,[74] or as co-deposited in a plasma polymerization process,[75] e.g. silver nanoparticles embedded in HMDS plasma-polymerized coating.[12] Alternatively, they can be embedded into the outer layer of the substrate, e.g. by Ag or Cu ion implantation,[76,77] F ion implantation,[78] or by Cu co-implantation with N_2 for regulation of the out diffusion.[79]

The possibility of photocatalytic disinfection in TiO_2 under UV light irradiation has also been extensively studied, e.g. magnetron sputtered or plasma ion implanted

TiO_2 thin film,[80] or Fe plasma ion implanted TiO_2 irradiated with visible light have shown antimicrobial activity.[81]

On the other hand, some coatings have shown to significantly reduce bacterial adhesion. Such is the case of plasma-polymerized coatings,[82] certain DLC coatings, e.g. acetylene PIII deposition of a-C:H films,[83] or PECVD produced a-SiO_x transparent coatings.[84] However, direct plasma treatment on the surface of the medical device in order to reduce bacterial adhesion on the polymers have often resulted in mixed results, e.g. poor results with CO_2, H_2O, and NH_3 plasma on polyurethane,[85] or improvement of the barrier against microbes with fluorocarbon-based plasma treatments on nonwoven fabrics for surgical gowns.[86]

Another promising option consists on grafting antibiotics onto the polymer surface by means of plasma processes. The plasma treatment is applied in order to activate the surface and promote a later fixing of the antibiotic, and can be applied during the deposition of the antibiotic, afterwards or in a combination of the previous, e.g. triclosan and bronopol on Ar PIII after activation by O PIII.[87] A polymeric film might also be deposited on the substrate as intermediate in order to make possible the grafting of an antibiotic, e.g. ethylene-based.[85] Conversely, a plasma-polymerized coating may be a suitable barrier for controlled release of antibiotics, eliminating the initial burst of release, e.g. RF plasma-deposited *n*-butyl methacrylate over drug loaded polyurethane.[88] It may also be deposited on top of polymeric resorbable[a] coatings loaded with antibacterial substances, which present the advantage of allowing a wide range of substances to be chosen and to tailor the release kinetics depending on its specific application. The use of plasma processes to co-deposit antibiotics while plasma polymerizing resorbable polymers should also be explored, as it might be able to overcome the shortcomings of current antibiotic loaded polymers, in terms of release control, adherence, etc.

5.8. *Interaction with Living Tissue*

The ability to control and elicit specific reactions between biomaterials and living tissues is a long pursued aim. This has led to several strategies to study, understand, and control the interactions occurring at the interface. It is known that cell processes such as adhesion, proliferation, differentiation, migration or apoptosis, depend on cell responses that are controlled by intracellular signaling pathways triggered by the interaction between the transmembrane proteins and the substrate's surface, where multiple surface properties play a role.

---

[a]Resorbable polymers are typically broken down and assimilated back into the body without any toxic side effect.

The possibility of producing surfaces, where the chemistry, pattern geometry and physical properties are spatially and temporally controlled, represents a powerful tool towards the aim of understanding fundamental aspects of the biological material–surface interactions in implantable materials, which additionally has found practical applications on biosensors and drug discovery.

Plasma-based processes offer interesting tools to produce chemically and spatially controlled surfaces. Actually, plasma-polymerized coatings, e.g. thin acetaldehyde and allylamine coatings,[89] often combined with photolithography[90] or laser ablation[91] have been used for producing two-dimensional patterns. An original approach to cell–biomaterial surface interaction investigation consists on producing chemical gradients by plasma polymerization process, e.g. assessing fibroblast adhesion on gradual allylamine and hexane plasma-polymerized surfaces,[92] so as to determine the most promising composition.

Gaseous plasma treatments have also been studied on several biopolymers for blood-contacting applications, where biomaterial-related thromboembolic complications remain a significant risk.[93]

Concerning bone tissue, surfaces that improve osseointegration are in high demand. Additionally to the commercially applied hydroxyapatite and titanium coatings deposited by Plasma Spraying, several plasma-based techniques, such as ion implantation, e.g. CO ions[23,94,95] (Figs. 11 and 12) or Ca ions,[96] PVD coatings,[97,98] IBAD,[99] plasma-polymerized coatings, e.g. allylamine plasma polymer layer to improve osteoblastic adhesion,[100] plasma-deposited acrylic acid coatings[101] or even plasma activated surfaces, e.g. H_2O_2 activation of stainless steel and subsequent treatment with simulated body fluids,[102] are being explored in order to accelerate and increase the osseointegration of implantable materials.

Ion implantation techniques allow new chemistries and topographies to be defined at the nanoscale, where the adsorption of proteins and interactions with integrins are regulated, which subsequently condition the signaling pathway that will eventually lead to bone formation around the implanted biomaterial.[103] On the other hand, PVD produces much thinner and controlled CaP-based coatings than conventional atmospheric plasma spraying techniques, while allowing for nanostructuring and doping, e.g. with bone inducing elements such as Sr.[104]

The study of the interaction of biomaterials with other tissues and cells, such as neuronal cells or stem cells, e.g. O plasmas on polyimide or polyurethane to attach adipose derived stem cells prior to implantation,[105] have also profited from plasma-based treatments.

On the other hand, sometimes the desired result is not to elicit a specific response, but avoiding eliciting a foreign body reaction and the consequent rejection. This is for example the case in Ni containing NiTi shape memory alloys used in dental devices

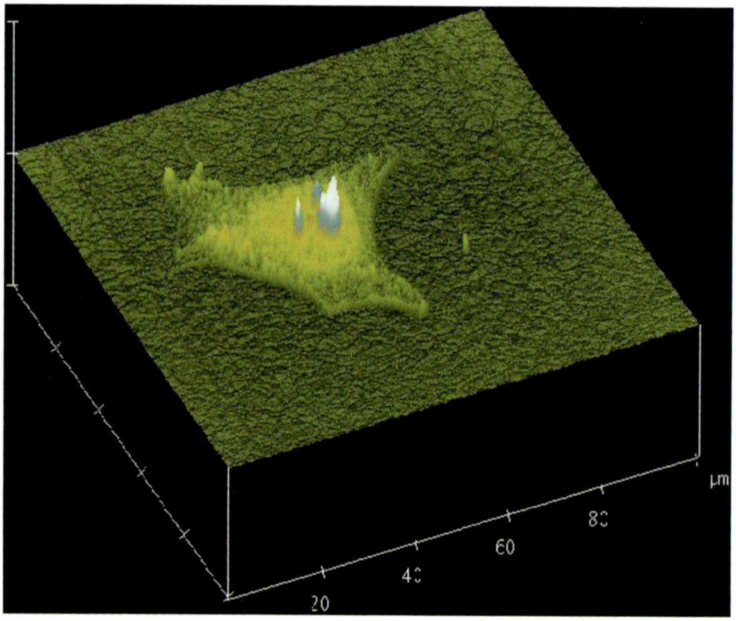

Fig. 11. AFM image of an osteoblast on a CO ion implanted nanostructured surface ($100 \times 100\,\mu m$). Reproduced with permission from Lifenova Biomedical.

such as archwires, which can be treated by PIII to avoid the leaching of Ni ions into the physiologic media,[106] or austenitic medical grade stainless steel, which can be surface oxidized in plasma.[107]

In brief, the application of plasma-based surface modified biomaterials that would originate controlled responses is very large, ranging from *in vitro* to *in vivo* applications.

5.9. *Therapies and Drug Release*

Localized and controlled release of drugs is much preferred than systemic treatments in many applications, especially when and where implantable medical devices are involved. Different approaches have been suggested, involving a wide variety of drugs, applications, and processes.

Among others, PECVD has been studied as a means to encapsulate drug molecules inside thin polymeric films, e.g. encapsulation of acetylsalicylic acid by plasma polymerization of allyl alcohol with controllable release rates and no undesirable changes in drug compositions.[108] More generally, plasma-based treatments have been described before (antimicrobial surfaces) as useful technologies for controlling the

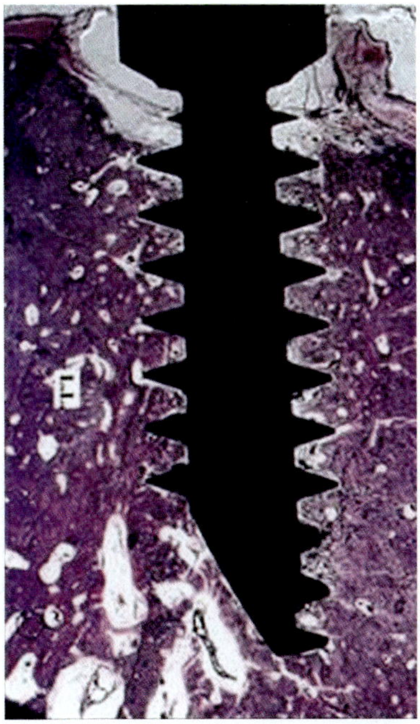

Fig. 12. Histological image of an ion implanted dental implant (×12). Reproduced with permission from Lifenova Biomedical.

release of antibiotics. One promising development may lay in the deposition of plasma-polymerized biodegradable coatings, e.g. RF plasma deposition of poly-*l*-lactic acid,[109] which could facilitate controlling release rates.

One relevant application involves vascular stents (endovascular prostheses), where thrombosis and restenosis remain a problem. So far, drug eluting coatings have been developed and applied with mixed results. One therapeutical approach involved ion implantation and cyclotron activation of stents to render them radioactive so as to inhibit neointimal formation and reduce restenosis, e.g. ^{31}P ion implantation to subsequently obtain radioactive ^{32}P or direct ^{32}P ion implantation.[110]

Furthermore, ion implantation of radioactive elements has been applied for cancer treatment, e.g. P ion implantation into glass microspheres[111] so as to avoid radiation been applied on healthy tissue.

Plasma treatments have also been applied to intraocular lenses (IOL) to prevent the formation of posterior capsular opacification, e.g. increasing their hydrophilicity and permitting cells to maintain their normal epithelial morphology,[112] or reduce the risk for endophthalimitis, e.g. by coating with poly(2-methacryloyloxyethyl phosphocholine-co-*n*-butyl-methacrylate) or methane gas plasma and thus inhibiting bacterial and fibroblast adhesion.[113] Inhibition of cell attachment has also been reported on F^+ ion implanted polymethylmethacrylate IOLs.[114]

5.10. *Treating Living Organisms*

So far, the applications described have basically consisted on modifying a biomaterial surface so as to elicit a specific response in biomolecules or the physiologic media. A quite different intriguing approach consists of applying the treatment to living tissue. The interaction of incident ions with biological material might induce radiation damage, sputtering, chemical reactions, fragmentation of biological building blocks, disruption of DNA, gene mutations, etc.[115]

Ion implantation has been used for mutation breeding to achieve higher yield rates, disease resistance, faster growth, higher quality, e.g. of rice, wheat or cotton, and other crops,[116] microorganisms for production of antibiotics,[117] bacteria,[118] and mushrooms.[119] It has proved to be a valuable method because of a lower damage rate induced, higher mutation rate, and wider mutational spectrum produced.

Last, but not least, C ion beams have actually been applied in human patients for cancer therapy.[120]

6. Conclusions

Plasma-based surface technologies offer plenty of possibilities for surface modification of almost any conceivable substrate material. Modifications range from relatively thick coatings of several μm to nanometrically modified layers and surface functionalization. The versatility of these technologies has found wide application in the biomedical field, both at industrial level and research stage, where new technologies, processes, and properties are continuously under development. A wide variety of treatments have been described here to achieve sterilization, biomolecule immobilization, hydrophobic and hydrophilic surfaces, hard and lubricious coatings, antireflective and antimicrobial surfaces, control of interaction with living tissue, biosensors and therapy. Medical devices for both *ex vivo* and *in vivo* usage are to profit from the potential of plasma-based technologies.

Acknowledgment

The authors thank the collaboration received from Esteban Broitman (Department of Chemical Engineering, Carnegie Mellon University), Marta Brizuela and Nerea Briz (INASMET-Tecnalia), Víctor Bellido (GENCOA SA), Gabriel López (Universidad del País Vasco), Belén Parodi and Leonardo Pazos (Instituto Nacional de Tecnología Industrial).

References

[1] Lieberman MA, Lichtenberg A. (2005) *Principles of Plasma Discharges and Materials Processing*, 2nd ed. Wiley-Interscience, John Wiley & Son, Inc Publication, USA.

[2] Brown SC. (1967) *Basic Data of Plasma Physics.* M.I.T. Press, Cambridge, Massachusetts.

[3] Denes FS, Manolache S. (2004) Macromolecular plasma-chemistry: An emerging field of polymer science. *Prog Polym Sci* **29**: 815–885.

[4] Vossen JL, Werner K. (1978) *Thin Film Processes.* Academic Press, New York.

[5] Tendero C, Tixier Ch, Tristant P, Desmaison J, Leprince P. (2006) Atmospheric pressure plasmas: A review. *Spectrochimica Acta Part* B **61**: 2–30.

[6] Zhang Z. (2003) Surface modification by plasma polymerization and application of plasma polymers as biomaterials, PhD Thesis, Johannes Gutenberg-Universität Mainz.

[7] Sioshansi P. (1987) Surface modification of industrial components by ion implantation. *Mater Sci Eng* **90**: 373–383.

[8] Biederman H. (1987) Polymer films prepared by plasma polymerization and their potential application. *Vacuum* **37**: 367–373.

[9] Biederman H, Slavinska D. (2000) Plasma polymer films and their future prospects. *Surf Coat Technol* **125**: 371–376.

[10] Shi FF. (1996) Developments in plasma-polymerized organic thin films with novel mechanical, electrical, and optical properties. *J Macromol Sci — Part C Rev Macromol Chem Phys* **36**: 795–826.

[11] Vossen J. (1979) Glow discharge phenomena in plasma etching and plasma deposition.*J Electrochem Soc.: Solid-State Sci Technol* **126**: 319–324.

[12] García-Luis A, Corengia P, González-Santamaría D, Brizuela M, Braceras I, Briz N, Azpiroz P, Bellido-González V, Powell S. (2007) Synthesis and characterization of plasma-polymerized HMDSO films using an ion gun inverse magnetron source. *Plasma Process Polym* **4**: S766–S770.

[13] Gombotz WR, Hoffman AS. (1987) Gas-discharge techniques for biomaterial modification. *Rev Biocompat* **4**: 1–42.

[14] Inagaki N. (1995) *Plasma Surface Modification and Plasma Polymerization.* Technomic Publishing, Basel. USA.

[15] Hynek B, Yoshihito O. (1992) *Plasma Polymerization Processes.* In L Holland (ed). Elsevier, Amsterdam.

[16] Grill A. (1994) *Cold Plasma Materials Fabrication: From Fundamentals to Applications, Cold Plasmas in Materials Fabrication.* In W Perkins (ed). Wiley-IEEE Press.

[17] Osada Y, Biederman H. (1990) Plasma chemistry of polymers. *Adv Polym Sci* **95**: 342–349.

[18] Shen M, Bell AT. (1979) *Plasma Polymerization*. In M Shen & AT Bell, ACS Symp. Series, Vol. 108, p. 1.

[19] Bell AT. (1980) Topics in current chemistry. In S Veprek & M Venugopalan (eds), *Plasma Chemistry* III, p. 43. Springer-Verlag, Berlin.

[20] Mattox DM. (2003) *The Foundations of Vacuum Coating Technology*. Noyes Publications/William Andrew Publishing, Norwich, New York.

[21] Rohde SL. (1999) *Sputter Deposition*. ASM International Handbook Committee (eds), ASM Handbook, Vol. 5, Surface Engineering, pp. 573–592. ASM International, USA.

[22] Braceras I, Briz N, García F, Muñoz R, Viviente JL, Oñate JI. (2007) Effects of ion implantation on nano-topographic properties. *Surf Coat Technol* **201**: 8511–8515.

[23] Braceras I, Alava JI, Goikoetxea L, De Maeztu MA, Oñate JI. (2007) Interaction of engineered surfaces with the living world: Ion implantation vs. osseointegration. *Surf Coat Technol* **201**: 8091–8098.

[24] Braceras I, Alava JI, Goikoetxea L, De Maeztu MA, Oñate JI. (2005) Bone cell adhesion on ion implanted titanium alloys. *Surf Coat Technol* **196**: 321–326.

[25] Nakajima A, Hashimoto K, Watanabe T, Takai, K, Yamauchi G, Fujishima A. (2000) Transparent superhydrophobic thin films with self-cleaning properties. *Langmuir* **16**: 7044–7047.

[26] Fujishima A, Rao Tata N, Tryk Donald A. (2000) Titanium dioxide photocatalysis. *J Photochem Photobiol. C: Photochem Rev.* **1**: 1–21.

[27] Elbert DL, Hubbell JA. (1996) Surface treatments of polymers for biocompatibility. *Ann Rev Mater Sci* **26**: 365–394.

[28] Park J, Kim DJ, Kim YK, Lee KH, Lee KH, Lee H, Ahn S. (2003) Improvement of the biocompatibility and mechanical properties of surgical tools with TiN coating by PACVD. *Thin Solid Films* **435**: 102–107.

[29] Raveh A, Zukerman I, Shneck R, Avni R, Fried I. (2007) Thermal stability of nanostructured superhard coatings: A review. *Surf Coat Technol* **201**: 6136–6142.

[30] Jozwik K, Karczemska A. (2007) The new generation Ti_6Al_4V artificial heart valve with nanocrystalline diamond coating on the ring and with Derlin disc after long-term mechanical fatigue examination. *Diamond Relat Mater* **16**: 1004–1009.

[31] Brizuela M, Garcia-Luis A, Viviente JL, Braceras I, Oñate JI. (2002) Tribological study of lubricious DLC biocompatible coatings. *J Mat Sci Mat Med* **13**: 1129–1133.

[32] Grill A. (2003) Diamond-like carbon coating as biocompatible materials — an overview. *Diamond Relat Mater* **12**: 166–170.

[33] Broitman E, Neidhardt J, Hultman L. (2007) Tribology of diamond-like carbon films: Fundamentals and applications. In C Donnet & A Erdemir (eds), p. 620. Springer, New York.

[34] Veprek S, Nesládek P, Niederhofer A, Glatza F, Jílekb M, Šímab M. (1998) Recent progress in the superhard nanocrystalline composites: Towards their industrialization and understanding of the origin of the superhardness. *Surf Coat Technol* **108–109**: 138–147.

[35] García-Luis A, Oñate JI, Alava JI, Braceras I, Brizuela M, Garagorri N. (2002) Tribological study of lubricious DLC biocompatible coatings. *Proceedings 7th European Congress on Biomaterials*, Barcelona, Spain, T191.

[36] Corengia P, Walther F, Ybarra G, Sommadossi S, Corbari R, Broitman E. (2006) Friction and rolling–sliding wear of DC-pulsed plasma nitrided AISI 410 martensitic stainless steel. *Wear* **260**: 479–485.

[37] Corengia P, Ybarra G, Moina C, Cabo A, Broitman E. (2004) Microstructure and corrosion behaviour of DC-pulsed plasma nitrided AISI 410 martensitic stainless steel. *Surf Coat Technol* **187**: 63–69.

[38] Corengia P, Ybarra G, Moina C, Cabo A, Broitman E. (2005) Microstructural and topographical studies of DC-pulsed plasma nitrided AISI 4140 low-alloy steel. *Surf Coat Technol* **200**: 2391–2397.

[39] De Las Heras E, González-Santamaría D, García-Luis A, Cabo A, Brizuela M, Ybarra G, Mingolo N, Brühl S, Corengia P. (2007) *Plasma Process Polym* 4: S741–S745.

[40] De Las Heras E, Egidi DA, Corengia P, González-Santamaría D, García-Luis A, Brizuela M., López GA, Flores Martínez M. (2008) Duplex surface treatment of an AISI 316L stainless steel: Microstructure and tribological behaviour. *Surf Coat Technol* **202**: 2945–2954.

[41] Lien TS, Wang SC, Chang JT, Davison A, Chen KC, He JL. (2007) In-service performance of components treated with plasma nitriding and arc ion plating duplex coatings. *J Mater Processing Technol* **184**: 401–406.

[42] Allan W, Williams ED, Kerawala CJ. (2005) Effects of repeated drill use on temperature of bone during preparation for osteosynthesis self-tapping screws. *Brit J Oral Maxillofacial Surgery* **43**: 314–319.

[43] Draenert FG, Mathys R, Jr., Ehrenfeld M, Draenert Y, Draenert K. (2007) Histological examination of drill sites in bovine rib bone after grinding in vitro with eight different devices. *Brit J Oral Maxillofacial Surgery* **45**: 548–552.

[44] Veli-Matti T. (2001) Amorphous carbon as a bio-mechanical coating-mechanical properties and biological applications. *Diamond Relat Mater* **10**: 153–160.

[45] Kaless A, Schulz U, Munzert P, Kaiser N. (2005) Nano-motheye antireflection pattern by plasma treatment of polymers. *Surf Coat Technol* **200**: 58–61.

[46] Burmeister F, Schäffer E, Kleer G, Döll W, Bläsi B, Gombert A. (2005) Investigation of structured TiAlN- and TiO_2-coatings with moth-eye-like surface morphologies. *Surf Coat Technol* **200**: 1088–1092.

[47] Xiaowen W, Weijia Z, Lanqin Y, Ruiying L. (2007) The deposition and optical properties of Ge1-xCx thin film and infrared multilayer antireflection coatings. *Thin Solid Films* **516**: 3189–3195.

[48] Stapinski T, Swatowska B. (2006) Amorphous hydrogenated silicon–carbon as new antireflective coating for solar cells. *J Non-Crystalline Solids* **352**: 1406–1409.

[49] Choi WS, Hong B. (2008) The effect of annealing on the properties of diamond-like carbon protective antireflection coatings. *Renewable Energy* **33**: 226–231.

[50] Demirgoz D, Elvira C, Mano JF, Cunha AM, Piskin E, Reis RL. (2000) Chemical modification of starch based biodegradable polymeric blends: Effects on water uptake, degradation behaviour and mechanical properties. *Poly Degrad Stab* **70**: 161–170.

[51] Kim YJ, Kang IK, Huh MW, Yoon SC. (2000) Surface characterization and in vitro blood compatibility of poly(ethylene terephthalate) immobilized with insulin and/or heparin using plasma glow discharge. *Biomaterials* **21**: 121–130.

[52] Baquey Ch, Palumbo F, Porte-Durrieu MC, Legeay G, Tressaud A, d'Agostino R. (1999) Plasma treatment of expanded PTFE offers a way to a biofunctionalization of its surface. *Nucl Instr Meth Phys Res B* **151**: 255–262.

[53] Muguruma H. (2007) Plasma-polymerized films for biosensors II. *Trends Anal Chem* **26**: 433–443.

[54] Janasek, D, Spohn U, Kiesow A, Heilmann A. (2001) Enzyme modified membranes protected by plasma polymer layers. *Sens Actuators B: Chemical* **78**: 228–231.

[55] Alp B, Mutlu S, Mutlu M. (2000) Glow-discharge-treated cellulose acetate (CA) membrane for a high linearity single-layer glucose electrode in the food industry. *Food Res Int* **33**: 107–112.

[56] Hiratsuka A, Kojima KI, Muguruma H, Lee KH, Suzuki H, Karube I. (2005) Electron transfer mediator micro-biosensor fabrication by organic plasma process. *Biosensors Bioelectr* **21**: 957–964.

[57] Miyachi H, Ikebukuro K, Yano K, Aburatani H, Karube I. (2004) Single nucleotide polymorphism typing on DNA array with hydrophobic surface fabricated by plasma-polymerization technique. *Biosensors Bioelectr* **20**: 184–189.

[58] Braceras I, Briz N, García-Luis A, Corengia P, González-Santamaría D, Brizuela M, Muñoz R. (2006). 10th International Conference on Plasma Surface Engineering, September 10–15, Proc. Cong, Garmisch-Partenkirchen, Germany, p. 502.

[59] Choi JH, Han I, Baik HK, Lee MH, Han DW, Park JC, Lee IS, Song KM, Lim YS. (2006) Analysis of sterilization effect by pulsed dielectric barrier discharge. *J Electrostatics* **64**: 17–22.

[60] Park BJ, Takatori K, Lee MH, Han DW, Woo YI, Son HJ, Kim JK, Chung KH, Hyun SO, Park JC. (2007) *Escherichia coli* sterilization and lipopolysaccharide inactivation using microwave-induced argon plasma at atmospheric pressure. *Surf Coat Technol* **201**: 5738–5741.

[61] Hongxia L, Jierong C, Liqing Y, Yuan Z. (2008) Long-distance oxygen plasma sterilization: Effects and mechanisms. *Appl Surf Sci* **254**: 1815–1821.

[62] Boscariol MR, Moreira AJ, Mansano RD, Kikuchi IS, Pinto TJA. (2008) Sterilization by pure oxygen plasma and by oxygen–hydrogen peroxide plasma: An efficacy study. *Int J Pharm* **353**: 170–175.

[63] Rutala WA, Gergen MF, Weber DJ. (1998) Comparative evaluation of the sporicidal activity of new low-temperature sterilization technologies: Ethylene oxide, 2 plasma sterilization systems, and liquid peracetic acid. *Am J Infection Control* **26**: 393–398.

[64] Ohkawa H, Akitsu T, Tsuji M, Kimura H, Kogoma M, Fukushima K. (2006) Pulse-modulated, high-frequency plasma sterilization at atmospheric-pressure. *Surf Coat Technol* **200**: 5829–5835.

[65] Batían Murali N, Mickelsen S, Brooks C, Jaramillo J, Hepton T, Kusumoto FM. (1998) Safety and efficacy of hydrogen peroxide plasma sterilization for repeated use of electrophysiology catheters. *J Am Coll Cardiol* **32**: 1384–1388.

[66] Kwon-Young L, Bong Joo P, Dong Hee L, In-Seop L, Soon OH, Kie-Hyung C, Jong-Chul P. (2005) Sterilization of *Escherichia coli* and MRSA using microwave-induced argon plasma at atmospheric pressure. *Surf Coat Technol* **193**: 35–38.

[67] Vassal S, Favennec L, Ballet JJ, Brasseur P. (1998) Hydrogen peroxide gas plasma sterilization is effective against *Cryptosporidium parvum* oocysts. *Am J Infect Control* **26**: 136–138.

[68] Pegueroles M, Gil FJ, Planell JA, Aparicio C. (2008) The influence of blasting and sterilization on static and time-related wettability and surface-energy properties of titanium surfaces. *Surf Coat Technol* **202**: 3470–3479.

[69] Baker DA, Hastings RS, Pruitt L. (2000) Compression and tension fatigue resistance of medical grade ultra high molecular weight polyethylene: The effect of morphology, sterilization, aging and temperature. *Polymer* **41**, 795–808.

[70] Serro AP, Saramago B. (2003) Influence of sterilization on the mineralization of titanium implants induced by incubation in various biological model fluids. *Biomaterials* **24**: 4749–4760.

[71] Campoccia D, Montanaroa L, Arciola CR. (2006) Microparticles developed by electrohydrodynamic atomization for the local delivery of anticancer drug to treat C6 glioma in vitro. *Biomaterials* **27**: 2331–3332.

[72] Montali A. (2006) Antibacterial coating systems. *Injury, Int J Care Injured*, **37**: S81–S86.

[73] Dowling DP, Betts AJ, Pope C, McConnell ML, Eloy R, Arnaud MN. (2003) Antibacterial silver coatings exhibiting enhanced activity through the addition of platinum. *Surf Coat Technol* **163–164**: 637–640.

[74] De los Arcos T, Oelhafen P, Aebi U, Hefti A, Düggelin M, Mathys D, Guggenheim R. (2002) Preparation and characterization of TiN–Ag nanocomposite films. *Vacuum* **67**: 463–470.

[75] Kampfrath G, Heilmann A, Hamann C. (1988) Plasma polymerized thin films containing small silver particles. *Vacuum* **38**: 1–3.

[76] Zhang W, Zhang YH, Ji JH, Zhao J, Yan Q, Chu, PK. (2006) Antimicrobial properties of copper plasma-modified polyethylene. *Polymer* **47**: 7441–7445.

[77] Wan YZ, Raman S, He F, Huang Y. (2007) Surface modification of medical metals by ion implantation of silver and copper. *Vacuum* **81**: 1114–1118.

[78] Yoshinari M, Oda Y, Kato T, Okuda K. (2001) Influence of surface modifications to titanium on antibacterial activity in vitro. *Biomaterials* **22**: 2043–2048.

[79] Zhang W, Ji J, Zhang Y, Yan Q, Kurmaev EZ, Moewes A, Zhao J, Chu PK. (2007) Effects of NH_3, O_2, and N_2 co-implantation on Cu out-diffusion and antimicrobial properties of copper plasma-implanted polyethylene. *Appl Surf Sci* **253**: 8981–8985.

[80] Liao JD, Chen HJ, Chang CW, Chiu SM, Chen ZS. (2006) Thin-film photo-catalytic TiO_2 phase prepared by magnetron sputtering deposition, plasma ion implantation and metal vapor vacuum arc source. *Thin Solid Films* **515**: 176–185.

[81] Yao KS, Wang DY, Yan JJ, Yang LY, Chen WS. (2007) Photocatalytic effects of TiO_2/Fe thin film irradiated with visible light on cellular surface ultrastructure and genomic DNA of bacteria. *Surf Coat Technol* **201**: 6882–6885.

[82] Çökeliler D, Caner H, Zemek J, Choukourov A, Biederman H, Mutlu M. (2007) A plasma polymerization technique to overcome cerebrospinal fluid shunt infections. *Biomed Mater* **2**: 39–47.

[83] Wang, N, Huang N, Pan CJ, Kwok SCH, Yang P, Leng YX, Chen JY, Sun H, Wan GJ, Liu, ZY, Chu PK. (2004) Bacterial repellence from polyethylene terephthalate surface modified by acetylene plasma immersion ion implantation–deposition. *Surf Coat Technol* **186**: 299–304.

[84] Mandracci P, Mussano F, Ricciardi C, Ceruti P, Pirri F, Carossa S. (2008) Low temperature growth of thin film coatings for the surface modification of dental prostheses. *Surf Coat Technol* **202**: 2477–2481.

[85] Sodhi RNS, Sahi VP, Mittelman MW. (2001) Application of electron spectroscopy and surface modification techniques in the development of anti-microbial coatings for medical devices. *J Electron Spectrosc Relat Phenom* **121**: 249–264.

[86] Virk RK, Ramaswamy GN, Bourham M, Bures BL. (2004) Plasma and antimicrobial treatment of nonwoven fabrics for surgical gowns. *Textile Res J* **74**: 1073–1079.

[87] Zhang W, Chu PK, Ji J, Zhang Y, Fu RKY, Yan Q. (2006) Antibacterial properties of plasma-modified and triclosan or bronopol coated polyethylene. *Polymer* **47**: 931–936.

[88] Kwok CS, Horbett TA, Ratner BD. (1999) Design of infection-resistant antibiotic-releasing polymers II. Controlled release of antibiotics through a plasma-deposited thin film barrier. *J Controlled Release* **62**: 301–311.

[89] Thissen H, Johnson G, Hartley PG, Kingshott P, Griesser HJ. (2006) Two-dimensional patterning of thin coatings for the control of tissue outgrowth. *Biomaterials* **27**: 35–43.

[90] Goessl A, Bowen-Pope DF, Hoffman AS. (2001) Control of shape and size of vascular smooth muscle cells in vitro by plasma lithography. *J Biomed Mater Res* **57**: 15–24.

[91] Thissen H, Hayes JP, Kingshott P, Johnson G, Harvey EC, Griesser HJ. (2002) Nanometer thickness laser ablation for spatial control of cell attachment, *Smart Mater Struct* **11**: 792–799.

[92] Zelzer M, Majani R, Bradley JW, Rose FRAJ, Davies MC, Alexander MR. (2008) Investigation of cell-surface interactions using chemical gradients formed from plasma polymers. *Biomaterials* **29**: 172–184.

[93] Rhodes NP, Wilson DJ, Williams RL. (2007) The effect of gas plasma modification on platelet and contact phase activation processes. *Biomaterials* **28**: 4561–4570.

[94] De Maeztu MA, Braceras I, Alava JI, Gay-Escoda C. (2008) Improvement of osseointegration of titanium dental implant surfaces modified with CO ions: A comparative histomorphometric study in beagle dogs. *Int J Oral Maxillofac Surg* **37**: 441–447.

[95] Braceras I, Alava JI, Onate JI, Brizuela M, Garcia-Luis A, Garagorri N, Viviente JL, de Maeztu MA. (2002) Improved osseointegration in ion implantation-treated dental implants. *Surf Coat Technol* **158–159**: 28–32.

[96] Nayab SN, Jones FH, Olsen I. (2005) Effects of calcium ion implantation on human bone cell interaction with titanium. *Biomaterials* **26**: 4717–4727.

[97] De Maeztu MA, Braceras I, Alava JI, Sánchez-Garcés MA, Gay-Escoda C. (2007) Histomorphometric study of ion implantation and diamond-like carbon as dental implant surface treatments in beagle dogs. *Int J Oral Maxillofac Implants* **22**: 273–279.

[98] Van der Wal E, Vredenberg AM, Ter Brugge PJ, Wolke JGC, Jansen JA. (2006) The in vitro behavior of as-prepared and pre-immersed RF-sputtered calcium phosphate thin films in a rat bone marrow cell model. *Biomaterials* **27**(8): 1333.

[99] Lee IS, Zhao B, Lee GH, Choi SH, Chung SM. (2007) Industrial application of ion beam assisted deposition on medical implants. *Surf Coat Technol* **201**: 5132–5137.

[100] Finke B, Luethen F, Schroeder K, Mueller PD, Bergemann C, Frant M, Ohl A, Nebe BJ. (2007) The effect of positively charged plasma polymerization on initial osteoblastic focal adhesion on titanium surfaces. *Biomaterials* **28**: 4521–4534.

[101] Mattioli-Belmonte M, Lucarini G, Virgili L, Biagini G, Detomaso L, Favia P, D'Agostino R, Gristina R, Gigante A, Bevilacqua C. (2005) Mesenchymal stem cells on plasma-deposited acrylic acid coatings: An in vitro investigation to improve biomaterial performance in bone reconstruction. *J Bioact Compat Polym* **20**: 343–360.

[102] Kumar S, Simpson D, Smart RSTC. (2007) Plasma processing for inducing bioactivity in stainless steel orthopaedic screws. *Surf Coat Technol* **202**: 1242–1246.

[103] Le Guéhennec L, Soueidan A, Layrolle P, Amouriq Y. (2007) Surface treatments of titanium dental implants for rapid osseointegration. *Dental Mater* **23**: 844–854.

[104] Ni GX, Lu WW, Chiu KY, Li ZY, Fong DYT, Luk, KDK. (2006) Strontium-containing hydroxyapatite (Sr-HA) bioactive cement for primary hip replacement: An in vivo study. *J Biomed Mate. Res B Appl Biomater* **77**: 409–415.

[105] Prichard HL, Reichert WM, Klitzman B. (2007) Adult adipose-derived stem cell attachment to biomaterials. *Biomaterials* **28**: 936–946.

[106] Chu PK. (2007) Enhancement of surface properties of biomaterials using plasma-based technologies. *Surf Coat Technol* **201**: 8076–8082.

[107] Vesel A, Drenik A, Mozetic M, Zalar A, Balat-Pichelinb M, Bele M. (2008) AES investigation of the stainless steel surface oxidized in plasma. *Vacuum* **82**: 228–231.

[108] Susut C, Timmons DF. (2005) Plasma enhanced chemical vapor depositions to encapsulate crystals in thin polymeric films: A new approach to controlling drug release rates. *Int J Pharm* **288**: 253–261.

[109] Favez D, Castner DG, Ratner BD, Mathieu HJ. (2002) RF-plasma deposition and surface characterization of a biodegradable thin film coating. *Eur Cells Mater* **3**: 20–21.

[110] Carter AJ, Fischell TA. (1998) Current status of radioactive stents for the prevention of In-stent restenosis. *Int J Radiat Oncol Biol Phys* **41**: 127–133.

[111] Kawashita M, Miyaji F, Kokubo T, Takaoka GH, Yamada I, Suzuki Y, Inoue M. (1999) Surface structure and chemical durability of P+-implanted Y_2O_3–Al_2O_3–SiO_2 glass for radiotherapy of cancer. *J Non-Crys Solids* **255**: 140–148.

[112] Yuen C, Williams R, Batterbury M, Grierson I. (2006) Modification of the surface properties of a lens material to influence posterior capsular opacification. *Clin Experiment Ophthalmol* **34**: 568–574.

[113] Shigeta M, Tanaka T, Koike N, Yamakawa N, Usui M. (2006) Suppression of fibroblast and bacterial adhesion by MPC coating on acrylic intraocular lenses. *J Cataract Refract Surg* **32**: 859–866.

[114] Li DJ, Cui FZ, Gu HQ. (1999) F^+ Ion implantation induced cell attachment on intraocular lens. *Biomaterials* **20**: 1889–1896.

[115] Feng H, Yu Z, Chu PK. (2006) Ion implantation of organisms. *Mat Sci Eng Reports* **54**: 49–120.

[116] Yu Z, Deng J, He J, Huo Y, Wu Y, Wang X, Lui G. (1991) Mutation breeding by ion implantation. *Nucl Instrum Meth B* **59–60**: 705–708.

[117] Yu C, Zhixin L, Zuyao Z, Feng Z, Duo L, Xianghuai L, Jianzhong T, Weimin Z, Bo H. (1998) High yield antibiotic producing mutants of *Streptomyces erythreus* induced by low energy ion implantation. *Nucl Instrum Meth B* **140**: 341–348.

[118] Mahadtanapuk S, Yu LD, Cutler R, Vilaithong T, Anuntalabhochai S. (2007) Mutation of *Bacillus licheniformis* using low-energy ion beam bombardment. *Surf Coat Technol* **201**: 8029–8033.

[119] Chen H, Wan H, Zhang J, Zeng X. (2008) Mutation induction of *Pleurotus ferulae* by low-energy N+ ion implantation and characters of the selected mutant. *Nucl Sci Tech* **19**: 13–16.

[120] Tinschert K, Iannucci R, Lang R. (2008) Electron cyclotron resonance ion sources in use for heavy ion cancer therapy. *Rev Sci Instrum* **79**: 02C505.

Index

About the Editors

Dr. Eduardo A. Favret is currently a research scientist at the Institute of Soils and the Botanical Garden "Arturo E. Ragonese", Natural Resources Research Centre (CIRN), National Institute for Agricultural Technology (INTA), Castelar, Argentina. He obtained his doctoral degree in Physics from the University of Buenos Aires and did a postdoctoral training in laser structuring of materials at the Institute of Functional Materials, University of Saarbrücken, Germany, with a German Science Foundation fellowship. He is also a research scientist of the National Council on Scientific and Technical Research (CONICET), Assistant Professor of the Institute of Technology "Prof. Jorge A. Sábato", UNSAM — CNEA, and director of postgraduate theses. He has published many articles and co-developed RIMAPS and ULOI techniques. He received international awards and participated in different conferences. He is a member of the Argentine Society of Microscopy (SAMIC), the American Society for Metals (ASM International), the International Metallographic Society (IMS) and the Microscopy Society of America (MSA). His research interests are in the field of micromorphology and mineralogy of soils, biomimetism (functional properties of biological surfaces and its development on artificial surfaces), metallography, microscopy and characterization of biological and technical surfaces.

Dr. Néstor O. Fuentes received his PhD in Physics from the University of Buenos Aires, where he later held a CONICET Research Fellowship. In 1985 he joined the National Atomic Energy Commission of Argentina (CNEA) where he is now Principal Investigator and Head of the Advanced Materials Division. He obtained postdoctoral, training and research fellowships from the Abdus Salam International Centre for Theoretical Physics (ICTP), the Italian Government and the International Atomic Energy Agency (IAEA). He is the Argentine Counterpart for International Cooperation Projects in the subject area of fluid flow and transport of contaminants. He is a member of the organizing committee of several international workshops and conferences. He also received national and international awards. He is Professor of the Institute of Technology "Prof. Jorge A. Sábato" (CNEA — University of San Martín), director of postgraduate theses and advisor of postgraduate fellows. He is co-founder and member of the Argentine Association on Nondestructive and Structural Testing (AAENDE). His research is focused on the development of the new imaging technique RIMAPS for surface characterization; the study, development and characterization of advanced nuclear materials; and the modelling of fluid flow and transport of contaminants.